Electronic Communication Systems

Electronic Communication Systems

**Edited by
Reuben Parker**

WILLFORD PRESS

www.willfordpress.com

Published by Willford Press,
118-35 Queens Blvd., Suite 400,
Forest Hills, NY 11375, USA

ISBN: 978-1-68285-770-0

Cataloging-in-Publication Data

Electronic communication systems / edited by Reuben Parker.
    p. cm.
Includes bibliographical references and index.
ISBN 978-1-68285-770-0
1. Telecommunication. 2. Telecommunication systems. 3. Digital communications. I. Parker, Reuben.
TK5101 .E44 2020
621.382--dc23

For information on all Willford Press publications
visit our website at www.willfordpress.com

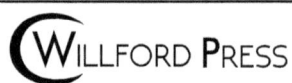

WILLFORD PRESS

Contents

Preface

The main aim of this book is to educate learners and enhance their research focus by presenting diverse topics covering this vast field. This is an advanced book which compiles significant studies by distinguished experts in the area of analysis. This book addresses successive solutions to the challenges arising in the area of application, along with it; the book provides scope for future developments.

An electronic communication system is a collection of communication networks, tributary stations, relay stations, transmission systems and data terminal equipment. These components are technologically compatible, respond to controls, use common procedures and operate in union. Electric communication systems are of different types, depending on the transmission media, such as optical communication system, power line communication system and radio communication system. Other classifications of communication systems, such as duplex communication system, tactical communications system, emergency communication system, etc. may be based on the technology used or the area of application. This book includes some of the vital pieces of work being conducted across the world, on various topics related to electronic communication systems. It attempts to understand the diverse aspects of electronic communication systems and how these have practical applications. This book is a complete source of knowledge on the present status of this important field.

It was a great honour to edit this book, though there were challenges, as it involved a lot of communication and networking between me and the editorial team. However, the end result was this all-inclusive book covering diverse themes in the field.

Finally, it is important to acknowledge the efforts of the contributors for their excellent chapters, through which a wide variety of issues have been addressed. I would also like to thank my colleagues for their valuable feedback during the making of this book.

Editor

Cooperative Cloud Service Aware Mobile Internet Coverage Connectivity Guarantee Protocol based on Sensor Opportunistic Coverage Mechanism

Qin Qin,[1] Yong-qiang He,[1] and Li-ming Nie[2]

[1]College of Computer, Henan Institute of Engineering, Zhengzhou 450007, China
[2]School of Software Technology, Dalian University of Technology, Dalian 116621, China

Correspondence should be addressed to Qin Qin; qq@haue.edu.cn

Academic Editor: James Nightingale

In order to improve the Internet coverage ratio and provide connectivity guarantee, based on sensor opportunistic coverage mechanism and cooperative cloud service, we proposed the coverage connectivity guarantee protocol for mobile Internet. In this scheme, based on the opportunistic covering rules, the network coverage algorithm of high reliability and real-time security was achieved by using the opportunity of sensor nodes and the Internet mobile node. Then, the cloud service business support platform is created based on the Internet application service management capabilities and wireless sensor network communication service capabilities, which is the architecture of the cloud support layer. The cooperative cloud service aware model was proposed. Finally, we proposed the mobile Internet coverage connectivity guarantee protocol. The results of experiments demonstrate that the proposed algorithm has excellent performance, in terms of the security of the Internet and the stability, as well as coverage connectivity ability.

1. Introduction

With the development and integration of wireless network, mobile communication [1], sensor [2], cloud platform, and so forth, the next generation of Internet has been widely used and developed in the field of real-time monitoring, full coverage of mobile network, and poor environment data acquisition and communication. The mapping of Internet nodes and sensor nodes establish the heterogeneous communication, and provide support for the monitoring of information storage and forwarding and real-time processing. However, because of the diversity of environmental monitoring needs and the forwarding of sensor information collection and the network coverage and connectivity [3] of all kinds of bad environment, it is still a hot and key issue of the next generation Internet.

There are some researches of Internet coverage connectivity. A coverage-based hybrid overlay was proposed in article [4], which disseminates messages to all subscribers without uninterested nodes involved in and increases the average number of node connections slowly with an increase in the number of subscribers and nodes. A new notion of intermittent coverage for mobile users was introduced in article [5], which provides worst-case guarantees on the interconnection gap, the distance, or expected delay between two consecutive mobile-AP contacts for a vehicle. Efficient node coverage scheme was proposed for addressing the scheduling issue of nodes of the underground space near surface [6]. The authors of article [7] proposed the supporting plant designer during wireless coverage prediction, virtual network deployment, and postlayout verification sensor opportunistic coverage mechanism. A low-cost way [8] was proposed for public transit operators to enhance quality of experience for passengers who access the Internet. The control flow graph and cyclomatic complexity of the example program [9] were used to find out the number of feasible paths present in the program and compared it with the actual number of paths covered by genetic algorithm.

About the coverage issue of wireless sensor networks (WSNs), Zhao et al. [10] analyzed the changes of opportunistic coverage ratios. The new cooperative opportunistic four-level model for IEEE 802.15.4 wireless personal area network was proposed by Rohokale et al. [11]. A localization scheme named Opportunistic Localization by Topology Control was proposed in [12] for sparse Underwater Sensor Networks. The use of an MAS as an appropriate mechanism was advocated by different stakeholders in [13].

For studying the relationship of cloud platform and coverage issue of Internet, a service aware location update mechanism was proposed in [14], which can detect the presence and location of the mobile device. A novel quality aware computational cloud selection service was proposed and evaluated in [15]. The cost and energy aware service-provisioning scheme [16] was presented for mobile client in mobile cloud, which includes two-stage optimization process. A novel disaster-aware service-provisioning scheme was proposed in [17], which multiplexes service over multiple paths destined to multiple servers/datacenters with many casting. A privacy-aware cross-cloud service composition method was proposed in [18] based on its previous basic version HireSome-I.

However, these research results ignored the resources management issue of cloud platform and the mobile Internet requirements and had a little research of coverage guaranteeing ability of WSNs. Then, we proposed the cooperative cloud service aware mobile Internet coverage connectivity guarantee protocol based on sensor opportunistic coverage mechanism.

The rest of the paper is organized as follows. Section 2 gave the sensor opportunistic coverage mechanism. In Section 3, we proposed the model of cooperative Cloud service aware. The mobile Internet coverage connectivity guarantee protocol was proposed in Section 4. The results of the mathematical analysis and simulation verification are given in Section 5. Finally, we conclude the paper in Section 6.

2. Sensor Opportunistic Coverage Mechanism

In the context of mobile Internet, how to deploy wireless sensor networks to achieve reliable and stable network coverage has become the key technology of wireless sensor networks and the Internet. Separate wireless sensor network coverage rule is specified based on different type of structure. Generally, the same wireless sensor network only includes a single cover rule or a geometric figure. This is not conducive to the combination of large scale sensor networks and mobile Internet. In order to meet the needs of the integration of the Internet and reduce the complexity and robustness of wireless sensor network coverage control, a mobile Internet based on sensor deployment and network coverage algorithm is proposed. The algorithm can ensure the distributed computing and the full connected routing of sensor nodes. The algorithm can be implemented based on the irregular coverage of mobility.

Based on the driving of Internet mobile node, the algorithm combines the opportunity dynamic coverage rule and

communication between the mobile node and the sensor node to achieve a high reliability and real-time security.

N denotes the number of mobile nodes which are used to connect the sensor network with the Internet. Wireless sensor networks (WSNs) deployment node number is M. The nearest distance between the mobile node and the sensor node is d_0. The minimum distance between the sensor nodes is d_1. In order to reduce the impact of the Internet node's mobility on sensor network coverage, the deployment distance parameters need to meet the following formula:

$$\frac{S(d_{\max})}{\sum_{i=1}^{M} d_1^i} \leq \frac{\text{MI}(d_{\max})}{\sum_{j=1}^{N} d_0^i}. \tag{1}$$

Here, $S(d_{\max})$ denotes the maximum connection distance of sensor nodes in WSNs. $\text{MI}(d_{\max})$ denotes the maximum connection distance between Internet and WSNs.

At the same time, in order to ensure the deployment of sensor nodes independently, the communication distance between adjacent sensor nodes and the Internet mobile node should be in accordance with formula (2), which is used to ensure that the mobile node and sensor nodes connected with independent characteristics

$$\frac{\text{MI}(d_{\max})}{v_{\max}} \leq t_0 \left[\frac{\mu \cos \alpha}{e^{-S(d_{\max})}} \right]^{N/M}. \tag{2}$$

Here, $v_{\max}$ is the maximum moving speed of Internet node. t_0 is the delay of transmitting signal from Internet node to sensor node. μ denotes the opportunistic connection ratio between Internet node and sensor node. α is the angle of the antenna direction of the Internet node and the sensor node direction.

Sensor node S_i can sense the Internet node MI_i. Internet nodes help sensor nodes to establish network coverage with strong connectivity. The opportunity coverage ratio is shown in the following formula:

$$C_{(S,\text{MI})} = \begin{cases} 0, & t \geq t_{\text{TH}} \\ 1, & t \leq t_{\text{TH}}, \end{cases} \tag{3}$$

$$t_{\text{TH}} = \mu \left(\frac{d_{(S,\text{MI})}}{v_{\text{MI}}} \right)^{\delta}.$$

Here, $C_{(S,\text{MI})}$ denotes the connection ratio between nodes. t_{TH} is the maximum delay of keeping communication. δ is the probability of occurrence and transformation of Internet node movement direction.

The mobile nodes and the sensor nodes would use signal level cooperation communication. The communication between Internet mobile node MI and the m sensor nodes would create the neighbor coverage relationship. They are common sensing coverage and control an area, which are shown in Figure 1. The opportunistic coverage strength OCI

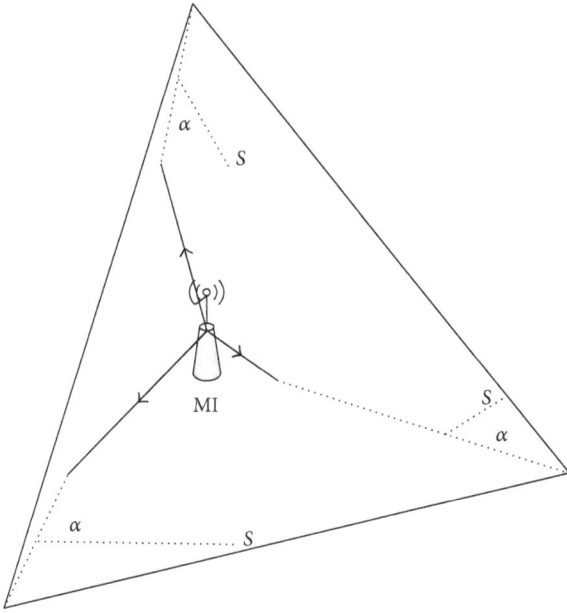

Figure 1: Common sense and coverage control area.

between sensor node and the mobile node can be obtained by the following formula:

$$OCI = |\gamma - \mu|^2 \sum_{i=1}^{m} C_{i,MI},$$

$$\gamma = \frac{\sum_{i=1}^{m} t_i}{t_{TH}}. \tag{4}$$

Here, γ is the probability of connecting time greater than t_{TH} between the Internet nodes and sensor nodes.

Through the combination of the Internet node deployment area and sensor network deployment area of the mobile Internet, the opportunity to cover the intensity MI-OCI can be obtained through the node between the opportunity connection and collaborate sensing, as shown in the following formula:

$$MI - OCI(N, M) = \frac{\sqrt{t_{TH} |v_{max} - v_{min}|^2} \sum_{i=1,j=1}^{N,M} d_{i,j}}{\mu |\gamma - \delta|^2 v_{TH}},$$

$$v_{TH} = \frac{\sqrt{\left| S(d_{max})^2 - MI(d_{max})^2 \right|}}{t_{TH}}. \tag{5}$$

Here, v_{TH} is the Internet node moving speed threshold.

3. Cooperative Cloud Service Aware Model

With the rapid development of the Internet and WSNs, the next generation network convergence and cloud platform business should be improved. So, the cloud platform service business platform is proposed. The service platform has the following characteristics.

(1) To guide the deployment, implementation, and management of WSNs is the center of service support and user requirements.

(2) Data services and forwarding services are capable of supporting multinetwork integration and heterogeneous services and are not independent of the front-end device and forwarding gateway for specific wireless sensor networks.

(3) The size of the cloud platform has a dynamic adjustment ability. According to different network traffic and forwarding data content, the cloud devices have the way of cooperation between the management and control.

(4) The information service quality and wireless sensor network resource utilization can be perceived as an open and reliable access to provide services to the front of the Internet and the sensor.

The architecture model of the cloud platform service business platform is shown in Figure 2. Cloud service business support platform of the Internet application service management capabilities and wireless sensor network communication service capabilities are abstracted as the architecture of the cloud support layer. The support layer is a business entity shared by all cloud services. Cloud service middleware platform provides the deployment and execution environment for all kinds of Internet applications and data services for wireless sensor networks. Cloud physical layer for the Internet and WSNs could provide the core support of various front-end equipment and services and cross layer interface.

The above architecture shows that the perception of cloud services has important significance for the Internet management and WSNs coverage, as well as topology management. Therefore, the application of the Internet could be encapsulated as cloud services. Various types of sensor front-end devices and cloud services could be considered as the perception of objects, through the aggregation of the Internet service and WSNs, which is used to provide a basis for the internet.

The cloud service perception model is shown in Figure 3. Based on the Internet mobility management business security mechanism and the WSNs connectivity guarantee mechanism, the perception model is proposed. The complexity of cloud computing through WSNs communication between different devices could be reduced. Cloud platform through cooperative cloud services for WSNs deployment and coverage could provide a unified positioning business. Cloud platform through the cooperative cloud service could be packaged in the cloud platform middleware layer. Web services of the Internet and WSNs data collection and forwarding services could be provided by the combination of cooperative cloud aggregation services. This service can provide reference information and decision basis for route maintenance and cooperative coverage in the WSNs coverage area. In particular, there are the perception pile between the Internet and WSNs, which is composed of wireless multiple word system bus. In order to improve the parallelism of the system, we designed the control feedback interaction

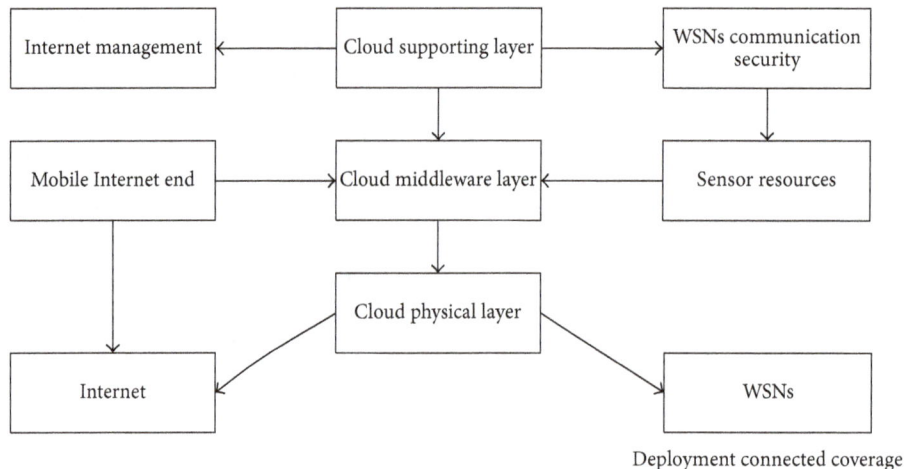

FIGURE 2: Architecture model of cloud service platform.

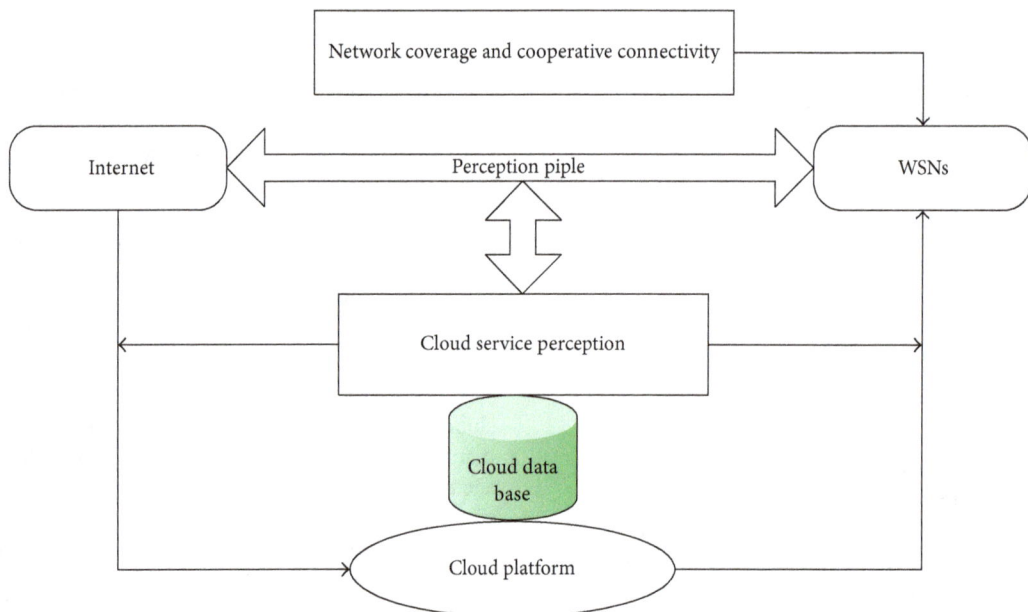

FIGURE 3: Cloud service perception model.

line, which is from cloud service perception platform to the Internet and WSNs.

4. Mobile Internet Coverage Connectivity Guarantee Protocol

Mobile Internet applications can achieve the application development through the integration of cloud platforms and WSNs. When the application is extended, the problem of mobile management and resource management and reliable information forwarding should be considered. The high coverage ratio could be maintained by deploying a large number of redundant devices or sensor nodes in the cloud platform and WSNs. But about the Internet node service life, WSNs sensor node resource management and cloud computing

scheduling, and so forth, the network connectivity could not kept by increasing the size of equipment.

Therefore, it is the key technology to make full use of the opportunity of each sensor node in wireless sensor network, and the effective driving ability of the cloud platform service, which is used to ensure the Internet connectivity. Through the WSNs, the sensor nodes and the Internet nodes are in the mobile state, which can reduce the resource consumption and realize the effective seamless coverage of the network. In order to reduce the impact of the mobile Internet connectivity coverage, the Internet and the cloud platform play a role in the root node of the cloud device. The diffusion of Internet nodes and sensor nodes coverage can be opportunistic, dynamic, and adjusted at the same time to achieve maximum coverage of connectivity and through collaborative optimization control to achieve global connectivity. This can reduce the resource consumption of each node and can reduce the probability of the occurrence of invalid overlap between nodes, as shown in the following formula:

$$C_{RC} = \sum_{k=1}^{N+M} R_k \frac{S(d_{max}) + MI(d_{max})}{\sum_{i=1}^{M} d_1^i d_0^i},$$

$$P_{IO} = \sum_{i=1}^{C_N} \rho_i \frac{\sqrt{|SC_N - SC_M|}}{SC_{N+M}}.$$

(6)

Here, C_{RC} denotes cover resource consumption. R_k denotes the resource consumption of node. P_{IO} is the probability of invalid overlap. C_N is the cloud number of the platforms. ρ denotes the active ratio of cloud. SC denotes the covering area.

Based on the Internet coverage mechanism and the cooperative security mechanism, the following problems should be considered.

(1) When the mobile node and sensor nodes are used in the model, the sensing distance of the node is related to the computing power of the cloud computing power. According to the node sensing distance, the Internet coverage is irregular geometry area composed by a number of mobile nodes and sensor nodes. Internet coverage area is not affected by node mobility.

(2) The sensor covers the sensing area opportunistically and deploys sensor nodes based on the remaining resources of the Internet node and cloud service needs.

(3) When the sensor network is deployed by the terrain or the impact of a large building, it can be through the cloud platform of collaborative cloud service sensing mechanism, the communication of all nodes to the same plane, as shown in Figure 4.

(4) The hierarchical communication architecture of sensor nodes and the hierarchical protocol of Internet nodes occur through the cloud platform to achieve cross layer interaction. The nodes and the cloud devices in the Internet are covered by the Internet, which can communicate directly with the sensor nodes and the cloud devices, and the heterogeneous communication protocols can be handled transparently.

The WSNs opportunity coverage role of Internet mobile coverage connectivity is reflected in the following aspects.

(1) When the sensor nodes have separated from the Internet, these nodes would broadcast a request of the subnetwork separation and reconstruct to the neighbor nodes. After receiving the request, the nodes are calculated by the cloud platform to maintain the network connectivity. If you cannot meet the needs of the Internet seamless coverage, the cloud platform issued a collaborative cloud service perception control in the Internet node and sensor nodes to

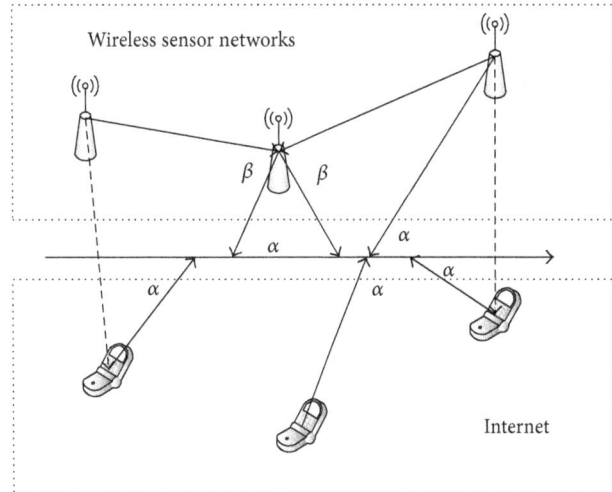

FIGURE 4: Communications mapping with cooperative cloud aware scheme.

search for collaborative gaps and opportunity to drive the region for ensuring the full connectivity of the Internet monitoring area and real-time coverage, as shown in the following formula:

$$d_{max(S,MI)} = \sqrt{|d_{max(S)} - d_{min(MI)}|^2} \sum_{i=1,j=1}^{N,M} d_{i,j},$$

$$SC_{crevice} = \frac{\sqrt{|SC^2_N - SC^2_M|}}{\tan(d_{max(S,MI)}/\theta)}.$$

(7)

Here, θ is the Internet node and sensor node antenna angle. $SC_{crevice}$ is the gap area.

(2) In order to avoid the loss caused by the network nodes and sensor nodes due to overlapping coverage area, the cloud platform and data forwarding provide seamless coverage and full connected routing. We must gradually reduce the communication delay between the nodes in the coverage area, so that the three-party interaction and coverage area of the cloud platform, the Internet, and WSN achieve the best equilibrium state.

(3) When the channel quality is good, it can increase the share of the three parties, the largest consumption of resources to achieve maximum coverage area. When the channel quality is poor, the sensor can increase the opportunity to cover the weight and the cloud service cooperative sensing coefficient and achieve the optimal control in the area of Internet coverage and connectivity.

5. Mathematical Analysis and Simulation Verification

The performance of the proposed algorithm is analyzed in this section, which includes the stability, connectivity, and

(a) $N = 15, M = 15$

(b) $N = 20, M = 10$

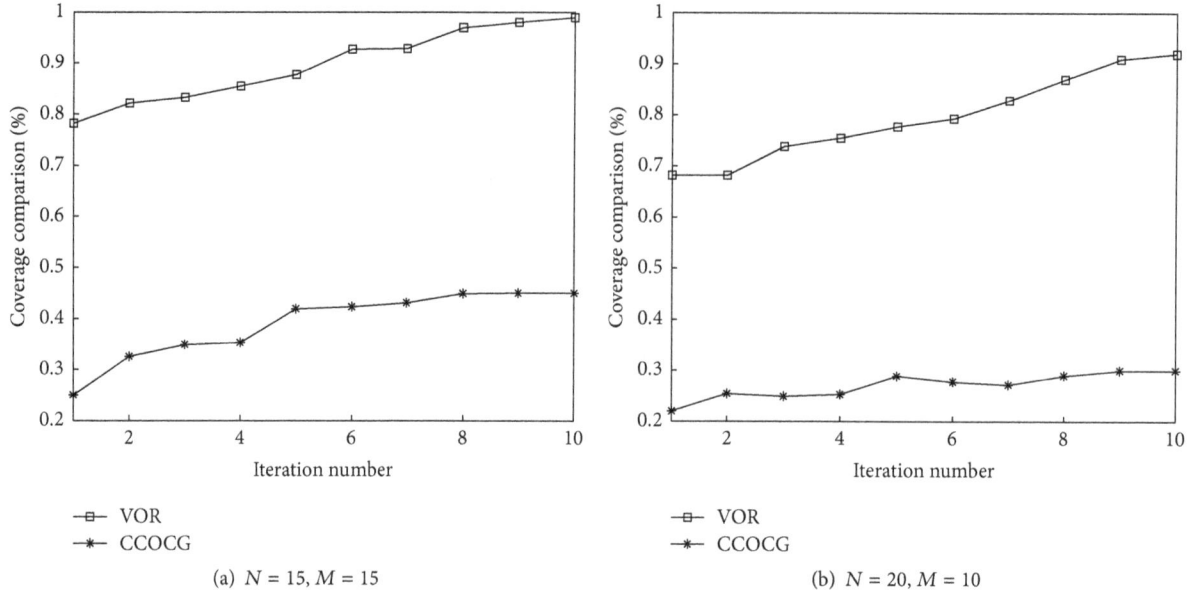

FIGURE 5: Coverage comparison.

security of the Internet. The coverage stability of the Internet is considered, both the Internet node and the sensor node sensing area coverage and the connectivity security capability. When the security capability is strong, the security time of the proposed algorithm is investigated, including the Internet node iteration time, the cloud computing time, and the forwarding time of the sensor. The time spent on the Internet is determined by the time of the node moving speed, cloud computing delay, and sensor node coverage.

The cost of Internet connectivity includes:

(1) the network nodes in order to maintain the loss of the mobile speed;

(2) the total number of nodes required by the Internet;

(3) the resource consumption of the cloud devices to maintain continuous calculation;

(4) the energy consumption of sensor nodes and the energy consumed in communication.

So, by comprehensive evaluation of coverage connectivity and security costs, we can better analyze the performance of the proposed algorithm (CCOCG) in the Internet node mobility management, cloud service perception, and WSNs opportunity communication and other environmental performance.

In the experiment, the Internet covering model of time domain, frequency domain, and spatial domain is considered. Random mobile model is adopted for the nodes and sensor nodes in the Internet. The cloud devices of cloud platform have the same configuration and deploy evenly. The basic parameters of the integrated network and WSN network are as follows: the maximum communication radius of nodes is 250 m, and the fusion network size is 800 m × 900 m. The average value of the results of the 50 experiments was used as the final experimental results.

In the experiment, we compared and analyzed the performance of the proposed algorithm and VOR algorithm [19].

In the comparison of the experimental results, the change of the distance between two extremes and the sensor is changed, and the effect of the algorithm is analyzed.

When the number of the Internet nodes and the number of sensor nodes are different or equal, the coverage performance of the proposed algorithm and the VOR algorithm have been shown in Figure 5. The experimental results are as follows.

(1) With the increase of the size of nodes, the coverage area of the two algorithms is increased.

(2) When the number of nodes and the number of sensor nodes are different, the coverage performance of the proposed algorithm is better than the VOR algorithm. Because the proposed algorithm uses the opportunity sensor coverage method, the local node coverage information and the global coverage of WSN can be deployed and connected with the Internet to achieve global optimal coverage.

(3) With the increase of the number of iterations, the proposed algorithm based on cloud service area has a stronger coverage restoration and stable full connectivity.

With the different node of the mobile speed of the Internet, which executed 30 iterations and recover update, the coverage of the 2 algorithms of the full connectivity ratio is shown in Figure 6. The proposed algorithm provides a transparent interactive platform through the cloud platform to build a middleware layer between the Internet and WSN. There is mapping relationship between the network nodes and sensor nodes, which weaken the influence of the node moving speed and communication distance on the whole network. And the nodes can adaptively adjust the moving speed with the help of the cooperative cloud service sensing strategy. The network algorithm of full connectivity ratio was significantly higher

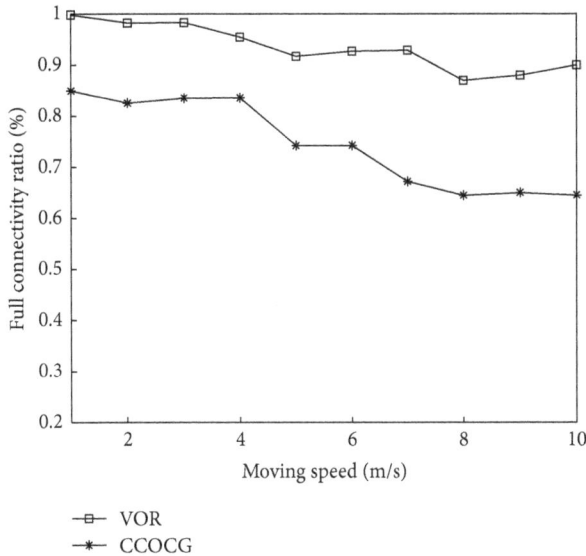

FIGURE 6: Full connectivity ratio.

than that in the VOR algorithm, and, in highly mobile nodes request, still has high communication rate.

6. Conclusions

How to improve the whole connectivity of the Internet through the integration of cloud platform and WSNs becomes an important technology of Internet application development. First, we studied the sensor opportunistic coverage mechanism based on the scale of Internet nodes, cloud, and sensor nodes. Second, according to the requirements of users and cloud platform computing ability, the cooperative cloud service aware model was given. Third, mobile Internet coverage connectivity guarantee protocol was proposed for resolving the mobility management, resource management and reliable information forwarding, and so forth. Mathematical analysis and simulation verification proved that the proposed scheme is superior to the VOR algorithm, such as coverage rate and full connectivity ratio, as well as connectivity guarantee price.

Conflict of Interests

The authors declare that there is no conflict of interests regarding the publication of this paper.

Acknowledgments

This work is supported in part by Key Scientific Research Project of Henan Province (15A520054) and Science and Technology Project of Henan Province (112102310550).

References

[1] B. Xie and A. Kumar, "A protocol for efficient Bi-directional connectivity between ad hoc networks and internet," *Journal of Internet Technology*, vol. 6, no. 1, pp. 101–108, 2005.

[2] N. Bartolini, G. Bongiovanni, T. F. La Porta, and S. Silvestri, "On the vulnerabilities of the virtual force approach to mobile sensor deployment," *IEEE Transactions on Mobile Computing*, vol. 13, no. 11, pp. 2592–2605, 2014.

[3] A. Sathiaseelan and J. Crowcroft, "Internet on the move: challenges and solutions," *Computer Communication Review*, vol. 43, no. 1, pp. 51–55, 2013.

[4] X. Ma, Y. Wang, and W. Sun, "Feverfew: a scalable coverage-based hybrid overlay for internet-scale pub/sub networks," *Science China Information Sciences*, vol. 57, no. 5, pp. 1–14, 2014.

[5] Z. Zheng, P. Sinha, and S. Kumar, "Sparse WiFi deployment for vehicular internet access with bounded interconnection gap," *IEEE/ACM Transactions on Networking*, vol. 20, no. 3, pp. 956–969, 2012.

[6] X. Huang and J. Chen, "Efficient node coverage scheme of wireless sensor network under multi-constraint conditions," *Journal of Networks*, vol. 8, no. 10, pp. 2277–2284, 2013.

[7] S. Savazzi, V. Rampa, and U. Spagnolini, "Wireless cloud networks for the factory of things: connectivity modeling and layout design," *IEEE Internet of Things Journal*, vol. 1, no. 2, pp. 180–195, 2014.

[8] S. G. Hong, S. Seo, H. Schulzrinne et al., "ICOW: internet access in public transit systems," *IEEE Communications Magazine*, vol. 53, no. 6, pp. 134–141, 2015.

[9] M. Panda and D. P. Mohapatra, "Generating test data for path coverage based testing using genetic algorithms," in *Proceedings of International Conference on Internet Computing and Information Communications*, vol. 216 of *Advances in Intelligent Systems and Computing*, pp. 367–379, Springer, New Delhi, India, 2014.

[10] D. Zhao, H. Ma, L. Liu, and X.-Y. Li, "Opportunistic coverage for urban vehicular sensing," *Computer Communications*, vol. 60, pp. 71–85, 2015.

[11] V. M. Rohokale, S. Inamdar, N. R. Prasad, and R. Prasad, "Energy efficient four level cooperative opportunistic communication for wireless personal area networks (WPAN)," *Wireless Personal Communications*, vol. 69, no. 3, pp. 1087–1096, 2013.

[12] S. Misra, T. Ojha, and A. Mondal, "Game-theoretic topology control for opportunistic localization in sparse underwater sensor networks," *IEEE Transactions on Mobile Computing*, vol. 14, no. 5, pp. 990–1003, 2015.

[13] R. Tynan, C. Muldoon, G. O'Hare, and M. O'Grady, "Coordinated intelligent power management and the heterogeneous sensing coverage problem," *Computer Journal*, vol. 54, no. 3, pp. 490–502, 2011.

[14] Q. Qi, J. Liao, and Y. Cao, "Cloud service-aware location update in mobile cloud computing," *IET Communications*, vol. 8, no. 8, pp. 1417–1424, 2014.

[15] S. Nizamani and A. Kumari, "A quality-aware computational cloud service for computational modellers," *Communications in Computer and Information Science*, vol. 414, pp. 184–194, 2014.

[16] L. Chunlin and L. LaYuan, "Cost and energy aware service provisioning for mobile client in cloud computing environment," *The Journal of Supercomputing*, vol. 71, no. 4, pp. 1196–1223, 2015.

[17] S. S. Savas, F. Dikbiyik, M. F. Habib, M. Tornatore, and B. Mukherjee, "Disaster-aware service provisioning with many-casting in cloud networks," *Photonic Network Communications*, vol. 28, no. 2, pp. 123–134, 2014.

[18] W. Dou, X. Zhang, J. Liu, and J. Chen, "HireSome-II: towards privacy-aware cross-cloud service composition for big data applications," *IEEE Transactions on Parallel and Distributed Systems*, vol. 26, no. 2, pp. 455–466, 2015.

[19] B. Wang, J. Qu, X. Wang, G. Wang, and M. Kitsuregawa, "VGQ-Vor: extending virtual grid quadtree with Voronoi diagram for mobile k nearest neighbor queries over mobile objects," *Frontiers of Computer Science*, vol. 7, no. 1, pp. 44–54, 2013.

A Chaos-Based Encryption Scheme for DCT Precoded OFDM-Based Visible Light Communication Systems

Zhongpeng Wang[1,2] and Shoufa Chen[1]

[1]*School of Information and Electronic Engineering, Zhejiang University of Science and Technology, Hang Zhou 310023, China*
[2]*State Key Laboratory of Millimeter Waves, Southeast University, Nanjing 210096, China*

Correspondence should be addressed to Zhongpeng Wang; wzp1966@sohu.com

Academic Editor: Maher Jridi

This paper proposes a physical encryption scheme for discrete cosine transform (DCT) precoded OFDM-based visible light communication systems by employing chaos scrambling. In the proposed encryption scheme, the Logistic map is adopted for the chaos mapping. The chaos scrambling strategy can allocate the two scrambling sequences to the real (I) and imaginary (Q) parts of OFDM frames according to the initial condition, which enhance the confidentiality of the physical layer. The simulation experimental results prove the efficiency of the proposed encryption method for DCT precoded OFDM-based VLC systems. The experimental results show that the proposed security scheme can protect the DCT precoded OFDM-based VLC from eavesdropper, while keeping the advantage of the DCT precoding technique, which can reduce the PAPR and improve the BER performance of OFDM-based VLC.

1. Introduction

Visible light communication (VLC) using light emitting diodes (LEDs), where the LEDs are used for both illumination and data wireless transmission, has received increasing attention among the researchers worldwide [1, 2]. Its distinct advantages are the license-free light spectrum, immunity to radio frequency (RF) interference, safety to human body, and the use of inexpensive light emitting diodes (LEDs). VLC can be viewed as a complement to RF in the face of the looming spectrum crunch [3].

Due to the greater immunity to multipath fading and reducing the complexity of equalizer, orthogonal frequency division multiplexing (OFDM) is one of the most popular techniques for high data rate communication. It has been accepted in the IEEE802.11a local area network (LAN), IEEE802.16 WiMax, digital audio broadcasting (DAB), and digital video broadcasting (DVB) and next generation mobile technologies 3GPP LTE [4]. To achieve higher spectral efficiency, OFDM is being considered as a crucial technique for indoor VLC systems. The most practical communication scheme for VLC systems is intensity modulation (IM) along with direct detection (IM/DD), of which the transmitted signal is usually modulated on the instantaneous power of light emitting diodes (LEDs) at the transmitters and photodiodes (PDs) are used at photoelectric converters at the receivers. Therefore, the generated OFDM time domain signal in OFDM-based IM/DD should be real-valued and nonnegative. This can be achieved by enforcing Hermitian symmetry constraint on the input data of the IFFT at the transmitters in order to generate real-valued signal. Depending on how the real-valued signal is converted to a nonnegative signal, three schemes have been proposed: (1) DC-biased optical OFDM (DCO-OFDM), (2) Asymmetrically-Clipped optical OFDM (ACO-OFDM), and (3) pulse-amplitude-modulated discrete multitone (PAM-DMT). For DCO-OFDM scheme, a DC offset is added to the real-valued signal to obtain a nonnegative signal [5]. For improving the throughput and capacity and/or the power efficiency, MIMO techniques, which offer a spatial gain, have been used in VLC. Meantime, in order to achieve higher bit error rate (BER) performance, precoding technique has been employed for OFDM systems, such as DCT precoding and DFT precoding [6, 7].

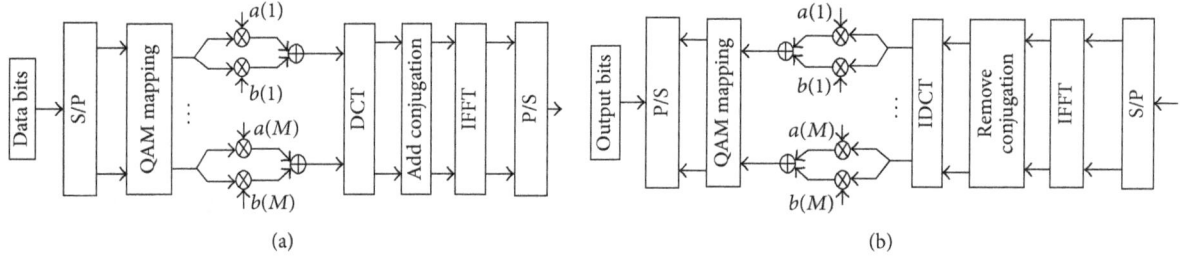

FIGURE 1: (a) Encryption principle for transmitter. (b) Decryption principle for receiver.

On the other hand, with the dramatic growth in wireless network capacity and accessibility of wireless network, data privacy and confidentiality are becoming a major concern for users. In some public areas such as classrooms, hallway, and planes, the transmitted signals in VLC link are susceptible to the eavesdropping. Many security measures can be adopted at upper layer of the network stack via access control and end-to-end encryption. For instance, the use of encryption end decryption scheme (DES, AES, etc.) can be implemented in the application layer of communication networks. During the past few years, however, the use of physical layer security techniques has attracted wide attention of scholars. Some physical secure strategies have been proposed in wireless communication and optical OFDM communication systems for fiber link [8–12]. Among these proposed schemes, chaos mapping techniques have usually been employed to enhance the security of physical layer. Recently, some related works in [13–16] considered improving the confidentiality of VLC links via physical security methods. However, to the best of our knowledge, the secure researching on precoded OFDM-based VLC systems has not been reported.

Without loss of generality, we will focus on DCO-OFDM-based VLC based in this paper. We firstly propose a novel secure DCT precoded OFDM-based VLC using chaos scrambling technique. In the proposed DCT precoded and encrypted OFDM-based VLC, the transmission security can be realized by the chaos scrambling sequence while the reliability can be improved via using DCT precoding. The simulation results show that the transmitted signal cannot be recovered at the eavesdropper due to the unknown secure key consisting of the initial value, the bifurcation parameter, and iteration step. Thus, the approach can provide scalable secure strategy in DCT precoded OFDM-based VLC application.

This paper is organized as follows. In Section 2, OFDM encryption scheme is introduced and described briefly. In Section 3, the system principle of encrypted and DCT precoded OFDM-based VLC is described. After that, simulation results and analysis are shown. Finally, Section 4 concludes this paper.

2. OFDM Encryption Scheme

In our proposed secure transmission scheme, the OFDM signal is encrypted using chaotic sequence. The chaos sequence is generated based on a Logistic map, which is controlled by the initial value and iteration parameter. A chaos model using Logistic map has the following iterative formula [17]:

$$x(n+1) = f(x(n)) = \mu x(n)(1 - x(n)), \quad (1)$$

where n is a time index, $x(0)$ is the initial value, $x(n)$ is the nth state value of (2), $x(n) \in (-1, 1)$, and $\mu \in [1, 4]$. μ is the bifurcation parameter or control parameter. When μ falls into the domain $3.569945 < \mu \leq 4$, the behavior changing of $x(n)$ will fall into chaos. To obtain a chaos sequence, a transform function is used to x_n; it is expressed as

$$s(n) = \begin{cases} -1, & 0 \leq x(n) \leq 0.5, \\ 1, & 0 \leq x(n) \leq 1, \end{cases} \quad (2)$$

where $s(n)$ is the nth element of the generated chaotic sequence. The chaos sequence can be obtained by (1) and (2). In the practical applications, some initial iterated values $\{x(n), n = 1, 2, \ldots, N\}$ are abandoned, where N is iteration step.

Based on (1) and (2), we can get two difference chaos sequences containing values form 1 and −1. The two chaos sequences are employed to encrypt the real and image parts of QAM symbol sequence, respectively. In this scheme, the secure key consists of the initial value x_0, the bifurcation parameter μ, and iteration step N. We assume that a and b are the generated chaos sequences where $a(m) \in \{-1, 1\}$ and $b(m) \in \{-1, 1\}$, respectively.

In the transmitter end, the transmitted data vector S with M length after the encryption can be represented as follows:

$$Z(m) = \text{real}(S(m)) \cdot a(m) + \text{imag}(S(m)) * b(m), \quad (3)$$

where $m = 1, 2, \ldots, M$. This is shown in Figure 1(a).

The receiver can decrypt the encrypted data by using its own key sequence. The decryption process can be written as follows:

$$S(m) = \text{real}(Z'(m)) \cdot a(m) + \text{imag}(Z'(m)) \cdot b(m), \quad (4)$$

where $Z'(m)$ is the output of the fast Fourier transform (FFT). The process is shown in Figure 1(b).

We assume that the M-order chaos scrambling sequence is generated to encrypt I and Q parts of frequency information signal. Here the randomness of Logistic chaos mapping

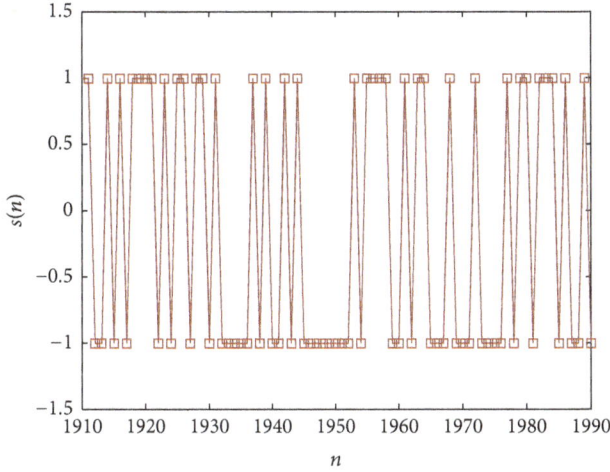

FIGURE 2: Chaos sequence for $\mu = 4$, $x_0 = 0.329999$, and $N = 2000$.

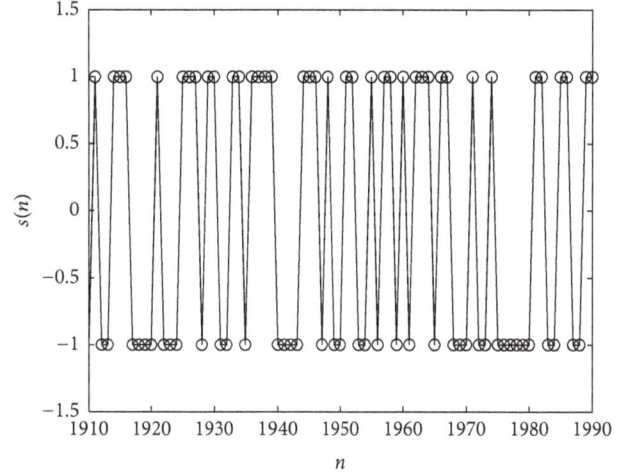

FIGURE 3: Chaos sequence for $\mu = 4$, $x_0 = 0.329998$, and $N = 2000$.

can be evaluated in terms of autocorrelation and cross-correlation functions of the chaos sequence. The normalized autocorrelation and cross-correlation functions of a random sequence with length M are defined as

$$R_{ac} = \frac{1}{M} \sum_{m=0}^{M-1} b_j(m) b_j(m+k),$$

$$-(M-1) \leq k \leq M-1, \quad (5)$$

$$R_{cc} = \frac{1}{M} \sum_{m=0}^{M-1} b_h(m) b_j(m+k),$$

$$-(M-1) \leq k \leq M-1.$$

Figures 2 and 3 show the chaos behavior of two different chaos sequences, which are generated by two different initial conditions, for example, with $\mu = 4$, $x_0 = 0.329999$, and $N = 2000$ and $\mu = 4$, $x_0 = 0.329998$, and $N = 2000$. We can see that the difference between the two waveforms of sequences is random. Figures 4 and 5 show the autocorrelation functions of the two chaos sequences, respectively. It can be seen that when lag $k \neq 0$ the value of the autocorrelation function of the chaos sequences is very small. Figure 6 shows the cross-correlation function of the chaos sequences for $x_0 = 0.329999$ and $x_0 = 0.329998$ and $N = 2000$. The values of cross-correlation functions are also around zero for all values of lag k. Therefore, the generated sequence by chaos mapping has very good random properties.

In our proposed scheme, the frequency information of OFDM signal is scrambled with chaos scrambling sequence to enhance the security of the physical layer of OFDM-based VLC. The scrambling sequence is generated from a Logistic mapping, in which the iteration parameters of Logistic mapping are used as security keys. Figure 7 illustrates the schematic of a DCT precoded OFDM-based visible light communications system with chaos scrambling sequence. The pseudorandom binary sequence (PRBS) information data is mapped into m-QAM data symbols and then goes through serial to parallel (S/P) transform. The generated

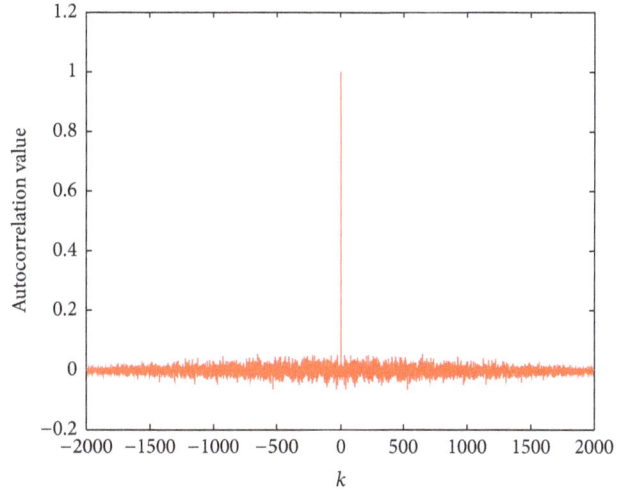

FIGURE 4: Autocorrelation of chaos sequence for $\mu = 4$, $x_0 = 0.329999$, and $N = 2000$.

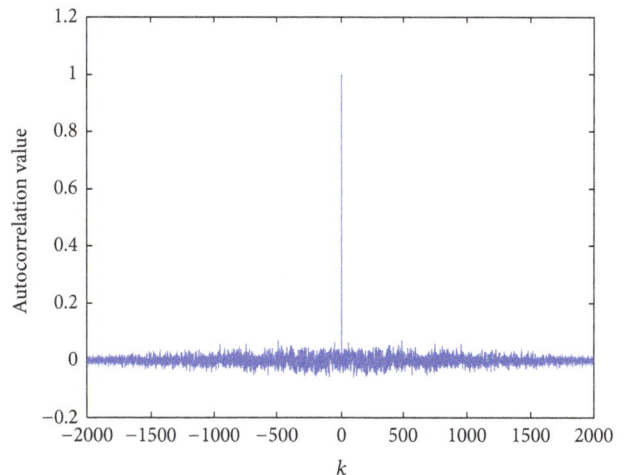

FIGURE 5: Autocorrelation of chaos sequence for $\mu = 4$, $x_0 = 0.329998$, and $N = 2000$.

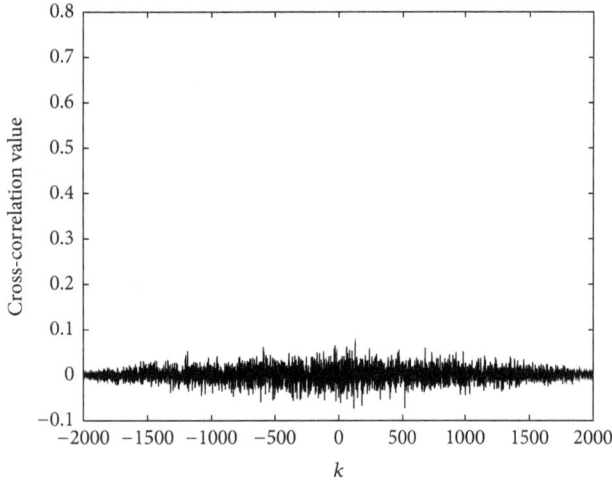

FIGURE 6: Cross-correlation of chaos sequences for $x_0 = 0.329998$ and $x_0 = 0.329999$, $\mu = 4$, and $N = 2000$.

complex vector of size M can be expressed as $S = [S(0) \; S(1) \; \cdots \; S(M-1)]^T$. After that the obtained m-QAM symbols are divided into I and Q parts. The encryption is performed by multiplying I and Q parts of the complex signal vector by a pair of chaos scrambling sequences separately according to (3).

The encrypted complex vector with size M can be written as $Z = [Z(0) \; Z(1) \; \cdots \; Z(M-1)]^T$. Then DCT precoding is applied to this complex vector which transforms this vector into new vector of length M that can be written as

$$Y = PZ = [Y(0) \; Y(1) \; \cdots \; Y(M-1)]^T, \qquad (6)$$

where $[\;]^T$ denotes the matrix transpose and P is DCT matrix with $M \times M$ dimension. The DCT precoded matrix can be stated as follows:

$$P = \begin{bmatrix} p_{00} & p_{01} & \cdots & p_{0(M-1)} \\ p_{10} & p_{11} & \cdots & p_{1(M-1)} \\ \vdots & \vdots & \ddots & \vdots \\ p_{(M-1)0} & p_{(M-1)1} & \cdots & p_{(M-1)(M-1)} \end{bmatrix}. \qquad (7)$$

The DCT kernel is real and can be expressed as

$$P_{lm} = \begin{cases} \dfrac{1}{\sqrt{M}}, & l = 0, \; 0 \le m \le M-1, \\[2ex] \sqrt{\left(\dfrac{2}{M}\right)} \cos\left[\dfrac{(2m+1)l\pi}{2M}\right], & 1 \le l \le M-1, \; 0 \le m \le M-1, \end{cases} \qquad (8)$$

where $l = 0, 1, \ldots, M-1$, $m = 0, 1, \ldots, M-1$ and p_{lm} means lth row and mth column of DCT precoding matrix.

In the receiver end, after FFT and equalization operation the inverse DCT precoding matrix is employed in receiver to recover the original data symbols. Assume that the chaotic map and secure key at the receiver are identical to those of the transmitter; they can provide sufficient information to generate identical chaotic scrambling sequences for decryption easily. Without knowledge of the secure key, the data cannot be recovered from the received signal by an eavesdropper.

In our proposed scheme, I and Q parts of QAM signal are both encrypted independently by chaos sequences. Therefore, there are two secret keys in our encryption algorithm. Every secret key of this chaos system contains three members, initial parameter x_0, control parameter μ, and iteration step N. By this, it can enhance secret key numbers and secret space. In our encrypted algorithm, the security keys can be expressed as $\{x_0^I, \mu^I, N^I, x_0^Q, \mu^Q, N^Q\}$, of which $\{x_0^I, \mu^I, N^I\}$ and $\{x_0^Q, \mu^Q, N^Q\}$ are the keys of I and Q parts of QAM signal, respectively.

The chaos state is highly sensitive to its initial values; only a slight change from the correct key value will fall into another absolute different chaotic state. This is beneficial for creating a huge key space which cannot be broken for illegal receiver. In our proposed system, there are four variables x_0^I, μ^I, x_0^Q, and μ^Q which are declared as Matlab type long. Every of the variables is scaled fixed point format with 15 digits precision for double. According to the IEEE floating-point standard [18], the computational precision of the 64-bit double-precision number is about 10^{-15}. After considering parameters involved, the key space size is approximately $10^{15 \times 4} \approx 2^{199}$, which is much larger than 2^{100}. Therefore, a sufficiently large key space is guaranteed in the proposed algorithm for application. The proposed encrypted scheme can efficiently resist the brute-force attack [19].

3. Simulation Results and Analysis

The chaos scrambling sequence is very import to ensure the security of the proposed encryption technique. In this work, we will evaluate the effect of scrambling sequence on the PAPR and BER performances of DCT precoded OFDM-VLC over multipath optical wireless channel by simulation. In the simulation setup, The OFDM frame structure has 192 data subcarriers and 8 pilot tones for channel estimation and equalization and 56 unused tones for the guard band. So the size of IFFT is 256. However, due to the Hermitian symmetry of input data of IFFT of DCO-OFDM systems, there are only 96 effective data subcarriers in OFDM frame. Therefore, the length of chaos scrambling sequence in the proposed scheme is set to 96. In the proposed encrypted scheme, the chaotic scrambling sequence can be generated based on Logistic map according to (1) and (2). In following simulation,

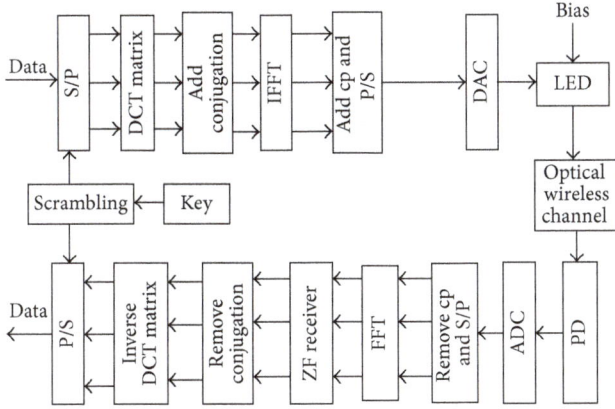

FIGURE 7: DCT precoded OFDM-based VLC system with encryption and decryption.

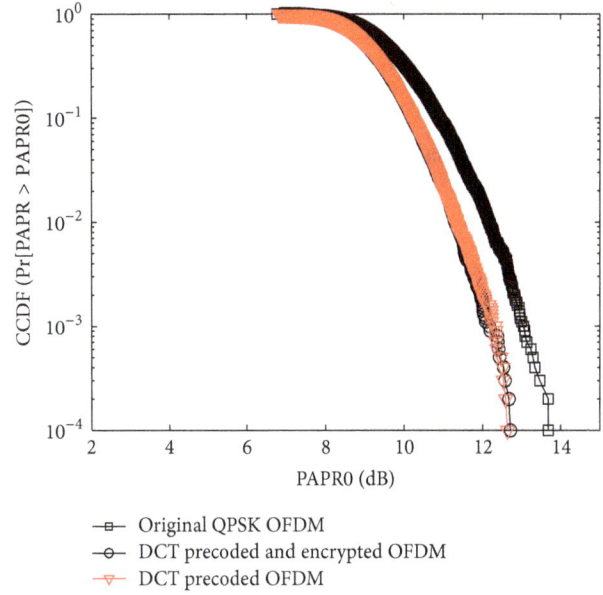

FIGURE 8: Comparison of the PAPRs of the precoded 16 QAM OFDM signals with and without chaos scrambling.

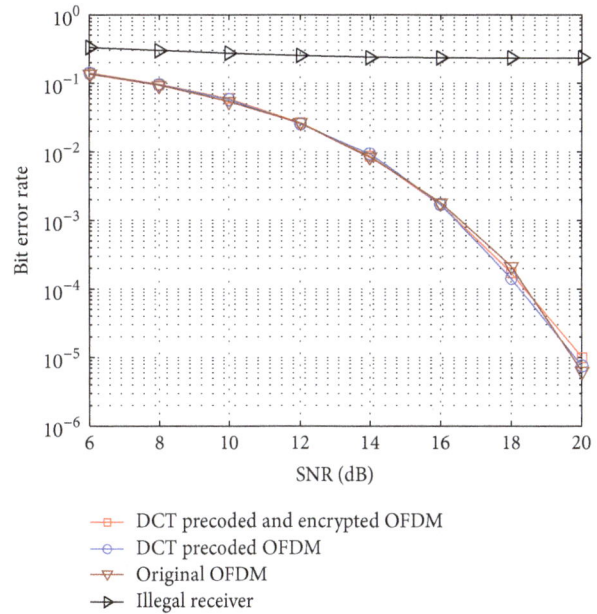

FIGURE 9: BER performance for AWGN channel.

the security keys of I or Q part of QAM signal can be fixed at $\{0.329999, 4, 2000\}$ and $\{0.329998, 4, 2000\}$.

3.1. PAPR Performance. One major drawback of OFDM is the peak-to-average power ratio (PAPR). It is verified that DCT precoding can reduce the PAPR and improve the BER performance of OFDM systems. In our proposed scheme, chaos scrambling is employed to improve security of physical layer of DCT precoded OFDM-based VLC. It is noteworthy to mention that the proposed encryption only changes the sign of the real and image parts of the transmitted symbols. Thus, the PAPR of the DCT precoded OFDM is kept unaffected. Therefore we mainly evaluated the effect of the chaos scrambling on the PAPR of OFDM signals in terms of the complementary cumulative distribution function (CCDF). Figure 8 shows the CCDF comparison of the PAPR of the scrambled and DCT precoded OFDM with that of conventional DCT precoded OFDM. From Figure 8 we can also see that the CCDF curves of the two cases are close to each other. Thus, the influence of the proposed encrypted scheme on the PAPR is negligible.

3.2. BER Performance. In following simulation experiment, the ceiling-bounce model developed by Carruthers and Kahn in [20] is chosen as the optical wireless channel model. This model is the most practical model and accurately represents the multipath dispersion of an indoor wireless optical channel. A single infinite-plane reflector with Lambertian reflectance is assumed. The continuous impulse response of an optical wireless link $h(t)$ is defined as

$$h_c(t) = H(0) \frac{6a^6}{(t+a)^7} u(t), \qquad (9)$$

where $H(0)$ is the channel DC gain, $u(t)$ is the step function, $a = 2H/c$, H is the ceiling height above the transmitter, and c is the velocity of light. The delay spread of a channel is a remarkably accurate predictor of ISI-induced signal-to-noise ratio (SNR) penalties, which is independent of the particular time dependence of the impulse response of that channel. This channel $h(n)$ was employed in the simulation.

The sample rate of the channel is represented by the symbol rate R_s. We mainly study the BER performance of DCT precoded 16 QAM OFDM-based system with and without chaos scrambling for AWGN and multipath optical wireless channel. In our simulation, the main parameters are shown in Table 1.

Figure 9 shows the BER performance demodulated by legitimate receiver and illegal receiver in the additive white Gaussian noise (AWGN) channel. The BER performance of the DCT precoded OFDM is almost the same as that of the conventional DCT precoded OFDM for the legitimate

TABLE 1: Simulation parameters.

R_s	125 M symbols/s
Modulation	16 QAM
FFT size	256
Number of pilot data	8
Length of CP	32
Scrambling size	96
H	3.5 m

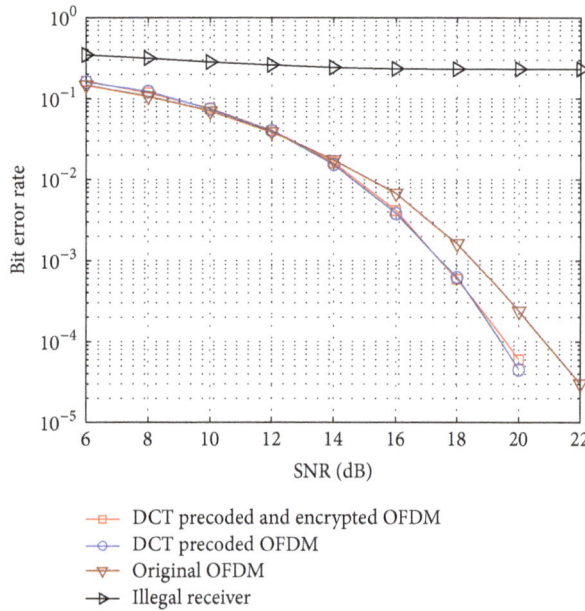

FIGURE 10: BER performance for multipath optical wireless channel.

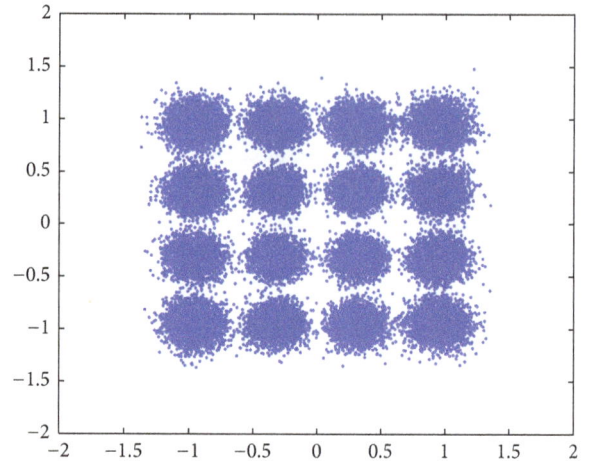

FIGURE 11: Received constellations of legitimate receiver for multipath optical wireless channel.

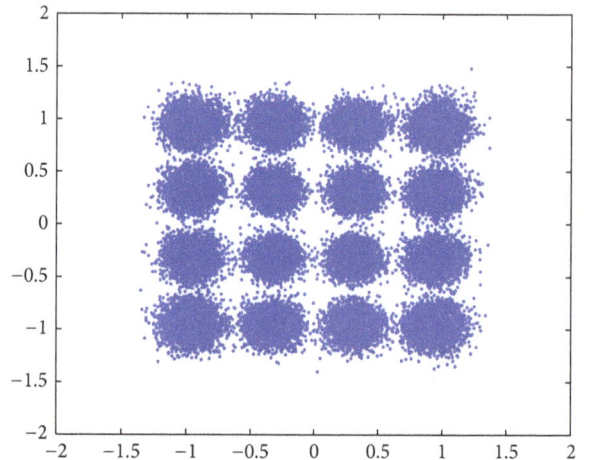

FIGURE 12: Received constellations of illegal receiver for multipath optical wireless channel.

receiver. Therefore, the encryption measure does not influence the BER performance of legal receiver. We also see that DCT precoding technique does not improve the BER of OFDM systems over AWGN channel. However, for the illegal receiver, the BER is around 0.5 because illegal receiver does not know the right secret key.

Figure 10 shows the BER performance comparison in a multipath VLC link. We can see that the BER performance of the encrypted and DCT precoded system is almost the same as that of the conventional DCT precoded system without encryption. For DCT precoded scheme with and without encryption, the improvement by DCT precoding can be clearly observed when the SNR is higher than 14 dB. At BER = 10^{-3}, the BER performance of the proposed DCT precoded system with chaos scrambling can be improved by an approximately 1 dB gain compared with that of the original OFDM system. The using of the chaos encryption maintains the advantage of DCT precoding technique, which can reduce the PAPR and improve the BER performance in OFDM systems. Form Figure 10, we can also see that illegal receiver cannot demodulate out the correct data if it does not know the right encryption keys. For the illegal receiver without right encryption keys, the BER is around 0.5. It is shown that

the proposed scheme can enhance both the reliability and security of OFDM-based VLC data transmission.

From Figure 10, it can be seen that the improvement of BER performance of the proposed DCT precoded and encrypted system is due to obtaining frequency diversity by using DCT precoding. Furthermore, in order to further improve the BER performance of the proposed system, channel coding such as convolution code and turbo-codes can be employed.

Figures 11 and 12 show the received constellations of legitimate receiver and illegal receiver for 16 QAM DCT precoded and encrypted OFDM signals, of which the signal-to-noise ratio is 18 dB, respectively. The legitimate receiver adopts chaos scrambling encryption technique while the illegal receiver does not adopt chaos scrambling encryption. Though the illegal receiver can obtain the right constellation it cannot demodulate the transmitted information due to have not the right encryption key.

4. Conclusions

In this paper, a physical layer encryption method is proposed to effectively enhance the security of a DCT precoded OFDM-based VLC system, where Logistic map is adopted to generate chaos scrambling sequences. The experimental results show that the chaos scrambling sequences can lead to a successful confidential data transmission in physical layer. Meanwhile, the proposed scheme does not influence the PAPR and BER performances of DCT precoded OFDM-based VLC. The advantage of DCT precoding technique can remain.

Competing Interests

The authors declare that they have no competing interests.

Acknowledgments

This work was supported in part by the Open Fund of the State Key Laboratory of Millimeter Waves (Southeast University, Ministry of Education, China) under K201214 and by the Zhejiang Provincial Natural Science Foundation of China under LY13F050005.

References

[1] D. Karunatilaka, F. Zafar, V. Kalavally, and R. Parthiban, "LED based indoor visible light communications: state of the art," *IEEE Communications Surveys and Tutorials*, vol. 17, no. 3, pp. 1649–1678, 2015.

[2] Y. Wang, N. Chi, Y. Wang, L. Tao, and J. Shi, "Network architecture of a high-speed visible light communication local area network," *IEEE Photonics Technology Letters*, vol. 27, no. 2, pp. 197–200, 2015.

[3] C. Rohner, S. Raza, D. Puccinlli, and T. Voigt, "Security in visible light communication: novel challenges and opportunities," *Sensors & Transducers*, vol. 192, no. 9, pp. 9–15, 2015.

[4] J. Liu, W. Noonpakdee, and S. Shimamoto, "Design and performance evaluation of OFDM-based wireless services employing radio over optical wireless link," *International Journal of Wireless & Mobile Networks*, vol. 3, no. 5, pp. 173–184, 2011.

[5] J. Tan, Z. Wang, Q. Wang, and L. Dai, "Near-optimal low-complexity sequence detection for clipped DCO-OFDM," *IEEE Photonics Technology Letters*, vol. 28, no. 3, pp. 233–236, 2016.

[6] B. Ranjha and M. Kavehrad, "Precoding techniques for PAPR reduction in asymmetrically clipped OFDM based optical wireless system," in *Broadband Access Communication Technologies VII, 86450R*, Proceedings of SPIE, International Society for Optics and Photonics, January 2013.

[7] L. Tao, J. Yu, Y. Fang, J. Zhang, Y. Shao, and N. Chi, "Analysis of noise spread in optical DFT-S OFDM systems," *Journal of Lightwave Technology*, vol. 30, no. 20, Article ID 6298919, pp. 3288–3294, 2012.

[8] T. Allen and N. Al-Dhahir, "Performance analysis of a secure STBC with coherent and differential detection," in *Proceedings of the IEEE Wireless Communications and Networking Conference (WCNC '15)*, pp. 522–527, IEEE, New Orleans, La, USA, March 2015.

[9] F. Huo and G. Gong, "A new efficient physical layer OFDM encryption scheme," in *Proceedings of the 33rd IEEE Conference on Computer Communications (IEEE INFOCOM '14)*, pp. 1024–1032, Toronto, Canada, May 2014.

[10] B. Liu, L. Zhang, X. Xin, and J. Yu, "Physical layer security in CO-OFDM transmission system using chaotic scrambling," *Optics Communications*, vol. 291, pp. 79–86, 2013.

[11] L. Deng, M. Cheng, X. Wang et al., "Secure OFDM-PON system based on chaos and fractional fourier transform techniques," *Journal of Lightwave Technology*, vol. 32, no. 15, pp. 2629–2635, 2014.

[12] W. Zhang, C. Zhang, W. Jin, C. Chen, N. Jiang, and K. Qiu, "Chaos coding-based QAM IQ-encryption for improved security in OFDMA-PON," *IEEE Photonics Technology Letters*, vol. 26, no. 19, pp. 1964–1967, 2014.

[13] A. Mostafa and L. Lampe, "Physical-layer security for indoor visible light communications," in *Proceedings of the 1st IEEE International Conference on Communications (ICC '14)*, pp. 3342–3347, Sydney, Australia, June 2014.

[14] H. Le Minh, A. T. Pham, Z. Ghassemlooy, and A. Burton, "Secured communications-zone multiple input multiple output visible light communications," in *Proceedings of the IEEE Globecom Workshops*, pp. 505–511, Austin, Tex, USA, December 2014.

[15] B. Zhang, K. Ren, G. Xing, X. Fu, and C. Wang, "SBVLC: secure barcode-based visible light communication for smartphones," in *Proceedings of the 33rd IEEE Conference on Computer Communications (IEEE INFOCOM '14)*, pp. 2661–2669, IEEE, Toronto, Canada, May 2014.

[16] A. Mostafa and L. Lampe, "Physical-layer security for MISO visible light communication channels," *IEEE Journal on Selected Areas in Communications*, vol. 33, no. 9, pp. 1806–1818, 2015.

[17] S.-L. Chen, T. T. Hwang, and W.-W. Lin, "Randomness enhancement using digitalized modified logistic map," *IEEE Transactions on Circuits and Systems II: Express Briefs*, vol. 57, no. 12, pp. 996–1000, 2010.

[18] IEEE Computer Society, *IEEE Standard for Binary Floating-Point Arithmetic*, ANSI/IEEE Standards 1985-754, 1985.

[19] G. Alvarez and S. Li, "Some basic cryptographic requirements for chaos-based cryptosystems," *International Journal of Bifurcation and Chaos in Applied Sciences and Engineering*, vol. 16, no. 8, pp. 2129–2151, 2006.

[20] J. B. Carruthers and J. M. Kahn, "Modeling of nondirected wireless infrared channels," *IEEE Transactions on Communications*, vol. 45, no. 10, pp. 1260–1268, 1997.

3

Adaptive Jamming Suppression in Coherent FFH System using Weighted Equal Gain Combining Receiver over Fading Channels with Imperfect CSI

Yishan He, Yufan Cheng, Gang Wu, Binhong Dong, and Shaoqian Li

National Key Laboratory of Science and Technology on Communications, University of Electronic Science and Technology of China, Chengdu 611731, China

Correspondence should be addressed to Yishan He; heyishan@live.com

Academic Editor: Tho Le-Ngoc

Fast frequency hopping (FFH) is commonly used as an antijamming communication method. In this paper, we propose efficient adaptive jamming suppression schemes for binary phase shift keying (BPSK) based coherent FFH system, namely, weighted equal gain combining (W-EGC) with the optimum and suboptimum weighting coefficient. We analyze the bit error ratio (BER) of EGC and W-EGC receivers with partial band noise jamming (PBNJ), frequency selective Rayleigh fading, and channel estimation errors. Particularly, closed-form BER expressions are presented with diversity order two. Our analysis is verified by simulations. It is shown that W-EGC receivers significantly outperform EGC. As compared to the maximum likelihood (ML) receiver in conventional noncoherent frequency shift keying (FSK) based FFH, coherent FFH/BPSK W-EGC receivers also show significant advantages in terms of BER. Moreover, W-EGC receivers greatly reduce the hostile jammers' jamming efficiency.

1. Introduction

As a powerful antijamming method, fast frequency hopping (FFH) is widely used in military applications. FFH employs a number of advantages including capability of antijamming, robustness against multipath fading, and low probability of interception [1, 2].

In the presence of jamming, various diversity combining schemes have been proposed for noncoherent frequency shift keying (FSK) based FFH, including maximum likelihood (ML) combining [1–7], FFT based combining schemes [8], linear combining (LC) [9], self-normalization combining [10], noise-normalization combining [11], product combining [12–17], and clipped combining [18, 19]. Among the noncoherent FFH/FSK combining schemes, ML combining yields the best BER performance in the presence of jamming.

In spite of the low complexity in implementation, noncoherent FFH systems have inevitable shortcomings, for example, performance loss due to noncoherent diversity combining. With the growing demand of better performance in antijamming communications, coherent phase shift keying

(PSK) based FFH system draws much attention. As indicated in [20] and the references therein, coherent reception has been made feasible by maintaining a continuous phase at the transmitter from hop to hop. Kang and Teh [20] studied the bit error ratio (BER) of coherent FFH/BPSK with partial band noise jamming (PBNJ) and AWGN channel. The authors considered coherent ML combining, LC combining, and hard-decision majority-vote combining, which significantly outperform various noncoherent FFH/FSK diversity combining schemes in terms of BER. However, the fading channels were not considered in [20]. In the presence of fading channels, we have proposed a novel FFH scheme [21], which enables reliable channel estimation for FFH signals. And we extended the study of [20] to the Rayleigh fading channels with imperfect channel state information (CSI) [22], where we analyzed the BER of FFH/BPSK with maximum ratio combining (MRC) and equal gain combining (EGC). It is illustrated that the two combining schemes have a close BER performance in the presence of PBNJ. However, the jamming suppression was not addressed in [21, 22].

This paper addresses the jamming suppression problem with coherent FFH/BPSK. In analysis, we consider PBNJ and frequency selective Rayleigh fading channels with imperfect CSI. Based on the studies of the EGC receiver [22], we give a further simplification on the BER expression. Then we propose adaptive jamming suppression schemes, namely, weighted EGC (W-EGC) with the optimum and suboptimum weighting coefficient, where the analytical BER expressions are also derived. Particularly, with diversity order $L = 2$, we work out closed-form BER expressions for the EGC and W-EGC. The theoretical results are validated by simulations. It is shown that the W-EGC receivers significantly outperform the noncoherent FFH/FSK ML receiver in terms of BER. It is also shown that with the increase of signal to jamming ratio (SJR), as compared with EGC, the optimum W-EGC lowers the error floor which is determined by the signal to noise ratio (SNR). Besides, W-EGC receivers reduce the hostile jammer's efficiency, by forcing the jammer to take full-band jamming to achieve the worst case jamming.

2. System Model

To guarantee reliable channel estimation, the so-called subset-based coherent FFH scheme [21] is adopted, where we partition the original hopping frequency set into a number of smaller subsets and choose only one of the frequency subsets as the hopping frequency set within a frame. The frame length T_f is designed to be shorter than the channel coherence time T_c. By controlling the subset size, the hopped frequencies are revisited within T_c, which makes channel estimation feasible.

In this paper, perfect synchronization and multipath fading channels are assumed. With a hopping rate sufficiently fast, the current hop received from the second path usually falls into a posterior hop. After dehopping and filtering, only the signal from the first path will be received. Note that each modulated symbol is L-fold hopped and the lth equivalent baseband-form received signal is given by

$$y_l = g_l s + n_l \quad \text{without PBNJ}$$
$$y_l' = g_l s + n_l + J_l \quad \text{with PBNJ,}$$

(1)

where y' denotes the received signal which is contaminated by PBNJ. The Rayleigh fading channel coefficient g_l is a zero mean complex Gaussian random variable (RV) with variance $2\sigma_g^2$. For the L hops of a modulated symbol, g_ls are independent and identically distributed (i.i.d.). The BPSK modulated symbol is denoted by s, $s = \pm\sqrt{P_d}$ with equal probability, where P_d is the instant power of s. The AWGN signal n_l is a zero mean complex Gaussian RV with variance $2\sigma_n^2$. The PBNJ signal J_l is also a zero mean complex Gaussian RV, with variance $2\sigma_J^2$. The jamming factor ρ_{PBNJ} is defined as the ratio of the jamming bandwidth to the entire hopping bandwidth, which is also the probability of a hop contaminated by PBNJ. Within a frame, if a hopped frequency f_i is disturbed by PBNJ, we assume that any hop with frequency f_i will be jammed.

Similar to [21, 22], the channel estimate is assumed to be disturbed by Gaussian errors, as

$$\hat{g}_l = g_l + \varepsilon_l \quad \text{without PBNJ}$$
$$\hat{g}_l' = g_l + \varepsilon_l' \quad \text{with PBNJ,}$$

(2)

where the estimation errors ε_l and ε_l' are zero mean complex Gaussian RVs with variances $2\sigma_\varepsilon^2$ and $2\sigma_{\varepsilon'}^2$, respectively, which both are independent of g_l. We have the following decomposition between $\hat{g}_l$ and g_l [23]:

$$g_l = (u_l + jv_l)\frac{\hat{g}_l}{|\hat{g}_l|} + \frac{R_c + jR_q}{\sigma_{\hat{g}}^2}\hat{g}_l,$$

(3)

where R_c and R_q are the second order moment between the real and imaginary part of $\hat{g}_l$ and g_l, as

$$R_c = \mathbb{E}\{\Re(\hat{g}_l)\,\Re(g_l)\} = \mathbb{E}\{\Im(\hat{g}_l)\,\Im(g_l)\} = \sigma_g^2,$$
$$R_q = \mathbb{E}\{\Re(\hat{g}_l)\,\Im(g_l)\} = -\mathbb{E}\{\Im(\hat{g}_l)\,\Re(g_l)\} = 0,$$

(4)

where $\mathbb{E}\{x\}$ is the expectation of x, $\Re(x)$ is the real part of x, and $\Im(x)$ is the imaginary part of x. u_l and v_l are i.i.d. zero mean Gaussian RVs, which are both independent of $\hat{g}_l$. The variance of u_l or v_l is $\sigma_e^2 = \sigma_g^2(1-|\rho|^2)$, where ρ is the complex correlation coefficient between $\hat{g}_l$ and g_l [23]:

$$\rho = \frac{\mathbb{E}\{g_l\hat{g}_l^*\}}{\sqrt{\mathbb{E}\{g_l g_l^*\}\,\mathbb{E}\{\hat{g}_l\hat{g}_l^*\}}} = \frac{R_c + jR_q}{\sigma_g\sigma_{\hat{g}}} = \frac{\sigma_g}{\sigma_{\hat{g}}}.$$

(5)

From (5), $|\rho| = \rho = \sigma_g/\sigma_{\hat{g}}$. Considering the similarity between PBNJ and AWGN, there is a similar decomposition between g_l and $\hat{g}_l'$, with $\rho' = \sigma_g/\sigma_{\hat{g}'}$ and $\sigma_{e'}^2 = \sigma_g^2(1-\rho'^2)$. For each single hop, we define the average SNR and the signal to jamming plus noise ratio (SJNR) as

$$\bar{\gamma} = \frac{\sigma_g^2 P_d}{\sigma_n^2},$$
$$\bar{\gamma}' = \frac{\sigma_g^2 P_d}{\sigma_n^2 + \sigma_J^2}.$$

(6)

Considering the influence of channels estimation error, we further define the effective SNR and the effective SJNR as

$$\bar{\gamma}_\rho = \frac{\rho^2\bar{\gamma}}{1 + \bar{\gamma}(1-\rho^2)},$$
$$\bar{\gamma}_{\rho'} = \frac{\rho'^2\bar{\gamma}'}{1 + \bar{\gamma}'(1-\rho'^2)}.$$

(7)

3. Performance Analysis of EGC Receiver

In this section, we first derive the BER of FFH/BPSK with EGC receiver, which further simplifies the results obtained in [22]. Then we calculate a closed-form BER expression for the case with $L = 2$.

3.1. BER for an Arbitrary L. In the presence of PBNJ, the EGC output is

$$r_{\text{EGC}} = \sum_{l=1}^{M} \frac{\widehat{g}_l'^*}{|\widehat{g}_l'|} y_l' + \sum_{l=M+1}^{L} \frac{\widehat{g}_l^*}{|\widehat{g}_l|} y_l, \tag{8}$$

where M is the number of jammed hops of a symbol.

With the BPSK constellation, the decision statistic is the real part of the combining output. Error occurs with $\Re(r_{\text{EGC}}) < 0$ when $s = \sqrt{P_d}$ is transmitted. Therefore, given M and the set $\mathbf{G} = \{g_1, \ldots, g_L, \widehat{g}_1', \ldots, \widehat{g}_M', \widehat{g}_{M+1}, \ldots, \widehat{g}_L\}$, the conditional error probability is

$$P_{\text{EGC}}(M, \mathbf{G}) = \Pr\left(\Re(r_{\text{EGC}}) < 0 \mid s = \sqrt{P_d}, M, \mathbf{G}\right). \tag{9}$$

Using (1)–(5), the decision statistic $\Re(r_{\text{EGC}})$ is expanded as

$$\Re(r_{\text{EGC}}) = s\left(\rho^2 \sum_{l=M+1}^{L} |\widehat{g}_l| + \rho'^2 \sum_{l=1}^{M} |\widehat{g}_l'|\right)$$

$$+ s\left(\sum_{l=M+1}^{L} u_l + \sum_{l=1}^{M} u_l'\right) \tag{10}$$

$$+ \Re\left(\sum_{l=M+1}^{L} \frac{\widehat{g}_l^*}{|\widehat{g}_l|} n_l + \sum_{l=1}^{M} \frac{\widehat{g}_l'^*}{|\widehat{g}_l'|}(J_l + n_l)\right).$$

According to (10), given s, M, and $\mathbf{G}$, $\Re(r_{\text{EGC}})$ is conditional Gaussian distributed. Hence, the $P_{\text{EGC}}(M, \mathbf{G})$ of (9) is calculated to be

$$P_{\text{EGC}}(M, \mathbf{G}) = Q\left(\frac{\mathbb{E}\left(\Re(r_{\text{EGC}}) \mid s = \sqrt{P_d}, M, \mathbf{G}\right)}{\sqrt{\text{var}\left(\Re(r_{\text{EGC}}) \mid s = \sqrt{P_d}, M, \mathbf{G}\right)}}\right), \tag{11}$$

where $\text{var}(x)$ is the variance of x and $Q(x)$ is the Gaussian Q function calculated by

$$Q(x) = \frac{1}{\pi} \int_0^{\pi/2} \exp\left(-\frac{x^2}{2\sin^2\psi}\right) d\psi, \quad x > 0. \tag{12}$$

Using (10) and (11), we simplify $P_{\text{EGC}}(M, \mathbf{G})$ as

$$P_{\text{EGC}}(M, \mathbf{G}) = Q\left(\alpha V + \beta V'\right), \tag{13}$$

where

$$V = \frac{1}{\sigma_{\widehat{g}}} \sum_{l=M+1}^{L} |\widehat{g}_l|, \qquad V' = \frac{1}{\sigma_{\widehat{g}'}} \sum_{l=1}^{M} |\widehat{g}_l'|,$$

$$\alpha = \frac{\rho}{\lambda}, \qquad \beta = \frac{\rho'}{\lambda}, \tag{14}$$

$$\lambda = \sqrt{\frac{(L-M)\rho^2}{\overline{\gamma}_\rho} + \frac{M\rho'^2}{\overline{\gamma}_{\rho'}}}.$$

By defining $Z = \alpha V + \beta V'$, the characteristic function (CHF) of Z is given by

$$\varphi_Z(t) = \left(1 - \sqrt{\frac{\pi}{2}} \alpha t e^{-\alpha^2 t^2/2}\left(-j + \text{erfi}\left(\frac{\alpha t}{\sqrt{2}}\right)\right)\right)^{L-M}$$

$$\times \left(1 - \sqrt{\frac{\pi}{2}} \beta t e^{-\beta^2 t^2/2}\left(-j + \text{erfi}\left(\frac{\beta t}{\sqrt{2}}\right)\right)\right)^M, \tag{15}$$

where $\text{erfi}(x)$ is the imaginary error function.

After averaging $P_{\text{EGC}}(M, \mathbf{G})$ over the distribution of Z, we obtain the $P_{\text{EGC}}(M)$ as

$$P_{\text{EGC}}(M) = \frac{1}{2\pi} \int_{-\infty}^{\infty} \int_0^{\infty} Q(z) e^{-jtz} \varphi_Z(t) \, dz \, dt, \tag{16}$$

where the internal integration of (16) can be calculated in a closed form, as

$$\Xi(t) = \int_0^{\infty} Q(z) e^{-jtz} dz$$

$$= \frac{1}{\pi} \int_0^{\pi/2} \int_0^{\infty} \exp\left(-\frac{z^2}{2\sin^2\psi}\right) e^{-jtz} dz \, d\psi$$

$$= \frac{1}{\sqrt{2\pi}} \int_0^{\pi/2} \exp\left(-\frac{t^2 \sin^2\psi}{2}\right) \tag{17}$$

$$\cdot \left(1 - j\, \text{erfi}\left(\frac{t \sin\psi}{\sqrt{2}}\right)\right) \sin\psi \, d\psi$$

$$= -j\frac{1}{2t} + \frac{1}{2t} \exp\left(-\frac{t^2}{2}\right)\left(j + \text{erfi}\left(\frac{t}{\sqrt{2}}\right)\right).$$

Then $P_{\text{EGC}}(M)$ is simplified to be

$$P_{\text{EGC}}(M) = \frac{1}{2\pi} \int_{-\infty}^{\infty} \Xi(t) \varphi_Z(t) \, dt. \tag{18}$$

Since $\varphi_Z(t)$ involves the CHF of a Rayleigh sum, a closed form for $P_{\text{EGC}}(M)$ with an arbitrary L has not yet been available so far as we know [24]. Compared with the quad-slope integration given by [22], we simplify $P_{\text{EGC}}(M)$ to be a 1-tuple integration, which reduces the complexity of numerical calculation.

Finally, the average error probability of EGC receiver is

$$P_{\text{EGC}} = \sum_{M=0}^{L} \binom{L}{M} \rho_{\text{PBNJ}}^M \left(1 - \rho_{\text{PBNJ}}\right)^{L-M} P_{\text{EGC}}(M). \tag{19}$$

3.2. Closed-Form BER Expression with L = 2. In the special case with $L = 2$, we work out a closed-form expression for $P_{\text{EGC}}(M)$. When $M = 0$, $P_{\text{EGC}}(M = 0)$ is given by [25]

$$P_{\text{EGC}}(M = 0) = \frac{1}{2}\left(1 - \frac{\sqrt{\overline{\gamma}_\rho(2 + \overline{\gamma}_\rho)}}{1 + \overline{\gamma}_\rho}\right). \tag{20}$$

Similarly, $P_{\text{EGC}}(M = 2)$ is calculated to be

$$P_{\text{EGC}}(M = 2) = \frac{1}{2}\left(1 - \frac{\sqrt{\overline{\gamma}_{\rho'}(2 + \overline{\gamma}_{\rho'})}}{1 + \overline{\gamma}_{\rho'}}\right). \tag{21}$$

For the case with $M = 1$, we have $V \sim \text{Rayleigh}(1)$, $V' \sim \text{Rayleigh}(1)$, and $\lambda = \sqrt{\rho^2/\overline{\gamma}_\rho + \rho'^2/\overline{\gamma}_{\rho'}}$. Then $P_{\text{EGC}}(M = 1)$ is calculated by

$$P_{\text{EGC}}(M = 1) = \int_0^\infty \int_0^\infty Q(\alpha x + \beta y) f_V(x) f_{V'}(y) \, dx \, dy, \tag{22}$$

which can be further expanded with (12), as

$$P_{\text{EGC}}(M = 1)$$
$$= \frac{1}{\pi} \int_0^{\pi/2} \int_0^\infty \int_0^\infty \exp\left(-\frac{(\alpha x + \beta y)^2}{2\sin^2\theta}\right) f_V(x) \tag{23}$$
$$\cdot f_{V'}(y) \, dx \, dy \, d\theta.$$

We first solve the internal integral with x, as

$$A(y, \theta) = \int_0^\infty \exp\left(-\frac{(\alpha x + \beta y)^2}{2\sin^2\theta}\right) f_V(x) f_{V'}(y) \, dx$$
$$= y\left(1 + \alpha^2 \csc^2\theta\right)^{3/2} e^{-y^2(1+\beta^2\csc^2\theta)/2}$$
$$\times \left(\sqrt{1 + \alpha^2 \csc^2\theta}\right.$$
$$- \sqrt{2\pi} y \alpha \beta \csc^2\theta e^{y^2\alpha^2\beta^2\csc^4\theta/(2+2\alpha^2\csc^2\theta)}$$
$$\left.\cdot Q\left(\frac{y\alpha\beta\csc^2\theta}{\sqrt{1 + \alpha^2\csc^2\theta}}\right)\right). \tag{24}$$

Then we solve the internal integral with y, as

$$B(\theta) = \int_0^\infty A(y, \theta) \, dy$$
$$= \left(1 + \frac{2\beta^2}{1 + 2\alpha^2 - \cos 2\theta}\right)^{-3/2} \left(1 + \alpha^2\csc^2\theta\right)^{-5/2}$$
$$\cdot \left(1 + \beta^2\csc^2\theta\right)^{-1} \left(\frac{\csc\theta^2}{\alpha^2 + \sin^2\theta}\right)^{-1/2}$$
$$\times \left\{\alpha^2\beta^2\sqrt{1 + \frac{2\beta^2}{1 + 2\alpha^2 - \cos 2\theta}}\csc^6\theta\right.$$
$$- \sqrt{\frac{2\csc^2\theta}{1 + 2\alpha^2 - \cos 2\theta}}$$
$$\times \left[\alpha\beta \arctan \frac{\sqrt{1 + 2\beta^2/(1 + 2\alpha^2 - \cos 2\theta)}}{\alpha\beta\sqrt{\csc\theta^2/(\alpha^2 + \sin\theta^2)}}\right.$$
$$\cdot \csc^2\theta\left(1 + \alpha^2\csc^2\theta\right)\left(1 + \beta^2\csc^2\theta\right)$$

$$- \sqrt{1 + \frac{2\beta^2}{1 + 2\alpha^2 - \cos 2\theta}}$$
$$\left.\left.\cdot \sqrt{1 + \alpha^2\csc^2\theta}\left(1 + (\alpha^2 + \beta^2)\csc^2\theta\right)\right]\right\}. \tag{25}$$

Finally, we calculate the internal integral with θ, as

$$P_{\text{EGC}}(M = 1) = \frac{1}{\pi} \int_0^{\pi/2} B(\theta) \, d\theta$$
$$= \frac{1}{2}\left(1 - \frac{\alpha\sqrt{1 + \alpha^2} + \beta\sqrt{1 + \beta^2}}{1 + \alpha^2 + \beta^2}\right). \tag{26}$$

By substituting (20), (21), and (26) into (19), we obtain the closed-form BER expression for $L = 2$.

4. Adaptive Jamming Suppression Schemes

4.1. W-EGC Receiver. In the W-EGC receiver, the jammed and unjammed received signal are, respectively, sent into an EGC receiver. Then the two EGC outputs are weighted and combined, with final output as

$$r_{\text{W-EGC}} = \eta \sum_{l=1}^M \frac{\widehat{g}_l'^*}{|\widehat{g}_l'|} y_l' + \sum_{l=M+1}^L \frac{\widehat{g}_l^*}{|\widehat{g}_l|} y_l, \tag{27}$$

where η is the weighting coefficient. In the following analysis, we will optimize η to minimize the BER.

Similar to (13), the conditional error probability of the W-EGC receiver can be written as

$$P_{\text{W-EGC}}(M, \mathbf{G}) = Q\left(aV + \eta b V'\right), \tag{28}$$

where

$$a = \frac{\rho}{\mu}, \qquad b = \frac{\rho'}{\mu},$$
$$\mu = \sqrt{\frac{(L - M)\rho^2}{\overline{\gamma}_\rho} + \frac{\eta^2 M \rho'^2}{\overline{\gamma}_{\rho'}}}. \tag{29}$$

Note that the Gaussian Q function is a monotone decreasing function; that is, minimizing $P_{\text{W-EGC}}$ is equivalent to maximizing $aV + \eta b V'$. By solving $\partial(aV + \eta b V')/\partial\eta = 0$, the optimum η can be obtained as

$$\eta_{\text{opt}} = \frac{L - M}{M} \frac{\overline{\gamma}_{\rho'}}{\overline{\gamma}_\rho} \frac{\sum_{l=1}^M |\widehat{g}_l'|}{\sum_{l=M+1}^L |\widehat{g}_l|}. \tag{30}$$

After substituting (30) into (28), we simplify $P_{\text{W-EGC}}$ as

$$P_{\text{W-EGC}} = Q\left(\sqrt{c_1 V^2 + c_2 V'^2}\right), \tag{31}$$

where $c_1 = \overline{\gamma}_\rho/(L - M)$ and $c_2 = \overline{\gamma}_{\rho'}/M$. Once again, we use the CHF method and calculate the $P_{\text{W-EGC}}$ as

$P_{\text{W-EGC}}$

$$= \frac{1}{4\pi^2} \int_0^\infty \int_0^\infty \int_{-\infty}^\infty \int_{-\infty}^\infty Q\left(\sqrt{c_1 x^2 + c_2 y^2}\right)$$
$$\cdot e^{-jt_1 x} e^{-jt_2 y} \varphi_V(t_1)$$
$$\cdot \varphi_{V'}(t_2)\, dt_1 dt_2 dx\, dy$$

$$= \frac{1}{4\pi^2} \int_0^{\pi/2} \int_0^\infty \int_0^\infty \int_{-\infty}^\infty \int_{-\infty}^\infty \frac{1}{\pi} \exp\left(-\frac{c_1 x^2 + c_2 y^2}{2\sin^2\psi}\right)$$
$$\cdot e^{-j(t_1 x + t_2 y)} \varphi_V(t_1)$$
$$\cdot \varphi_{V'}(t_2)\, dt_1 dt_2 dx\, dy$$

$$= \frac{1}{4\pi^2} \int_0^{\pi/2} \int_{-\infty}^\infty \int_{-\infty}^\infty f(t_1, t_2, \psi)$$
$$\varphi_V(t_1) \varphi_{V'}(t_2)\, dt_1 dt_2 d\psi, \tag{32}$$

where

$$f(t_1, t_2, \psi) = \frac{1}{\pi}\left(\int_0^\infty \exp\left(-\frac{c_1 x^2}{2\sin^2\psi}\right) e^{-jt_1 x} dx\right)$$
$$\cdot \left(\int_0^\infty \exp\left(-\frac{c_2 y^2}{2\sin^2\psi}\right) e^{-jt_2 y} dy\right)$$
$$= -\frac{\sin^2\psi}{2\sqrt{c_1 c_2}} \exp\left(-\frac{1}{2}\left(\frac{t_1^2}{c_1} + \frac{t_2^2}{c_2}\right)\sin^2\psi\right)$$
$$\cdot \left(j + \text{erfi}\left(\frac{t_1 \sin\psi}{\sqrt{2c_1}}\right)\right) \tag{33}$$
$$\cdot \left(j + \text{erfi}\left(\frac{t_2 \sin\psi}{\sqrt{2c_2}}\right)\right),$$

$$\varphi_V(t_1) = \left(1 - e^{-t_1^2/2}\sqrt{\frac{\pi}{2}} t_1 \left(-j + \text{erfi}\left(\frac{t_1}{\sqrt{2}}\right)\right)\right)^{L-M},$$

$$\varphi_{V'}(t_2) = \left(1 - e^{-t_2^2/2}\sqrt{\frac{\pi}{2}} t_2 \left(-j + \text{erfi}\left(\frac{t_2}{\sqrt{2}}\right)\right)\right)^{M}.$$

Note that η is eliminated in (28) with $M = 0$ or $M = L$, which indicates that η_{opt} is used only for $1 \le M \le L - 1$. Due to the Rayleigh sum involved, a closed-form expression for $P_{\text{W-EGC}}$ is not available for an arbitrary L. Similar to (19), the BER for W-EGC is

$$P_{\text{W-EGC}} = \sum_{M=0}^L \binom{L}{M} \rho_{\text{PBNJ}}^M (1 - \rho_{\text{PBNJ}})^{L-M} P_{\text{W-EGC}}(M). \tag{34}$$

We would like to compare the BER between EGC and W-EGC. Due to the monotonicity of the Gaussian Q function,

we only need to compare the internal fraction of (13) and (31), which is calculated to be

$$\vartheta(V, V') = c_1 V^2 + c_2 V'^2 - (\alpha V + \beta V')^2$$
$$= \frac{M}{L-M} \frac{\rho^2}{(L-M)\rho^2 \overline{\gamma}_{\rho'} + M\rho'^2 \overline{\gamma}_\rho} \tag{35}$$
$$\cdot \left(\frac{\rho'}{\rho}\overline{\gamma}_\rho V - \frac{L-M}{M}\overline{\gamma}_{\rho'} V'\right)^2.$$

Take the expectation of $\vartheta(V, V')$ with regard to V and V', as

$$\mathbb{E}\{\vartheta(V, V')\} = \frac{M}{(L-M)\overline{\gamma}_{\rho'} + (\rho'^2/\rho^2)M\overline{\gamma}_\rho}$$
$$\cdot \left(\left(2 - \frac{\pi}{2}\right)\left[\left(\frac{L}{M} - 1\right)\overline{\gamma}_{\rho'}^2 + \overline{\gamma}_\rho^2 \frac{\rho'^4}{\rho^4}\right]\right.$$
$$\left. + \frac{\pi}{2}(L-M)\left(\overline{\gamma}_{\rho'} - \overline{\gamma}_\rho \frac{\rho'^2}{\rho^2}\right)^2\right) > 0. \tag{36}$$

From (36), it is shown that, as compared with EGC, W-EGC has a lower conditional error probability. Besides, with the increase of SJR, we have

$$\lim_{\sigma_J^2 \to 0} \mathbb{E}\{\vartheta(V, V')\} = \left(2 - \frac{\pi}{2}\right)\overline{\gamma}_\rho > 0. \tag{37}$$

From (37), it is indicated that, in a high SJR region, W-EGC will lower the error floor, which is determined by the SNR.

In the special case with $L = 2$, $P_{\text{W-EGC}}$ has a closed-form expression. For $M = 0$ and $M = 2$, we have $P_{\text{W-EGC}}(M = 0, 2) = P_{\text{EGC}}(M = 0, 2)$. For $M = 1$, $P_{\text{W-EGC}}(M = 1)$ is calculated to be

$P_{\text{W-EGC}}(M = 1)$

$$= \int_0^\infty \int_0^\infty Q\left(\sqrt{\overline{\gamma}_\rho x + \overline{\gamma}_{\rho'} y}\right) f_X(x) f_Y(y)\, dx\, dy$$
$$= \frac{1}{4\pi} \int_0^{\pi/2} \int_0^\infty \int_0^\infty \exp\left(-\frac{\overline{\gamma}_\rho x + \overline{\gamma}_{\rho'} y}{2\sin^2\psi} - \frac{x+y}{2}\right) dx\, dy\, d\psi$$
$$= \frac{1}{\pi} \int_0^{\pi/2} \frac{1}{(1 + \overline{\gamma}_\rho \csc^2\psi)(1 + \overline{\gamma}_{\rho'} \csc^2\psi)} d\psi$$
$$= \frac{1}{2}\left(1 - \sqrt{\frac{\overline{\gamma}_{\rho'}}{1 + \overline{\gamma}_{\rho'}}} \frac{\overline{\gamma}_{\rho'}}{\overline{\gamma}_{\rho'} - \overline{\gamma}_\rho} + \sqrt{\frac{\overline{\gamma}_\rho}{1 + \overline{\gamma}_\rho}} \frac{\overline{\gamma}_\rho}{\overline{\gamma}_{\rho'} - \overline{\gamma}_\rho}\right). \tag{38}$$

4.2. Suboptimum W-EGC Receiver. Note that the optimum weighting coefficient of (30) contains the instantaneous channel estimates. To achieve a simpler result, we replace the $|\hat{g}_l|$ and $|\hat{g}_l'|$ with the corresponding mathematical expectations $\sqrt{\pi/2}\sigma_{\hat{g}}$ and $\sqrt{\pi/2}\sigma_{\hat{g}'}$, respectively. The suboptimum weighting coefficient η_{Sopt} is then calculated to be

$$\eta_{\text{Sopt}} = \frac{\overline{\gamma}_{\rho'}}{\rho'\overline{\gamma}_\rho}. \tag{39}$$

The conditional error probability of the suboptimum W-EGC (SW-EGC) receiver is calculated to be

$$P_{\text{SW-EGC}}(M, \mathbf{G}) = Q\left(\alpha_S V + \beta_S V'\right), \tag{40}$$

where

$$\alpha_S = \frac{\rho\overline{\gamma}_\rho}{\lambda_S}, \qquad \beta_S = \frac{\overline{\gamma}_{\rho'}}{\lambda_S},$$
$$\lambda_S = \sqrt{\rho^2(L-M)\overline{\gamma}_\rho + M\overline{\gamma}_{\rho'}}. \tag{41}$$

Since $P_{\text{SW-EGC}}(M, \mathbf{G})$ and $P_{\text{EGC}}(M, \mathbf{G})$ are in similar form, $P_{\text{SW-EGC}}(M)$ can be simply obtained by substituting (41) into (18). Similarly, the closed-form expressions for $P_{\text{SW-EGC}}(M, \mathbf{G})$ with $L = 2$ are given by $P_{\text{SW-EGC}}(M = 0, 2) = P_{\text{EGC}}(M = 0, 2)$ and

$$P_{\text{SW-EGC}}(M = 1) = \frac{1}{2}\left(1 - \frac{\alpha_S\sqrt{1+\alpha_S^2} + \beta_S\sqrt{1+\beta_S^2}}{1+\alpha_S^2+\beta_S^2}\right). \tag{42}$$

Similar to (35), the comparison between the SW-EGC and EGC can be denoted as $\vartheta_S(V, V') = (\alpha_S - \alpha)V + (\beta_S - \beta)V'$. In the high SJR region, we have

$$\lim_{\sigma_J^2 \to 0} \mathbb{E}\left\{\vartheta_S\left(V, V'\right)\right\} = \left(\frac{\rho(L-M)+M}{\sqrt{\rho^2(L-M)+M}} - \sqrt{L}\right)\sqrt{\frac{\pi}{2}\overline{\gamma}_\rho}. \tag{43}$$

For a moderate large SNR, that is, $\rho^2 \to 1$, we can see

$$\lim_{\sigma_J^2 \to 0, \rho^2 \to 1} \mathbb{E}\left\{\vartheta_S\left(V, V'\right)\right\} = 0. \tag{44}$$

From (44), it is indicated that, with a moderate large SNR, the SW-EGC and EGC receivers approach exactly the same error floor.

5. Numerical Results and Discussions

In this section, we present some numerical results and corresponding discussions. In simulation, the channel coefficient for each frequency and the location of jammed bandwidth are assumed to be unchanged within a frame. The simulation parameters are given in Table 1.

In Figure 1, the analytical results match the simulation very well. Both W-EGC and SW-EGC outperform EGC in terms of BER, especially in the low E_b/J_0 region. When

TABLE 1: Simulation parameters.

Type	Value
Frequency subset size	15
Frame length	0.9 ms
Hopping rate	1×10^5 hops/s
Number of pilot groups	2
Pilot power P_p	$P_p = P_d$
Fading type	Frequency selective Rayleigh fading
$2\sigma_g^2$	1
Jamming type	PBNJ
Channel estimation method	Pilot-based LS estimation

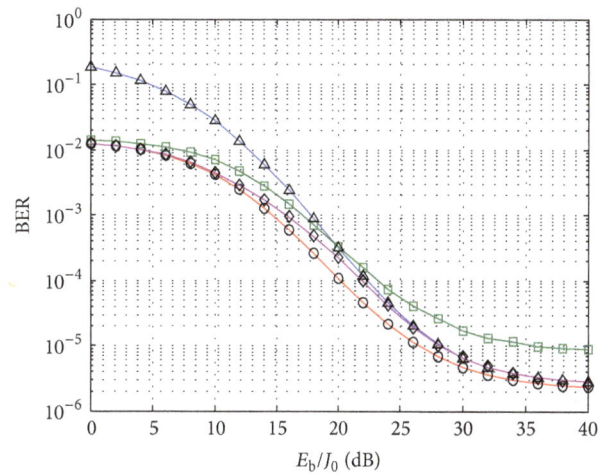

	FFH/BPSK W-EGC	Analytical
○	FFH/BPSK W-EGC	Simulation
	FFH/BPSK SW-EGC	Analytical
◇	FFH/BPSK SW-EGC	Simulation
	FFH/BPSK EGC	Analytical
△	FFH/BPSK EGC	Simulation
⊟	FFH/BFSK ML	Simulation

FIGURE 1: BER comparison, with $E_b/N_0 = 25$ dB, $L = 3$, and $\rho_{\text{PBNJ}} = 0.3$.

BER $= 1 \times 10^{-3}$, for example, W-EGC and SW-EGC show, respectively, 3.5 dB and 2 dB gain over EGC. In a high E_b/J_0 region, W-EGC shows a lower error floor than that of EGC, while the SW-EGC receiver approaches the same error floor, which have been explained in Section 4. As compared to the noncoherent FFH/BFSK with ML receiver, which is the optimum receiver for the noncoherent FFH in the presence of jamming, coherent FFH/BPSK with W-EGC and SW-EGC receivers shows performance gain even with imperfect CSI. As seen in Figure 1, the performance gain is 2.5 dB and 1 dB, respectively, when BER $= 1 \times 10^{-3}$. And this performance improvement increases with the increase of E_b/J_0.

Figure 2 shows the influence of the jamming factor ρ_{PBNJ} on the performance of EGC, W-EGC, and SW-EGC, respectively. For the EGC receiver, there is a worst case jamming factor, which is less than 1. With the worst case jamming factor, the hostile jammer achieves the worst case jamming effect by jamming only a small fraction of the bandwidth. In

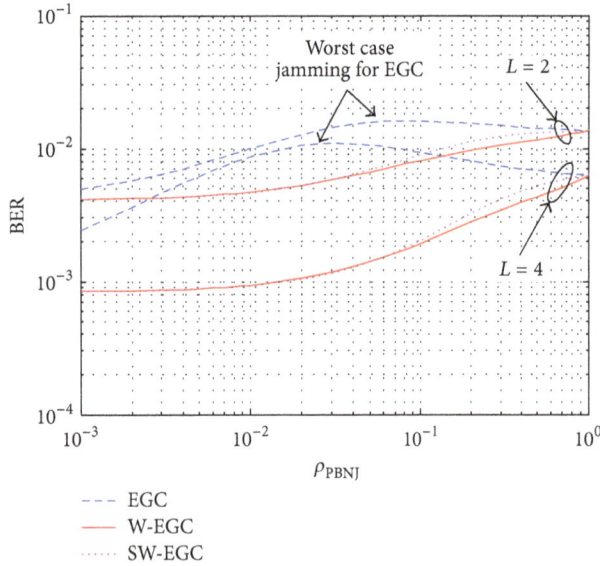

FIGURE 2: BER of FFH/BPSK with W-EGC and EGC versus ρ_{PBNJ} for $L = 2$ and 4, $E_b/N_0 = 15$ dB, and $E_b/J_0 = 15$ dB.

contrast, as seen in Figure 2, the worst case jamming factor for the W-EGC and SW-EGC receiver is 1. Then the jammer has to take full-band jamming to achieve the worst case jamming, whose jamming effectiveness is greatly reduced.

6. Conclusion

We have proposed jamming suppression schemes for coherent FFH/BPSK system, which is based on the weighted equal gain combining. And we analyzed the BER performance in the presence of PBNJ, frequency selective Rayleigh fading channel, and imperfect CSI. From theoretical analysis and simulation validation, it is shown that the proposed schemes significantly outperform EGC and noncoherent ML receiver in terms of BER. Besides, the proposed schemes greatly reduce the jammer's efficiency, where the jammer has to implement full-band jamming to achieve the worst case jamming.

Conflict of Interests

The authors declare that there is no conflict of interests regarding the publication of this paper.

Acknowledgments

This work is supported in part by Natural Science Foundation of China (NSFC) Grant no. 61201126, Ministry Sponsored Project 4010103020201-2, Program for New Century Excellent Talents in University (no. NCET-11-0058), and Program for Sichuan Youth Science and Technology Fund (no. 2012JQ0020).

References

[1] L.-M.-D. Le, K. C. Teh, and K. H. Li, "Jamming rejection using FFH/MFSK ML receiver over fading channels with the presence of timing and frequency offsets," *IEEE Transactions on Information Forensics and Security*, vol. 8, no. 7, pp. 1195–1200, 2013.

[2] F. Yang and L.-L. Yang, "A single-user noncoherent combining scheme achieving multiuser interference mitigation for FFH/MFSK systems," *IEEE Transactions on Wireless Communications*, vol. 12, no. 9, pp. 4306–4314, 2013.

[3] G. Li, Q. Wang, V. K. Bhargava, and L. J. Mason, "Maximum-likelihood diversity combining in partial-band noise," *IEEE Transactions on Communications*, vol. 46, no. 12, pp. 1569–1574, 1998.

[4] J. Zhang, K. C. Teh, and K. H. Li, "Performance analysis of a maximum-likelihood FFH/MFSK receiver with partial-band-noise jamming over frequency-selective fading channels," *IEEE Communications Letters*, vol. 12, no. 6, pp. 401–403, 2008.

[5] J. Zhang, K. C. Teh, and K. H. Li, "Error probability analysis of FFH/MFSK receivers over frequency-selective Rician-fading channels with partial-band-noise jamming," *IEEE Transactions on Communications*, vol. 57, no. 10, pp. 2880–2885, 2009.

[6] J. Zhang, K. C. Teh, and K. H. Li, "Maximum-likelihood FFH/MFSK receiver over rayleigh-fading channels with composite effects of MTJ and PBNJ," *IEEE Transactions on Communications*, vol. 59, no. 3, pp. 675–679, 2011.

[7] L.-M. Le, K. C. Teh, and K. H. Li, "Performance analysis of a suboptimum fast frequency-hopped/*M*-ary frequency-shift-keying maximum-likelihood receiver over Rician-fading channels with composite effects of partial-band noise jamming and multitone jamming," *IET Communications*, vol. 6, no. 13, pp. 1903–1911, 2012.

[8] K. C. Teh, A. Kot, and K. Li, "Partial-band jammer suppression in FFH spread-spectrum system using FFT," *IEEE Transactions on Vehicular Technology*, vol. 48, no. 2, pp. 478–486, 1999.

[9] R. Robertson and K. Y. Lee, "Performance of fast frequency-hopped mfsk receivers with linear and self-normalization combining in a rician fading channel with partial-band interference," *IEEE Journal on Selected Areas in Communications*, vol. 10, no. 4, pp. 731–741, 1992.

[10] R. Robertson and T. Ha, "Error probabilities of fast frequency-hopped fsk with self-normalization combining in a fading channel with partial-band interference," *IEEE Journal on Selected Areas in Communications*, vol. 10, no. 4, pp. 714–723, 1992.

[11] C.-D. Chung and P.-C. Huang, "Effects of fading and partial-band noise jamming on a fast FH/BFSK acquisition receiver with noise-normalization combination," *IEEE Transactions on Communications*, vol. 44, no. 1, pp. 94–104, 1996.

[12] K. C. Teh, A. C. Kot, and K. H. Li, "Partial-band jamming rejection of FFH/BFSK with product combining receiver over a Rayleigh-Fading channel," *IEEE Communications Letters*, vol. 1, no. 3, pp. 64–66, 1997.

[13] T. C. Lim, W. He, and K. H. Li, "Rejection of partial-band noise jamming with FFH/BFSK product combining receiver over Nakagami-fading channel," *Electronics Letters*, vol. 34, no. 10, pp. 960–961, 1998.

[14] G. Huo and M.-S. Alouini, "Another look at the BER performance of FFH/BFSK with product combining over partial-band jammed Rayleigh-fading channels," *IEEE Transactions on Vehicular Technology*, vol. 50, no. 5, pp. 1203–1215, 2001.

[15] C.-L. Chang and T.-M. Tu, "Performance analysis of FFH/BFSK product-combining receiver with partial-band jamming over independent rician fading channels," *IEEE Transactions on Wireless Communications*, vol. 4, no. 6, pp. 2629–2635, 2005.

[16] S. Ahmed, L.-L. Yang, and L. Hanzo, "Mellin-transform-based performance analysis of FFH M-ary FSK using product combining for combatting partial-band noise jamming," *IEEE Transactions on Vehicular Technology*, vol. 57, no. 5, pp. 2757–2765, 2008.

[17] J. Zhang, K. C. Teh, and K. H. Li, "Performance study of fast frequency-hopped/M-ary frequency-shift keying systems with timing and frequency offsets over Rician-fading channels with both multitone jamming and partial-band noise jamming," *IET Communications*, vol. 4, no. 10, pp. 1153–1163, 2010.

[18] C. M. Keller and M. B. Pursley, "Clipped diversity combining for channels with partial-band interference—part I: clipped-linear combining," *IEEE Transactions on Communications*, vol. 35, no. 12, pp. 1320–1328, 1988.

[19] L. Xiao, X. Xu, and Y. Yao, "Adaptive threshold clipper combining receiver for fast frequency hopping systems during partial-band noise jamming," *Tsinghua Science & Technology*, vol. 6, no. 1, pp. 42–44, 2001.

[20] J. J. Kang and K. Teh, "Performance of coherent fast frequency-hopped spread-spectrum receivers with partial-band noise jamming and AWGN," *IEE Proceedings: Communications*, vol. 152, no. 5, pp. 679–685, 2005.

[21] Y. He, Y. Cheng, Y. Yang, G. Wu, B. Dong, and S. Li, "A subset-based coherent FFH system," *IEEE Communications Letters*, vol. 19, no. 2, pp. 199–202, 2015.

[22] Y. He, Y. Cheng, G. Wu, and S. Li, "Performance analysis of FFH/BPSK system with partial band noise jamming and channel estimation error in high-mobility wireless communication scenarios," *Chinese Science Bulletin*, vol. 59, no. 35, pp. 5011–5018, 2014.

[23] R. Annavajjala and L. B. Milstein, "Performance analysis of linear diversity-combining schemes on Rayleigh fading channels with binary signaling and Gaussian weighting errors," *IEEE Transactions on Wireless Communications*, vol. 4, no. 5, pp. 2267–2278, 2005.

[24] J. Hu and N. C. Beaulieu, "Accurate simple closed-form approximations to rayleigh sum distributions and densities," *IEEE Communications Letters*, vol. 9, no. 2, pp. 109–111, 2005.

[25] A. Annamalai, C. Tellambura, and V. K. Bhargava, "Equal-gain diversity receiver performance in wireless channels," *IEEE Transactions on Communications*, vol. 48, no. 10, pp. 1732–1745, 2000.

4

Performance Analysis of Homogeneous On-Chip Large-Scale Parallel Computing Architectures for Data-Parallel Applications

Xiaowen Chen,[1,2] Zhonghai Lu,[2] Axel Jantsch,[3] Shuming Chen,[1] Yang Guo,[1] Shenggang Chen,[1] and Hu Chen[1]

[1]*College of Computer, National University of Defense Technology, Changsha, Hunan 410073, China*
[2]*Department of Electronic Systems, KTH-Royal Institute of Technology, Kista, 16440 Stockholm, Sweden*
[3]*Institute of Computer Technology, Vienna University of Technology, 1040 Vienna, Austria*

Correspondence should be addressed to Xiaowen Chen; xiaowenc@kth.se

Academic Editor: Dimitrios Soudris

On-chip computing platforms are evolving from single-core bus-based systems to many-core network-based systems, which are referred to as *On-chip Large-scale Parallel Computing Architectures (OLPCs)* in the paper. Homogenous OLPCs feature strong regularity and scalability due to its identical cores and routers. Data-parallel applications have their parallel data subsets that are handled individually by the same program running in different cores. Therefore, data-parallel applications are able to obtain good speedup in homogenous OLPCs. The paper addresses modeling the speedup performance of homogenous OLPCs for data-parallel applications. When establishing the speedup performance model, the network communication latency and the ways of storing data of data-parallel applications are modeled and analyzed in detail. Two abstract concepts (*equivalent serial packet and equivalent serial communication*) are proposed to construct the network communication latency model. The uniform and hotspot traffic models are adopted to reflect the ways of storing data. Some useful suggestions are presented during the performance model's analysis. Finally, three data-parallel applications are performed on our cycle-accurate homogenous OLPC experimental platform to validate the analytic results and demonstrate that our study provides a feasible way to estimate and evaluate the performance of data-parallel applications onto homogenous OLPCs.

1. Introduction and Motivation

As technology advances, on-chip computing platforms are evolving from single-core bus-based systems to many-core network-based systems, which feature integrating a number of computing cores that run in parallel and adopting an on-chip network that provides concurrent pipelined communication. The many-core network-based systems are referred to as *On-chip Large-scale Parallel Computing Architectures (OLPCs)* in the paper. OLPCs can be highly homogeneous or irregular and heterogeneous. Homogenous OLPC owns its characteristics of strong regularity and scalability, since processor cores and routers in it are the same. Each processor core has the same computation capability. As one way of parallel processing, data parallelism partitions data into several blocks that are mapped to different processors and processors

work in SPMD (Single Program Multiple Data) mode, that is, they handle their own data blocks by running the same program. Data-parallel applications have the parallel data set that can be partitioned in parallel into data subsets and each data subset can be handled individually by the same program and has marginal synchronization overhead, so they are well scalable and can be used to exploit the potential of multiple computing cores. Therefore, homogenous OLPCs and data-parallel applications match each other well. Data-parallel applications are able to obtain good speedup on homogenous OLPCs. Therefore, the focus of the paper is to provide a workable way to estimate and evaluate the performance of homogenous OLPCs with data-parallel applications.

Scalability is one of the important features of homogenous OLPCs. In homogenous OLPCs, as the network size is scaled up, the network communication latency is increasing and becoming one of the most significant factors affecting the system performance. Therefore, we firstly propose two abstract concepts: *equivalent serial packet* and *equivalent serial communication*, and then we construct a detailed network communication latency model. Then, based on Amdahl's Law, we propose a performance model including the detailed network communication latency. Two traffic models (*uniform* and *hotspot*) are used to reflect the two ways of storing data of data-parallel applications. The uniform traffic model matches the *distributed* way that data are equally distributed into all nodes, while the hotspot traffic model matches the *centralized* way that data are only maintained in the central node. Our models also analyze the performance impact of the noncommunication/communication ratio. Some useful suggestions are presented during the performance model's analysis. Finally, we map three data-parallel applications (Wavefront Computation, Vector Norm, and Block Matching Algorithm in Motion Estimation) on our cycle-accurate homogenous OLPC experimental platform to validate and demonstrate our performance analysis.

The contributions of the paper are summarized as follows.

(1) Since homogenous OLPCs match data-parallel applications well and vice versa, our study exhibits a workable way to formulate and evaluate the speedup performance of data-parallel applications onto homogenous OLPCs before application programming and hardware design.

(2) Two abstract concepts, *equivalent serial packet* and *equivalent serial communication*, are proposed and then used to construct the detailed network communication latency model (see Section 4.3).

(3) Based on Amdahl's Law, we propose a performance model of homogeneous OLPCs for data-parallel applications (see Section 4.4). The proposed performance includes the proposed network communication latency model and adopts two traffic models (*uniform* and *hotspot*) so as to have two forms (see Sections 4.4.1 and 4.4.2). They, respectively, reflect the distributed way and the centralized way of storing data of data-parallel applications.

(4) A cycle-accurate homogenous OLPC experimental platform is built up and three real data-parallel applications are mapped to validate the effectiveness of the proposed performance model.

The rest of the paper is organized as follows. Section 2 presents the background and related work. Section 3 discusses the characteristics of homogenous OLPCs and data-parallel applications and their relationship. Section 4 proposes the communication latency model and the performance model of homogenous OLPCs and details the analysis. Section 5 maps three data-parallel applications on our homogenous OLPC platform to validate the effectiveness of the performance model. Section 6 discusses the applicabilities and the limitations of our performance model. Finally, we conclude in Section 7.

2. Background and Related Work

The development of on-chip computation presents two trends. One is towards a growing number of processors integrated on a chip [1, 2]. it is moving away from a sequential to a parallel paradigm leading to tens, dozens, hundreds, and soon even thousands of computing cores on a single chip. A number of computing cores are potential to cooperate in parallel to obtain higher performance of parallel applications. The other trend is about the interconnection of on-chip resources. The communication infrastructure is developing into a similarly parallel structure, which is often called a Network-on-Chip (NoC) [3–5]. Shared, serial buses are replaced by pipelined communication networks that allow hundreds or thousands of communications going on concurrently at any time. Combining the two trends, on-chip computing platforms are evolving from single-core bus-based systems to many-core network-based systems, which are referred to as *On-chip Large-scale Parallel Computing Architectures (OLPCs)* in the paper. Understanding the speedup potential that OLPC computing platforms can offer is a fundamental question to continually pursuing higher performance.

With respect to performance analysis, Amdahl's Law [6] provides a simple, yet very useful method to evaluate the performance of a parallel system. Its fundamental hypothesis is that the computation problem size does not change when running on enhanced parallel systems. Its main result shows that the percentage of the serial portion dominates the speedup limit. Amdahl's Law is a pessimistic view that the speedup does not increase infinitely along with the increase of the number of paralleled processor cores. Based on Amdahl's Law, many researchers discussed their variants for different purposes. In [7], Li and Malek discussed the effect of noncommunication/communication ratio on the speedup based on Amdahl's Law, but his communication delay model is simple without considering the detail of interconnects. In [8], Paul revisited Amdahl's Law on the single chip heterogeneous multiprocessor. His focus was on the performance impact induced by different types of processor cores with different processing capability. In [9], Cho and Melhem presented the corollaries to Amdahl's Law in order to study the interaction between parallelization and energy consumption. In [10], Hill and Marty offered a corollary of a simple model of multicore hardware resources based on Amdahl's Law. He proved that an enhanced core is necessary for the high system performance but the parallelism supported by systems with such cores suffers. In [11], Loh extended Hill's work to study the performance impact of uncore function units on the multicore system's throughput. In both Hill's and Loh's discussions, the effect of network communication latency is omitted. In OLPCs, the enhancement of application performance may be restricted by the increasing network communication latency, even though the number of cores increases. We note that less work aforementioned discusses the effect of network communication latency on the performance of OLPCs. In the paper, we detail the network communication latency by proposing two abstract concepts, *equivalent serial*

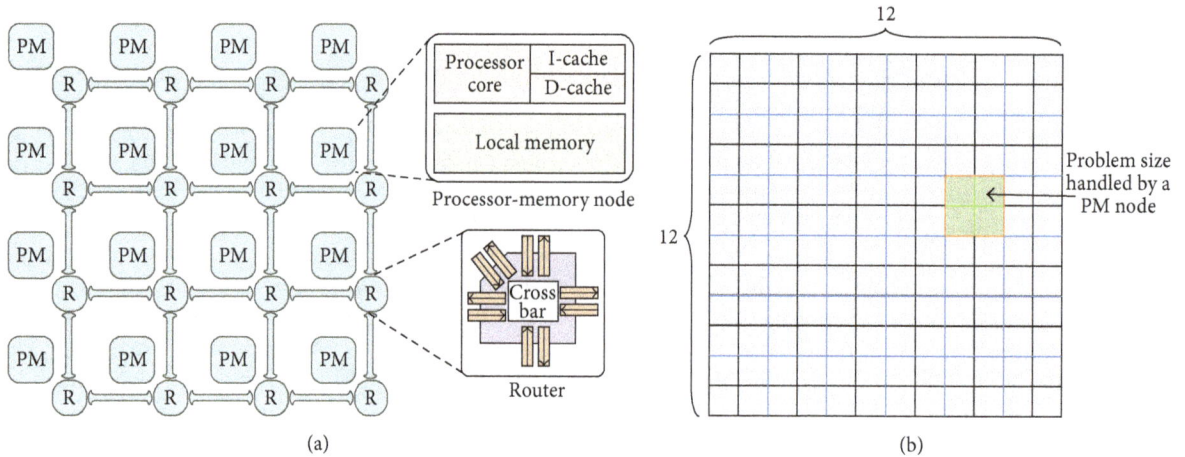

FIGURE 1: (a) Sketch map of homogenous OLPCs and (b) an example of data partitioning of data-parallel applications.

packet and *equivalent serial communication*, and establish the performance model of homogenous OLPCs. Our model, verified by real data-parallel applications, exhibits a workable way to estimate and evaluate the performance of homogenous OLPCs.

3. Homogenous OLPCs and Data-Parallel Applications

Homogenous OLPCs are a suitable architecture for data-parallel applications and vice versa. Regularity and scalability are the key features of homogenous OLPCs. Figure 1(a) shows an example of homogenous OLPCs. The communication infrastructure is a regular 2D-mesh NoC, which is the most popular NoC topology proposed today [12]. As we can see, the processor type and the local memory volume in each Processor-Memory (PM) node is the same so that each PM node has the same computation capability. All PM nodes are networked by routers. The network size is scalable. As one way of parallel processing, data parallelism partitions data into several blocks that are mapped to different processors. Processors handles their own data blocks by running the same program. Data parallelism is efficient for applications with high computation complexity (e.g., image processing, hydrodynamics computing). These data-parallel applications are well scalable and their data are regular. They are easily parallelized by partitioning their data. Figure 1(b) illustrates a data partitioning way of data-parallel applications. Assuming that there are 144 (12 × 12) data to be processed by a data-parallel application and the homogenous OLPC is with the network size of 36 (6 × 6). Since the computation ability of each PM node is the same, it is obvious to partition the 144 data into 36 equivalent parts. Each equivalent part contains 4 sets of data and is handled by a PM node. As the network size is scaled up and hence more PM nodes are included, we can repartition the data to suit the number of PM nodes in order to gain higher performance. However, the network communication limits the performance. We consider two

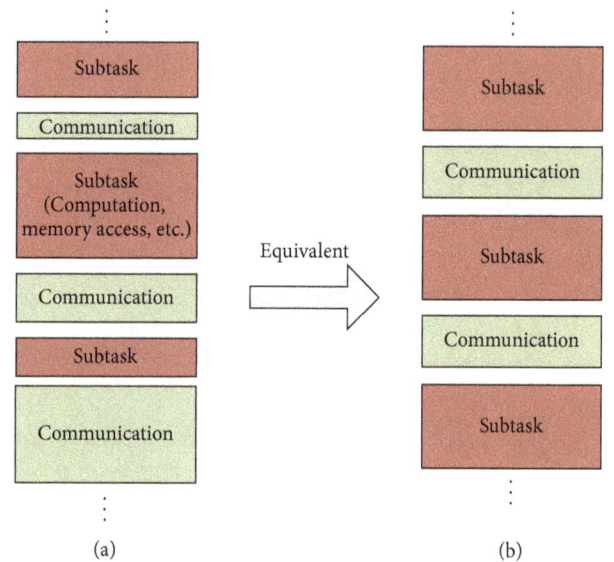

FIGURE 2: The subprogram running on a processor node is abstracted as a set of subtasks and communications.

traffic models which reflect two ways of storing data of data-parallel applications. The uniform traffic model matches the *distributed* way that data are distributed equally into all local memories of all nodes. The hotspot traffic model matches the *centralized* way that data are only maintained in the central node.

4. Models and Analysis

4.1. Problem Definition. The problem we consider is the performance in the context of homogeneous OLPCs for data parallel applications. We give detailed analysis on communication latency. The program running on OLPCs are divided into several subprograms running on different processor nodes. The subprogram can be abstracted as a set of subtasks and communications (see Figure 2(a)). The communication

TABLE 1: Calculation of Hop Count in k-ary-2-mesh.

Uniform	Hotspot
$H = 2\left(\dfrac{k}{3} - \dfrac{1}{3k}\right)$	$H = \begin{cases} \dfrac{k}{2} + \dfrac{k}{2\left(k^2-1\right)k} & k\text{: even} \\[2ex] \dfrac{k}{2} & k\text{: odd} \end{cases}$

denotes the interaction between two communicating processor nodes. A communication contains one or more packets transmitted in the network. The subtask denotes the noncommunication processing (e.g., computation, memory access, etc.) between two successive communications. To facilitate constructing the models of communication latency and the performance, we make the following three assumptions.

(1) The noncommunication time and communication time of the subprogram assigned to each node is equal to each other. That is, the subprogram in each node contains the same number of subtasks and communications.

(2) The execution time of each subtask is also equal to each other.

(3) The time of each communication is also equal to that of others.

Figure 2(b) is the reabstracted subprogram based on assumption (2) and (3). The sum of the subtasks and communications of Figure 2(b) is equal to that of Figure 2(a).

4.2. Notations. To facilitate the analysis, we first define a set of symbols in Notaions section.

4.3. Communication Latency Model. Communication latency contains two parts: minimal (noncontention) latency and contention latency.

The minimal latency is determined by the distance of the two communicating nodes. We use hop count to calculate the latency. Table 1 lists our calculated hop count following [13]. We consider two representative traffic models (Uniform and Hotspot) in 2D-mesh networks. For hotspot traffic, the central node is chosen as the hotspot node.

The contention latency mainly depends on the behavior of parallel applications running on OLPCs. In general, it is difficult to quantify the contention latency exactly. "When to communicate," "which processor core starts a message passing" and "where the destination is" lead to different contention latency. If no contention occurs, transmitting a packet in one hop takes 1 cycle ($\tau_{1hop} = 1$) in our experimental platform shown in Figure 8. However, network contention makes τ_{1hop} uncertain. Hence, in order to facilitate constructing the performance models, we consider the contention latency from another angle. Since network contention occurs only when multiple communications issued by different processor nodes appear simultaneously in the on-chip network, we introduce an abstract concept: *equivalent serial communication.* The equivalent serial communications are sequential when the program is running so that network contention

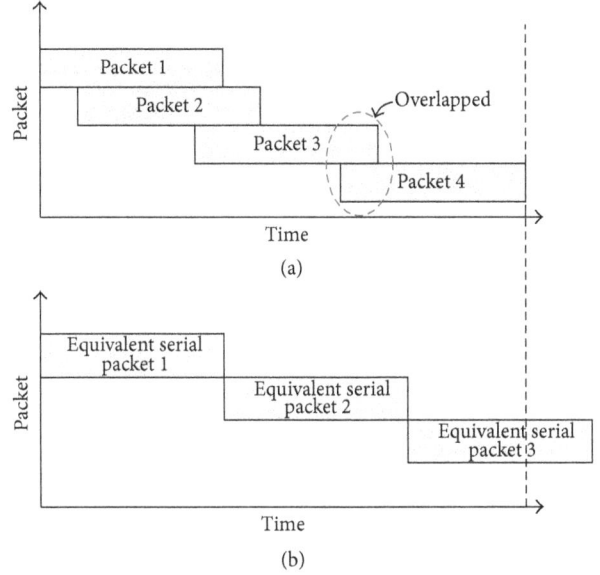

FIGURE 3: Packets in a communication issued by a processor node.

does not exist at all. To a certain extent, the number of equivalent serial communication (ω) reflects the network contention. Equivalent serial communication is discussed in detail in Step 3 below.

In the next, we use three steps to establish the communication latency model.

Step 1 (calculating the time of transmitting a packet). With packet switching, the average time of transmitting a packet in the network is

$$\tau_t = H \cdot \tau_{1hop}, \tag{1}$$

where H reflects the distance and τ_{1hop} reflects the architectural latency without contention.

Step 2 (calculating the time of a communication). In general, a communication issued by a processor node contains one or more packets. These packets are launched by the same processor node. Packet transmissions may overlap. In the best case, a packet in a communication is launched one cycle after the preceding packet. A packet transmits in the on-chip network without need of waiting for the completion of its preceding packet transmission. The packet transmissions are overlapping. For the worst case, all packets are transmitted serially; that is, a packet will not be transmitted until the previous one is finished. The overlap among packet transmissions improves the performance by shortening the network communication latency.

To measure the time of a communication, we define an abstract concept: *equivalent serial packet.* The equivalent serial packet is considered to be transmitted sequentially. A communication is abstracted to consist of several equivalent serial packets. As shown in Figure 3(a), assuming that the communication contains four packets, the program behavior determines the concurrent degree of packets' transmission.

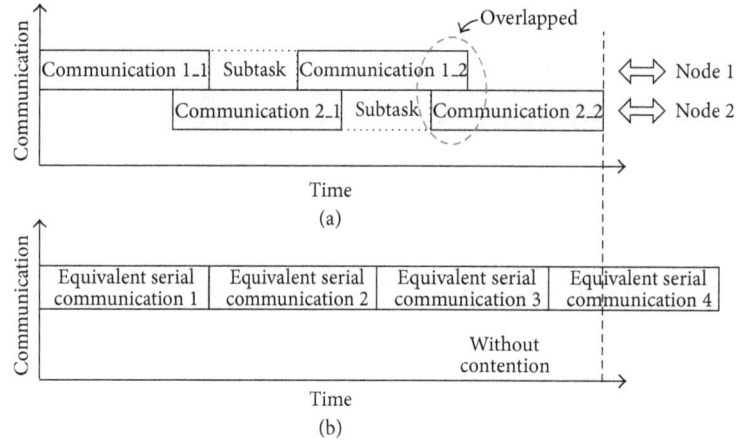

FIGURE 4: Communications in the entire program.

For example, Packet 1 and Packet 2 are almost fully overlapped, while small portion of Packet 3 and 4 are overlapped. For ease of measuring the communication time, the communication is abstracted to be composed of several equivalent serial packets. In Figure 3(b), the number of equivalent serial packets (γ) is about 2.67, which is less than the packet number: 4. γ meets the inequation below:

$$1 < \gamma \leqslant M. \tag{2}$$

γ describes the concurrent degree of packet transmission in a communication. The ideal best case is that all packets is transmitted concurrently. However, it cannot be reached, because there is only one physical channel from the node to the router. The best case is that packets in a communication are launched one cycle by one cycle, so γ is close to, but not equal to, 1. For the worst case that all packets transmit sequentially, $\gamma = M$. That means the number of equivalent serial packets is equal to the number of real packets (M) in a communication.

From (1) and (2), we can obtain the time of a communication:

$$\tau_c = \gamma \cdot \tau_t = \gamma \cdot H \cdot \tau_{1\text{hop}} \quad (1 < \gamma \leqslant M). \tag{3}$$

Step 3 (calculating the communication overhead of a program). The program is parallelized on N nodes, so the subprogram in each node contains p/N communications. Communications issued by the same node are sequential, because the subprogram is sequentially executed in the processor node. Communications issued by different nodes may exist in the network at the same time. For the best case, the program is fully parallelized. The communication overhead of the entire program is equal to communication latency of the subprogram distributed in each node. For the worst case, communications from different nodes do not overlap one another. The communication overhead of the entire program is equal to the sum of communication latency of each node. In this case, there is no network contention. However, in general, communications are partially overlapped and network

contention always exists. Moreover, the existence of multiple communications in the network leads to the occurrence of network contention. The behavior of parallel programs (e.g., "when a communication is generated" and "which node sends or receives packets in the communication") determines the concurrent degree of communications and the network contention latency.

Therefore, in order to quantify the network contention and measure the communication overhead of the entire program, we define an abstract concept: *equivalent serial communication*. The equivalent serial communication is considered to be sequential so that there is no network contention. A program is abstracted to contain several equivalent serial communications. As shown in Figure 4(a), assuming that the program is mapped on two nodes: Node 1 and Node 2. There are four communications. Communication 1_1 and Communication 1_2 are generated by Node 1, while Communication 2_1 and Communication 2_2 are generated by Node 2. Communications generated by different nodes may be overlapped due to the program behavior. For example, Communication 1_2 is overlapped with Communication 2_2. There are network contention between Communication 1_2 and Communication 2_2. The number of equivalent serial communications (ω) is about 3.33, which is less than the communication number: 4. ω meets the inequation below:

$$\frac{p}{N} \leqslant \omega \leqslant \frac{p}{N} \cdot N = p. \tag{4}$$

ω describes the concurrent degree of communications as well as the network contention. The equivalent serial communications are sequential when the program is running so that no network contention occurs. Therefore, the contention latency is removed and fused into the ω when calculating the network communication latency. The network contention and the concurrent degree of communications together determine the value of ω.

(i) If communications are concurrent and they all exist in the same local area resulting in a *hotspot*,

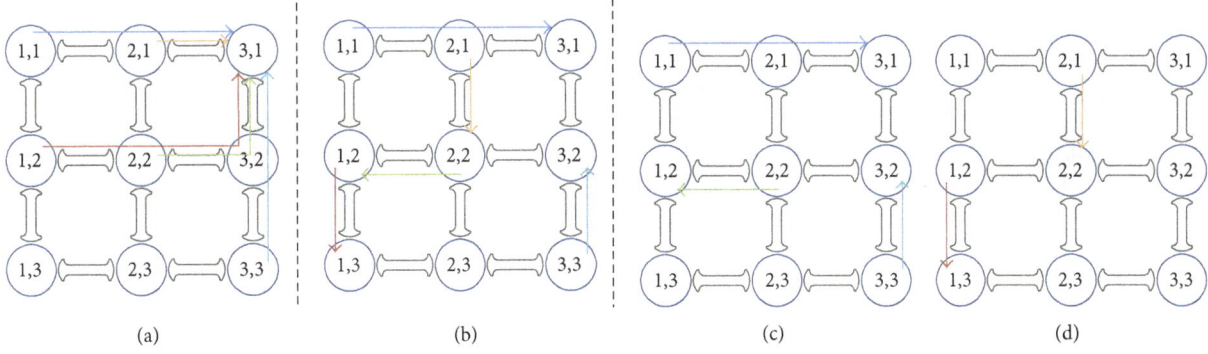

FIGURE 5: Examples of communications in a 3×3 2D-mesh network.

the network contention is heavy. In this case, the total communication time of the program is longer and hence ω is larger, close to $(p/N) \cdot N = p$. For instance, as illustrated in Figure 5(a), Node (1,1), Node (2,1), Node (1,2), Node (2,2), and Node (3,3) communicate with Node (3,1) concurrently. A hotspot is formed near Node (3,1) and network contention is heavy there. Although the five communications are issued concurrently, the network contention serializes them.

(ii) If communications are concurrent and they are uniformly distributed in the entire on-chip network, the network contention becomes light. In this case, the total communication time of the program is shorter and hence ω is smaller, close to p/N. For instance, as shown in Figure 5(b), there are also five communications occurring concurrently in the network. However, they belong to different source nodes and destination nodes and their routing tracks do not overlap, so there is no network contention. Therefore, its ω is smaller than that in Figure 5(a).

(iii) If communications are sequential, although the network contention is not heavy, the total communication time of the program is always long and hence ω is large and close to $(p/N) \cdot N = p$. For instance, as shown in Figure 5(c), Node (1,1) communicates with Node (3,1), Node (2,2) communicates with Node (1,2), and Node (3,3) communicates with Node (3,2). After that, Node (2,1) communicates with Node (2,2) and Node (1,2) communicates with Node (1,3) (see Figure 5(d)). Although there is no network contention, the five communications are not issued concurrently. Therefore, its ω is bigger than that in Figure 5(b).

(iv) For the best case that all nodes are fully concurrent and there is no network contention, the number of equivalent serial communication is equal to the number of real communication in each node ($\omega = p/N$). For the worst case that communications from all nodes occur sequentially, the number of equivalent serial communication is equal to the sum of

the number of real communication in each node ($\omega = (p/N) \cdot N = p$).

From (1), (2), and (4), we can calculate the communication overhead of a program running on homogenous OLPCs:

$$T_T = \omega \cdot \tau_c = \omega \cdot \gamma \cdot H \cdot \tau_{1hop} \quad \begin{pmatrix} 1 < \gamma \leqslant M \\ \dfrac{p}{N} \leqslant \omega \leqslant p \end{pmatrix}. \quad (5)$$

From (5), we can observe that (i) when $\gamma = M$ and $\omega = p$, $T_T = p \cdot M \cdot H \cdot \tau_{1hop}$ is the maximal communication overhead of the program for the worst case that all packets are transmitted in the network in a sequential way and (ii) when $\gamma \rightarrow 1$ and $\omega = p/N$, $T_T \rightarrow (p/N) \cdot H \cdot \tau_{1hop}$ is the minimal communication overhead of the program for the ideal best case that all packets in a communication are transmitted concurrently, all communications from different nodes are concurrent and no network contention occurs and (iii) when the network size is scaled up, H and T_T increases due to the longer communication distance.

Network contention is hard to quantify exactly. The concrete behavior of parallel applications leads to different traffic patterns, packet generation rate, and other factors. These factors influence the network contention. In this section, by introducing two abstract concepts, *equivalent serial packet* and *equivalent serial communication*, we could quantify the network contention and formulate the network communication latency. The equivalent serial packets and equivalent serial communications are sequential so that the network contention does not exist. To a certain extent, the effect of network contention is fused into the number of equivalent serial packet (γ) and the number of equivalent serial communication (ω). With the two extremes of traffic patterns (*Uniform* and *Hotspot* traffic models), we obtain the upper and lower bounds of γ and ω (see Formulas (2) and (4)). The bounds are determined by the number of packets in a communication (M), parallel part of the program (p) and the total processor number (N). M reflects the packet generation rate. Our model offers a feasible way to evaluate the network communication latency of homogenous OLPCs, but here comes a question: how do we determine or estimate

N, p, M, γ, and ω? The network size of OLPCs decides N. Different applications have their own p. Data-parallel applications are scalable and their data are regular. Their programs generally consist of a set of identical subtasks. Analyzing computation and communication behavior of the subtask, we could determine M and estimate γ and ω. Section 5.3 exemplifies the way of estimating γ and ω. Based on the analysis in this subsection, we could have a piece of implication.

Implication 1. Network communication latency has significant influence on the system's performance. The basic three threads to reduce the latency are (1) decreasing the number of communications in the program and the number of packets in a communication, (2) improving the concurrency of communications and packets, and (3) avoiding hotspot traffic. Architects or programmers can try their best to achieve these three threads by optimizing hardware design and application mapping, for instance by offering support for outstanding transactions or caching remote data in the local memory.

4.4. Performance Model. In this subsection, inspired by Amdahl's Law, we establish the performance model for homogenous OLPCs, incorporating the network communication latency. We elaborate the performance model under both *uniform* and *hotspot* traffic patterns. Under the two traffic models, we discuss and analyze the performance's trend, limitation, minimum, and maximum. The impact of network size (N), the ratio of the serial part and the parallel part in a program (α), the number of equivalent serial packets in a communication (γ), and the execution time of a subtask (τ_{nc}) on the performance are also discussed in detail. γ reflects the influence of network contention and congestion, while τ_{nc} reflects the influence of noncommunication/communication ratio.

The same as with Amdahl's Law, we assume that the total problem size is fixed as the number of computing nodes increases. The parallel part in the program is speeded up. The parallel part assigned to each processor node decreases with the increase of the system size. So we can get the performance model as the formula below shows:

$$S = \frac{(s + p) \cdot \tau_{nc}}{s \cdot \tau_{nc} + (p/N) \cdot \tau_{nc} + T_T}. \tag{6}$$

By including (5), we can get

$$S = \frac{(s + p) \cdot \tau_{nc}}{s \cdot \tau_{nc} + (p/N) \cdot \tau_{nc} + \omega \cdot \gamma \cdot H \cdot \tau_{1hop}}$$

$$\begin{pmatrix} 1 < \gamma \leqslant M \\ \dfrac{p}{N} \leqslant \omega \leqslant p \end{pmatrix}. \tag{7}$$

The last product item in the denominator describes the communication overhead. If this item is ignored, (6) can be simplified to

$$S = \frac{s + p}{s + p/N} \tag{8}$$

which is Amdahl's Law [6].

The behavior of parallel programs determines the communication patterns, affecting the value of γ and ω. Uniform traffic model is a well-distributed traffic model, while hotspot traffic model is a centralized traffic model. They are two extremes, representing the upper bound and the lower bound of the communication patterns, respectively. Hence, we consider both uniform and hotspot traffic models below to analyze the speedup in detail. Although hotspot traffic has smaller average hop count and hence less minimal latency, hotspot traffic causes much heavier network contention than uniform traffic. For uniform traffic, it has lower network contention and ω is closer to p/N. For hotspot traffic, it has higher network contention and ω is closer to p, because of the serialization effect in the destination node. Therefore, to facilitate the formula transformation and analysis, we consider $\omega = p/N$ for uniform traffic, while $\omega = p$ for hotspot traffic. This assumption is thought to be reasonable without the loss of analyzing the performance trend.

4.4.1. Uniform Traffic Model. Assuming $\omega = p/N$, we can refine (7) as

$$S = \frac{(s + p) \cdot \tau_{nc}}{s \cdot \tau_{nc} + (p/N) \cdot \tau_{nc} + p \cdot \gamma \cdot (2/3) \cdot (1/N^{1/2} - 1/N^{3/2}) \cdot \tau_{1hop}}$$

$$= \frac{(\alpha + 1) \cdot \tau_{nc}}{(\alpha + 1/N) \cdot \tau_{nc} + (2/3) \cdot \gamma \cdot (1/N^{1/2} - 1/N^{3/2}) \cdot \tau_{1hop}}. \tag{9}$$

Since τ_{1hop} reflects the architectural latency without contention, it is a constant for a given homogenous OLPC architecture. Therefore, The speedup (S) is a quaternion function: $S = S(N, \alpha, \tau_{nc}, \gamma)$. Its value is determined by N, α, τ_{nc}, and γ. To obtain the variation trend of S, we conduct two steps below.

Step 1 (calculating the speedup's limitation). We have the limitation of S as below:

$$\lim_{N \to \infty} S = \frac{\alpha + 1}{\alpha} = 1 + \frac{1}{\alpha}. \tag{10}$$

Step 2 (calculating the value of N related to the extreme minimal value of S). Let $\partial S / \partial N = 0$; then, we can get

$$N_{\partial S/\partial N=0} = \frac{6 \cdot \gamma^2 \cdot \tau_{1hop}^2 + 9 \cdot \tau_{nc}^2 - 3 \cdot \sqrt{12 \cdot \gamma^2 \cdot \tau_{1hop}^2 \cdot \tau_{nc}^2 + 9 \cdot \tau_{nc}^4}}{2 \cdot \gamma^2 \cdot \tau_{1hop}^2}. \tag{11}$$

Let $\beta = \tau_{nc}/(\gamma \cdot \tau_{1hop})$; Formula (11) is refined as

$$N_{\partial S/\partial N=0} = \frac{9 \cdot \beta^2 + 6 - \sqrt{9 \cdot \beta^2} \cdot \sqrt{9 \cdot \beta^2 + 12}}{2}. \tag{12}$$

From formula (12), we can get

$$N_{\partial S/\partial N=0} > \frac{9 \cdot \beta^2 + 6 - \left(9 \cdot \beta^2 + 9 \cdot \beta^2 + 12\right)/2}{2} = 0,$$

$$N_{\partial S/\partial N=0} < \frac{9 \cdot \beta^2 + 6 - \sqrt{9 \cdot \beta^2} \cdot \sqrt{9 \cdot \beta^2}}{2} = 3. \tag{13}$$

The extreme minimal value of S exists; its related N is defined as N_{opt}. Because N is a positive integer, we have

$$N_{opt} \in \{1, 2, 3\}. \tag{14}$$

The OLPC hosts at least one processor core, so $N \geqslant 1$. Combining the two steps, we can obtain that

(1) when $N_{opt} = 1$,

    (i) S monotonically increases with the increase of N; parallelization enables the performance improvement; however, S is bounded by $1 + 1/\alpha$ for $N \to \infty$; the ratio of the serial part in a program limits the performance improvement;

(2) when $N_{opt} = 2$ or $N_{opt} = 3$,

    (i) when $N < N_{opt}$, S decreases with the increase of N; parallelization degrades the performance rather than improves it, because the negative effect of network communication latency on the performance surpasses the positive effect of cooperation of multiple processor cores on the performance;

    (ii) when $N = N_{opt}$, S reaches its minimum (S_{min});

    (iii) when $N > N_{opt}$, S increases when N is increasing; the positive effect of parallelization surpasses the negative effect of network communication latency, thus improving the performance;

(3) the ratio between the serial part and the parallel part in a program determines the upper limit of S. The limit is inversely proportional to α. It indicates that reducing the serial part or enlarging the parallel part in a program is good for improving the performance limit.

As we can see, S reaches its minimum when N is very small. The OLPC hosts a number of processor cores. Therefore, for a lager range of N, S keeps going up when N increases. To further discuss the effect of N, α, τ_{nc}, and γ on S, Figure 6 shows performance trends of S under uniform traffic model. Without loss of trend analysis, we consider

    (a) $N \in \{n \mid 1 \leqslant n \leqslant 256, n \in \mathbb{N}\}$; the network size is scaled up from 1 to 256; the increase of the network size makes more processor cores involved;

    (b) $\alpha \in \{x \mid 0 \leqslant x \leqslant 1, x \in \mathbb{R}\}$; with the increase of α, the serial part takes more proportion in a program; the performance limit $(1 + 1/\alpha)$ becomes less;

    (c) $\gamma \in \{1, 16, 256\}$; the number of equivalent serial packets in a communication increases from 1, 16 to 256; more packets lead to larger network communication latency, causing negative effect on the performance;

    (d) $\tau_{nc} \in \{10, 100, 1000\}$; the execution time of a subtask increases from 10, 100 to 1000; increasing noncommunication time can bring positive effect on the performance.

From aforementioned formula transformation and Figure 6, we can have results regarding the performance under uniform traffic model.

    (I) The increase of the network size (N) makes more processor cores, exploiting larger parallelism. As shown in Figure 6, as N increases, the speedup (S) firstly decreases and soon reaches its minimum when $N = N_{opt}$ (this situation is shown in Figure 6(g). It also exists in other subfigures, but it is not obvious, since S is much larger); then, S increases. However, S increases more and more slowly; it is limited by $1 + 1/\alpha$ finally.

    (II) Both the incremental ratio and the limit of S are deeply influenced by α. As shown in all subfigures, as α increases, the incremental ratio of S becomes very low and the limit of S is very small. Even if the network

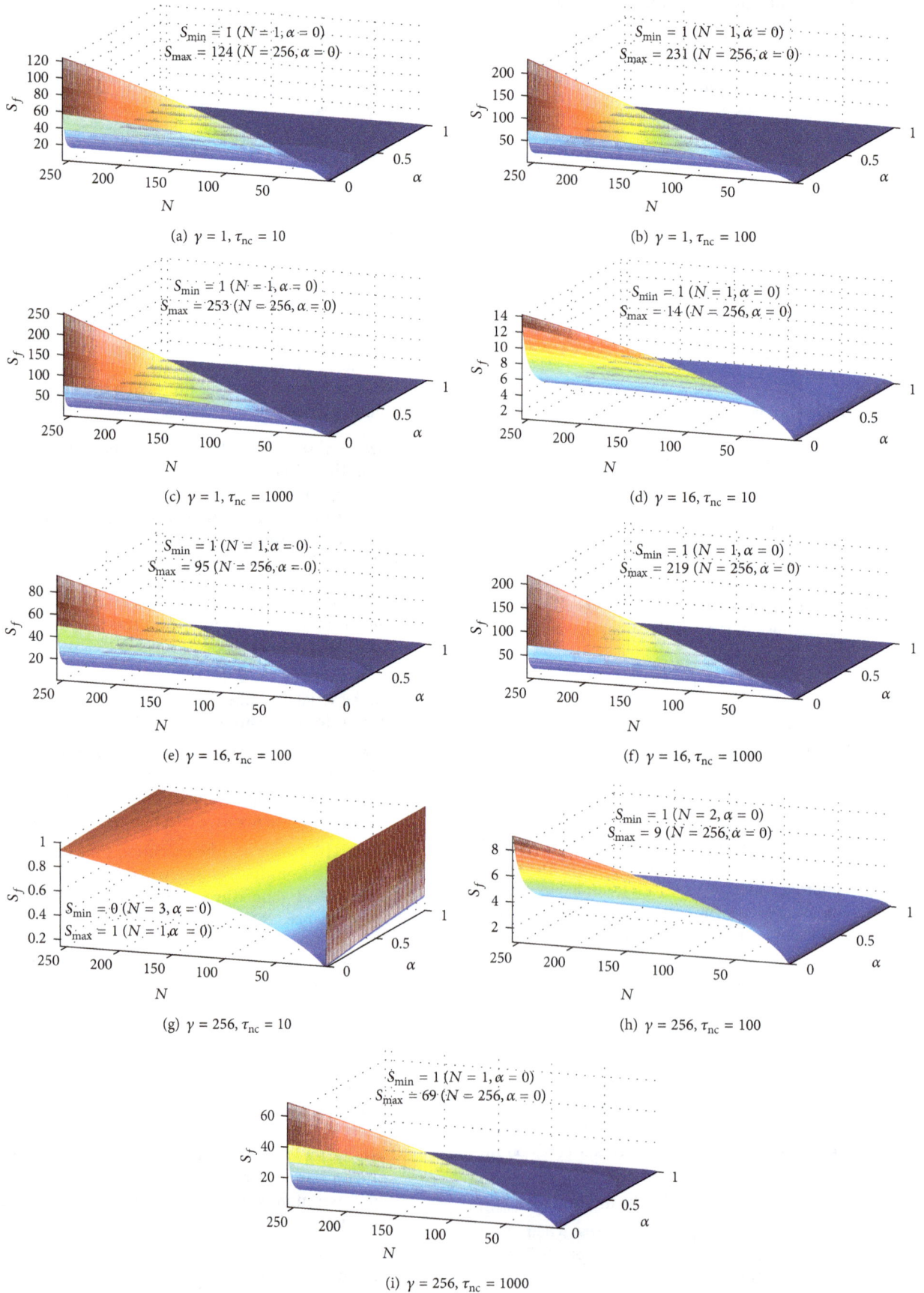

(a) $\gamma = 1, \tau_{nc} = 10$

(b) $\gamma = 1, \tau_{nc} = 100$

(c) $\gamma = 1, \tau_{nc} = 1000$

(d) $\gamma = 16, \tau_{nc} = 10$

(e) $\gamma = 16, \tau_{nc} = 100$

(f) $\gamma = 16, \tau_{nc} = 1000$

(g) $\gamma = 256, \tau_{nc} = 10$

(h) $\gamma = 256, \tau_{nc} = 100$

(i) $\gamma = 256, \tau_{nc} = 1000$

FIGURE 6: Performance trends under uniform traffic model.

size (N) is scaled up, the performance improvement is very little.

(III) As γ increases, network communication hosts more packets, worsening network congestion or contention and thus generating larger network communication latency. Larger network communication latency brings negative effect on the performance. Frequent network communication and huge latency makes the performance very bad. For instance, for $\tau_{nc} = 10$, $\alpha = 0$, and $N = 256$ (see Figures 6(a), 6(d), and 6(g)), (i) when $\gamma = 1$, S can reach its maximum ($S_{max} = 124$); (ii) when $\gamma = 16$, the maximal speedup becomes small ($S_{max} = 14$); (iii) when $\gamma = 256$, the heavy network communication makes the performance even not improved.

(IV) The increase of τ_{nc} can improve the performance, alleviating and making up the negative effect of network communication latency. For instance, for $\gamma = 16$, $\alpha = 0$, and $N = 256$ (see Figures 6(d), 6(e), and 6(f)), (i) when $\tau_{nc} = 10$, S reaches its maximum ($S_{max} = 14$); (ii) when $\tau_{nc} = 100$, the maximal speedup becomes large ($S_{max} = 95$); (iii) As τ_{nc} rises up to 1000, the maximal speedup ($S_{max} = 219$) is close to the ideal maximal value (256).

4.4.2. Hotspot Traffic Model.

Assuming $\omega = p$ and k is odd, we can refine (7) as

$$
S = \frac{(s + p) \cdot \tau_{nc}}{s \cdot \tau_{nc} + (p/N) \cdot \tau_{nc} + p \cdot \gamma \cdot \left(\sqrt{N}/2\right) \cdot \tau_{1hop}}
$$

$$
= \frac{(\alpha + 1) \cdot \tau_{nc}}{(\alpha + 1/N) \cdot \tau_{nc} + (1/2) \cdot \gamma \cdot \sqrt{N} \cdot \tau_{1hop}}. \tag{15}
$$

The same as with Section 4.4.1, the speedup (S) is also a quaternion function: $S = S(N, \alpha, \tau_{nc}, \gamma)$. Its value is decided by N, α, τ_{nc}, and γ. In (15), when N becomes larger, $(1/N) \cdot \tau_{nc}$ decreases but $(1/2) \cdot \gamma \cdot \sqrt{N} \cdot \tau_{1hop}$ increases, so S may increase or decrease. To obtain the variation trend of S, we also conduct two steps below.

Step 1 (calculating the speedup's limitation). We have the limitation of S as below:

$$
\lim_{N \to \infty} S = 0. \tag{16}
$$

Step 2 (calculating the value of N related to the extreme maximal value of S). Let $\partial S/\partial N = 0$; then, we can get

$$
N_{\partial S/\partial N = 0} = \sqrt[3]{\left(\frac{4 \cdot \tau_{nc}}{\gamma \cdot \tau_{1hop}}\right)^2} > 0. \tag{17}
$$

The extreme maximal value of S exists; its related N_{opt} is obtained by the formula below:

$$
N_{opt} = \begin{cases} 1, & \text{when } N_{\partial S/\partial N = 0} < 1 \\ \lfloor N_{\partial S/\partial N = 0} \rfloor, & \text{when } N_{\partial S/\partial N = 0} \geq 1, \\ & S(\lfloor N_{\partial S/\partial N = 0} \rfloor) \geq S(\lceil N_{\partial S/\partial N = 0} \rceil) \\ \lceil N_{\partial S/\partial N = 0} \rceil, & \text{when } N_{\partial S/\partial N = 0} \geq 1, \\ & S(\lfloor N_{\partial S/\partial N = 0} \rfloor) < S(\lceil N_{\partial S/\partial N = 0} \rceil). \end{cases} \tag{18}
$$

With Formulas (15) and (17), we can have the extreme maximal value of S:

$$
S_{max} = S(N_{opt}) \approx S(N_{\partial S/\partial N = 0})
$$

$$
= \frac{\alpha + 1}{\alpha + 3 \cdot \sqrt[3]{\left((\gamma \cdot \tau_{1hop})/(4 \cdot \tau_{nc})\right)^2}}. \tag{19}
$$

Let $\beta = 3 \cdot \sqrt[3]{((\gamma \cdot \tau_{1hop})/(4 \cdot \tau_{nc}))^2}$; Formula (19) is refined as

$$
S_{max} \simeq \frac{\alpha + 1}{\alpha + \beta}. \tag{20}
$$

Because N is a positive integer, combining the two steps, we can obtain that

(1) when $N_{opt} = 1$,

   (i) S monotonically decreases with the increase of N; parallelization degrades the performance rather than improves it, because the negative effect of network communication latency on the performance surpasses the positive effect of cooperation of multiple processor cores on the performance; S tends to zero when $N \to \infty$;

(2) when $N_{opt} \geq 2$,

   (i) when $N < N_{opt}$, S increases with the increase of N; within this condition, the network communication latency is not much and parallelization is able to improve the performance;

   (ii) when $N = N_{opt}$, S reaches its maximum (S_{max});

   (iii) when $N > N_{opt}$, S becomes decreasing when N keeps going up; performance degrades because the network communication latency dominates;

(3) $N_{\partial S/\partial N = 0} \sim \tau_{nc}^{2/3}$ and $N_{\partial S/\partial N = 0} \sim \gamma^{-2/3}$; when $\tau_{nc} \uparrow$ and $\gamma \downarrow$, $N_{\partial S/\partial N = 0} \uparrow$, resulting in $N_{opt} \uparrow$; it indicates that increasing noncommunication time and improving packet concurrency can increase the extreme value of S and the performance improves further covers a larger system size.

To further discuss the effect of N, α, τ_{nc}, and γ on S, Figure 7 shows performance trends of S under uniform traffic model. We consider the values of N, α, τ_{nc}, and γ as the same with Section 4.4.1. From aforementioned formula transformation and Figure 7, we can have results regarding the performance under hotspot traffic model.

(I) Although the increase of the network size (N) could make more processor cores involved to cooperation together so as to seek higher parallel performance, it also induces network communication latency, limiting the performance improvement and even worsen-ing the performance. As shown in Figure 7, as N increases, in some cases (see Figures 7(a), 7(b), 7(c), 7(e), 7(f), and 7(i)), the speedup (S) firstly increases and then becomes decreasing after reaching its maximum; in other cases (see Figures 7(d), 7(g), and 7(h)), it monotonically decreases. For all cases, as N increases, S finally tends to zero.

(II) Both the incremental/decremental ratio and the maximal value of S are influenced by α. As shown in all subfigures, as α increases, the incremental/decremental ratio becomes very low. With the increase of α, the maximal value of S may increase or decrease: (i) if $1 > \beta$ in Formula (20), S_{max} decreases (see Figures 7(a), 7(b), 7(c), 7(e), 7(f), and 7(i)); (ii) if $1 = \beta$ in Formula (20), $S_{max} \equiv 1$; (iii) if $1 < \beta$ in Formula (20), S_{max} increases (see Figures 7(d), 7(g), and 7(h)).

(III) As γ increases, network communication hosts more packets. Larger network communication latency makes the performance goes bad. The maximal value of S reached by parallelism declines. For instance, for $\tau_{nc} = 1000$ and $\alpha = 0$ (see Figures 7(c), 7(f), and 7(i)), (i) when $\gamma = 1$, S can reach its maximum ($S_{max} = 84$ with $N = 252$); (ii) when $\gamma = 16$, the maximal speedup becomes small ($S_{max} = 13$ with $N = 40$); (iii) when $\gamma = 256$, the heavy network communication makes the speedup soon reach its little maximum ($S_{max} = 2$ with $N = 6$).

(IV) The increase of τ_{nc} can improve the performance, alleviating and making up the negative effect of network communication latency. For instance, for $\gamma = 1$ and $\alpha = 0$ (see Figures 7(a), 7(b), and 7(c)), (i) when $\tau_{nc} = 10$, the maximal value of S is very small ($S_{max} = 4$ with $N = 12$); (ii) when $\tau_{nc} = 100$, the maximal speedup becomes big ($S_{max} = 18$ with $N = 54$); (iii) as τ_{nc} rises up to 1000, the maximal speedup further becomes large ($S_{max} = 84$ with $N = 252$).

In all, performance under hotspot traffic model is worse than that under uniform traffic model.

With the performance analysis in this subsection, we could have the following.

Implication 2. With the uniform traffic model, the communication overhead is modest, assuming that there is limited contention, so the performance can still keep improving.

Under the uniform traffic model, the concurrent degree of communications are usually high. Architects or programmers need to pay more attention to improve the concurrent degree of packets in a communication. The performance improvement can benefit more from the improvement of packet concurrency.

Implication 3. With the hotspot traffic model, parallelization cannot always improve the system's performance, when the network communication latency dominates. To alleviate the impact of network communication latency on the performance and hence keep the performance's improvement continuous, designers need to address increasing the non-communication time and improving packet concurrency.

Implication 4. Exploiting the parallelism of multiple processor cores well is potential to make up the negative effect of network communication latency and even obtains the continuous improvement of the performance. Following this view, architects or programmers need to pay more attention to exploit the parallelism of processor cores.

Implication 5. Besides, increasing the noncommunication time is a viable way to alleviate the negative effect induced by the network communication latency.

5. Experiments and Results

In this section, we apply three real data-parallel applications on our cycle-accurate homogenous OLPC experimental platform to validate and demonstrate the effectiveness of our performance analysis.

5.1. Experimental Platform. Figure 8 shows our homogenous OLPC experimental platform. The platform uses the LEON3 [14] as the processor in each PM node and uses the Nostrum NoC [15] as the on-chip network. Each Processor-Memory (PM) node has a LEON3 processor, an enhanced memory controller plus a local memory. The enhanced memory controller extends the function of LEON3's own memory control module to support memory accesses from/to remote nodes via the network. The LEON3 processor core is a synthesizable VHDL model of a 32-bit processor compatible with the SPARC V8 architecture. The Nostrum NoC is a 2D-mesh packet-switched network with configurable size. Besides, moving one hop in the network takes one cycle.

5.2. Application Examples. We use Wavefront Computation, Vector Norm, and Block Matching Algorithm in Motion Estimation as application examples and perform experiments on various instances of the three applications. Wavefront Computation and Vector Norm are mostly used in wireless communication, computer vision, and image/video processing. And Block Matching Algorithm in Motion Estimation is one of the basic components in image/video processing.

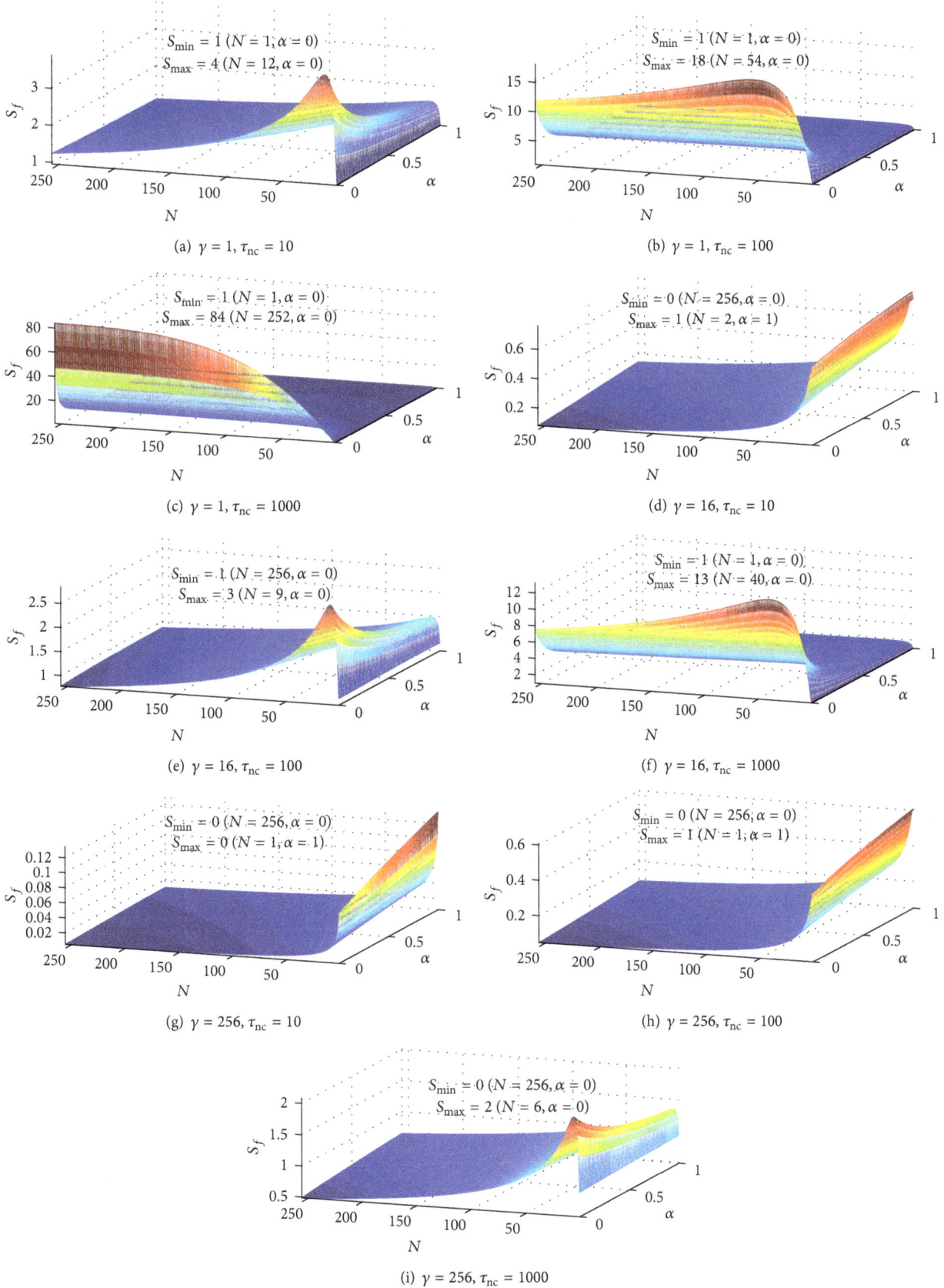

FIGURE 7: Performance trends under hotspot traffic model.

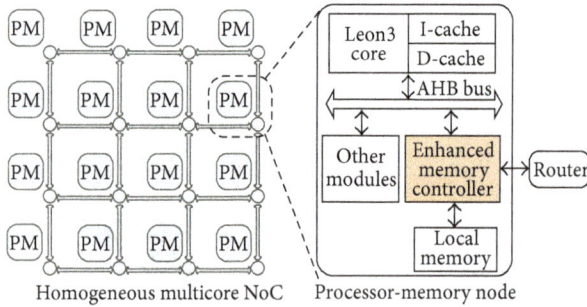

Homogeneous multicore NoC   Processor-memory node

FIGURE 8: The homogenous OLPC experimental platform.

5.2.1. Wavefront Computation.

5.2.1. Wavefront Computation. Wavefront Computations are common in scientific applications. Given a matrix (see Figure 9(a)), the left and top edges of which are all a constant, the computation of each remaining element depends on its neighbors to the left, above, and above-left. If the solution is computed in parallel, the computation at any instant forms a wavefront propagating toward in the solution space. Therefore, this form of computation gets its name as wavefront. We use the same method as [16] to parallelize the Wavefront Computation, the rows of the matrix are assigned to PM nodes in a round-robin fashion (see Figure 9(b)). With this static scheduling policy, to compute an element, only the availability of its above neighbor needs to be checked (synchronized). For instance, PM node 0 computes the elements in row 1. PM node 1 cannot compute the elements in row 2 until the corresponding elements in row 1 has been figured out by PM node 0. After finishing the computation in row 1, PM node 0 goes on to compute the elements in row 3 according to the round-robin scheduling policy. In our experiment, we conduct various instances of Wavefront Computation described below.

(1) Two ways of data storing are realized to reflect the two traffic models. One is "*Uniform*" meaning that the matrix data are uniformly distributed over all nodes. The other is "*Hotspot*" meaning that the matrix data are only located in the central node.

(2) Both integer matrix and floating point matrix are implemented to vary the noncommunication time: τ_{nc}. For the same problem size and algorithm, floating point computation needs more time than integer computation and hence has bigger τ_{nc}.

(3) The Wavefront Computation conducts a matrix with the size of 256×256, on the homogenous OLPC with the network size varying from 1×1 (1), 1×2 (2), 2×2 (4), 2×4 (8), 4×4 (16), 4×8 (32), 8×8 (64), 8×16 (128), to 16×16 (256). The total problem size is fixed and the problem size assigned to each node varies from 256 rows, 128 rows, 64 rows, 32 rows, 16 rows, 8 rows, 4 rows, 2 rows to 1 row.

5.2.2. Vector Norm. Vector Norm is used to compute the magnitude (length) of the vector. Figure 10(a) shows the formula of Vector Norm. When $p = 2$, the Vector Norm is also called L^2-*Norm* or *Euclidean Norm*, which is common in operations of 2D/3D computer graphics. In the paper, we choose to parallelize and compute L^2-*Norm*. Figure 10(b) illustrates the parallelization of L^2-*Norm* on our OLPC platform. Different from Matrix Multiplication and Wavefront Computation, Vector Norm only can be partially parallelized. Its computation contains two steps. Step 1 is parallel. In Step 1, PM nodes are responsible for computing the square (t_i) of x_i ($i = 1, \ldots, n$). x_i are assigned to PM nodes in a round-robin fashion. Step 2 is sequential. In Step 2, a central PM node takes charge of computing the square root of the sum of all t_i. For instance, as shown in Figure 10(b), there are two PM nodes computing the L^2-*Norm* of a vector with four elements. In Step 1, PM node 0 computes the square (t_1) of x_1, while PM node 1 computes the square (t_2) of x_2. After finishing the computation of t_1, PM node 0 goes on to compute the square (t_3) of x_3 according to the round-robin scheduling policy. In Step 2, PM node 1 (the central PM node) computes $\sqrt{t_1 + t_2 + t_3 + t_4}$. In our experiment, we apply various instances of L^2-*Norm* described below.

(1) Two ways of data storing are realized to reflect the two traffic models. One is "*Uniform*" meaning that the data in Step 1 are uniformly distributed over all nodes. The other is "*Hotspot*" meaning that all data in both Steps 1 and 2 are only located in the central node.

(2) Both integer data type and floating point data type are implemented to vary the noncommunication time: τ_{nc}. For the same problem size and algorithm, floating point computation needs more time than integer computation and hence has bigger τ_{nc}.

(3) The L^2-*Norm* conducts a vector with 1024 elements, on the homogenous OLPC with the network size varying from 1×1 (1), 1×2 (2), 2×2 (4), 2×4 (8), 4×4 (16), 4×8 (32), 8×8 (64), 8×16 (128), to 16×16 (256). The total problem size is fixed and the problem size assigned to each node varies from 1024 elements, 512 elements, 256 elements, 128 elements, 64 elements, 32 elements, 16 elements, 8 elements, to 4 elements.

5.2.3. Block Matching Algorithm in Motion Estimation. Motion Estimation is one of important parts in H.264/AVC standard, which addresses obtaining high coding efficiency and good picture quality [17]. It is of importance to find the best Motion Vector in Motion Estimation. The Block Matching Algorithm in Motion Estimation aims at looking for the best matching block with the best Motion Vector in Reference Frame. Figure 11(a) illustrates the Block Matching Algorithm. As shown in the figure, there is a Current Block (**C**) in the Current Frame. For a Reference Frame, the Block Matching Algorithm first predicts a Search Center (**SC**) according to the position of the Current Block (**C**). Then, it exhaustively checks all search points (i.e., candidate Reference Blocks, e.g., **R**) in the Search Window (**SW**) of the Reference Frame to find the best matching block (**R**$_{opt}$) with the best Motion Vector (**MV**). The position of the Search Window (**SW**) is decided by the Search Center (**SC**), while its size is decided by the Search Range (**SR**). It is obvious that larger Search

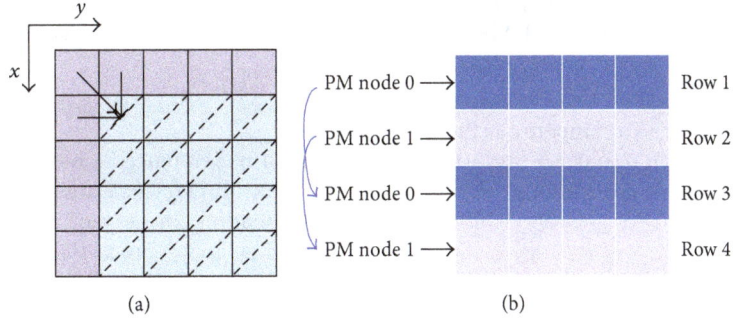

FIGURE 9: (a) Wavefront Computation; (b) its parallelization.

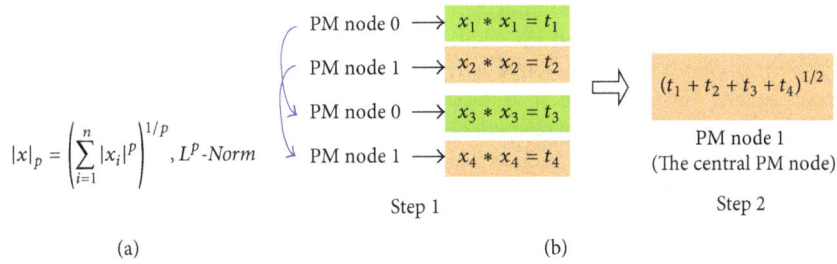

FIGURE 10: (a) Vector Norm and (b) its parallelization.

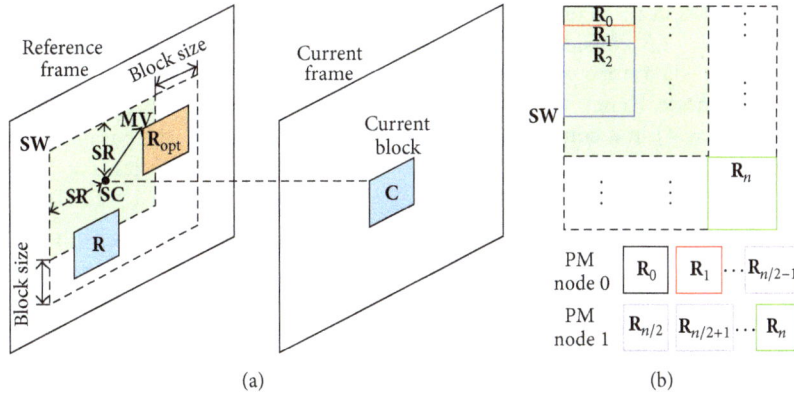

FIGURE 11: (a) Block Matching Algorithm in Motion Estimation and (b) its parallelization.

Window (**SW**) leads to more accurate prediction of the best matching block with the best Motion Vector but consumes more amount of computation time. Figure 11(b) shows how the Block Matching Algorithm is parallelized on our OLPC platform. We uniformly assign candidate Reference Blocks (**R_i**) into each PM node so that each PM node handles the same number of candidate Reference Blocks. For instance, assume that there are n search points in the Search Window (**SW**) and two PM nodes take charge of obtaining the best matching block. The PM node 0 is responsible for comparing **$R_0, R_1, \ldots, R_{n/2-1}$** with the Current Block (**C**), while PM node 1 takes charge of comparing **$R_{n/2}, R_{n/2+1}, \ldots, R_n$** with the Current Block (**C**). In our experiment, we perform a various instances described below.

(1) We also realize two ways of data storing to reflect the two traffic models. One is "*Uniform*" meaning that the candidate reference blocks are uniformly distributed over all nodes. The other is "*Hotspot*" meaning that all candidate reference blocks are located in the central node.

(2) Only integer data type is considered, since the data in image processing are "integer."

(3) We conduct a Search Window with the size of 128 × 128 (i.e., 16384 candidate reference blocks), on the homogenous OLPC with the network size varying from 1 × 1 (1), 1 × 2 (2), 2 × 2 (4), 2 × 4 (8), 4 × 4 (16), 4 × 8 (32), 8 × 8 (64), 8 × 16 (128), to 16 × 16 (256).

The total problem size is fixed and the problem size assigned to each node varies from 16384, 8192, 4096, 2048, 1024, 512, 256, 128, to 64 reference blocks.

5.3. Theoretical Speedup Estimation.

To compare our theoretical analysis with the real simulation results, we first estimate the theoretical speedups of the three applications.

5.3.1. Wavefront Computation

(1) The program of Wavefront Computation can be fully parallelized, thus $s = 0$.

(2) The subtask on each node is $E(i, j) = E(i - 1, j - 1) * E(i - 1, j - 1) - E(i, j - 1) * E(i - 1, j)$. Here, $E(i, j)$ represents the current element in the matrix in Figure 9, while $E(i - 1, j - 1)$, $E(i, j - 1)$ and $E(i - 1, j)$ are $E(i, j)$'s neighboring element to the above-left, left and above, respectively. The time of such subtask (including computation and local memory reference) is collected in our experiment: $\tau_{nc} = 176$ clock cycles for integer data type; $\tau_{nc} = 2032$ clock cycles for floating point data type.

(3) For "*Uniform*" data storing, the elements computed by a PM node are located on the local memory of that PM node. Hence, $E(i, j)$ and $E(i, j - 1)$ are local, while $E(i - 1, j - 1)$ and $E(i - 1, j)$ are remote. There are two packets ($M = 2$) transmission in a communication and we assume that $\gamma = 1.5$ considering packet concurrency. For "*Hotspot*" data storing, all elements are located on the central node. Hence, there are four packet transmissions ($M = 4$) in a communication. Considering packet transmissions are overlapped, we assume that $\gamma = 2$.

5.3.2. Vector Norm

(1) The program of Vector Norm is partially parallelized. The serial part consumes much time.

(2) Step 1 is parallel. In Step 1, the subtask on each node is $t_i = x_i \times x_i$. The time of such subtask (including computation and local memory reference) is collected in our experiment: $\tau_{nc} = 110$ clock cycles for integer data type; $\tau_{nc} = 1270$ clock cycles for floating point data type. Because the vector contains 1024 elements, $p = 1024$. Step 2 is sequential. in our experiment, the computation of $\sqrt{t_1 + t_2 + \cdots + t_{1024}}$ takes 32700 cycles for integer data type and 337560 cycles for floating point data type. So $s = 32700/110 \approx 297$ for integer data type and $s = 337560/1270 \approx 266$ for floating point data type.

(3) For "*Uniform*" data storing, x_i in Step 1 used by a PM node are located on the local memory of that PM node. t_i is stored in the central PM node. Hence, there is one packet ($M = 1$) transmission in a communication and we assume that $\gamma = 1$. For "*Hotspot*" data storing, all data are located on the central node. Hence, there are two packet transmissions ($M = 2$) in

a communication. Considering packet transmissions are overlapped, we assume that $\gamma = 1.5$.

5.3.3. Block Matching Algorithm in Motion Estimation

(1) The Reference Frame has been computed and stored in on-chip local memories in the last Motion Estimation. In current Motion Estimation, the "Block Matching" processing starts until the Current Block in the Current Frame is transferred from the off-chip DRAM into the on-chip memory. The elapsed time of transferring the Current Block from the off-chip DRAM memory into the on-chip memory is the serial part of the Block Matching Algorithm. In our OLPC platform, the central PM node features an External Memory Interface connecting with the off-chip DRAM. The External Memory Interface reads a datum from the DRAM in 20 cycles and the size of the Current Block is 16×16. Hence, for "*Hotspot*" data storing that all data are stored in the central PM node, the data transfer takes 5120 ($=16 \times 16 \times 20$) cycles. For "*Uniform*" data storing that data are uniformly stored in each PM node, the Current Block is transferred from the DRAM to the External Memory Interface and routed to all PM nodes in a broadcast way, so the time of the Current Block's transfer is $5120 + N/2$ cycles (a packet from the central node to the corner one takes $N/2$ hops), approximately equal to 5120 cycles. The subtask on each node is the comparison of the Current Block and a candidate Reference Block, consuming 7680 cycles. Therefore, the problem size is 128×128, so the parallel part takes 125829120 ($=7680 \times 128 \times 128$) cycles. $s = 5120/(5120 + 125829120) \approx 0.00\%$, and $p = 1 - s \approx 100.00\%$.

(2) The subtask on each node is the comparison of the Current Block and a candidate Reference Block. The time of such subtask (including computation and local memory reference) is collected in our experiment: $\tau_{nc} = 7680$ cycles.

(3) For "*Uniform*" data storing, the Current Block and the candidate blocks are located in each PM node, so there is no network communication and $\gamma = 0$. For "*Hotspot*" data storing, the Current Block and the candidate blocks are in the central PM node, Hence, there are 512 ($=16 \times 16 \times 2$) packet transmissions ($M = 512$) in a communication. Considering that such many packets are routed to the central node, the network contention is extremely heavy and we assume that $\gamma = 512$.

Then, using the Formula (9) and (15) estimates the theoretical speedups of the three applications.

5.4. Simulation Results.

The real speedups of the three applications are calculated based on the simulation results on our homogenous OLPC experimental platform (because the sequential part in the program of Vector Norm dominates, the performance improvement is limited).

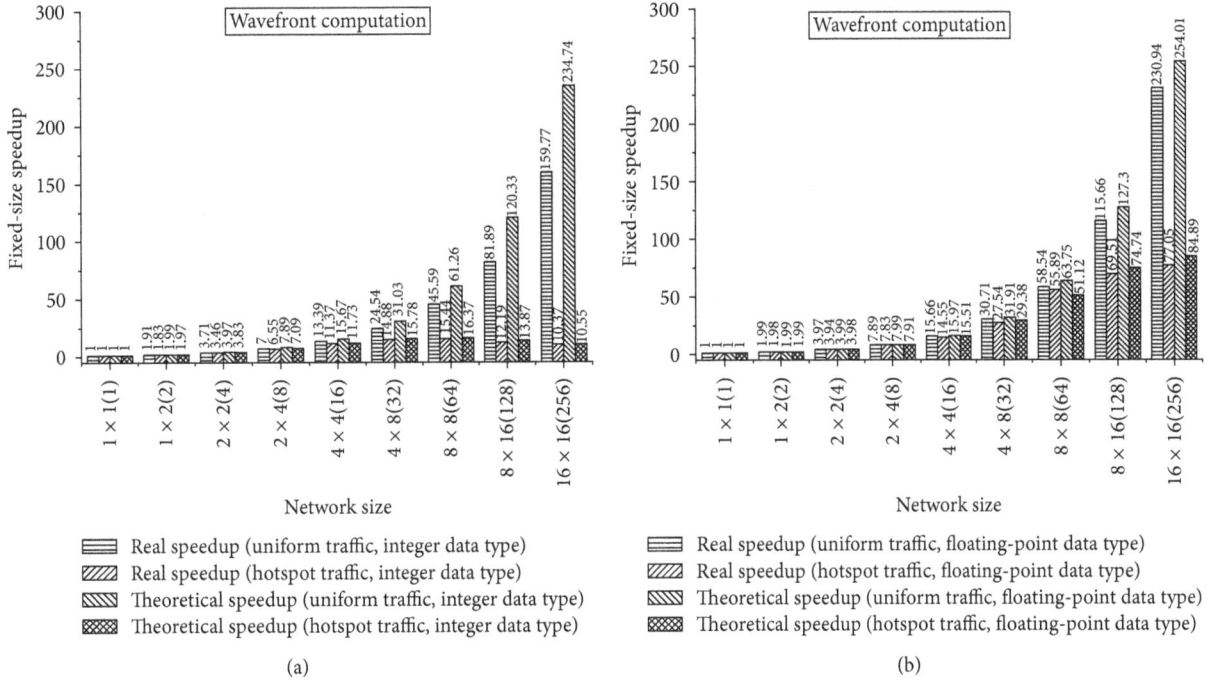

FIGURE 12: Effect of traffic models: Wavefront Computation with (a) integer data type and (b) floating-point data type.

5.5. Analysis and Discussion

5.5.1. Effect of Network Size.
The effect of network size on the performance reflects the scalability of homogenous OLPCs. Figures 12, 13, 14, 15, and 16 plot the real and theoretical speedups versus the size of the homogenous OLPC from 1×1 (1), 1×2 (2), 2×2 (4), 2×4 (8), 4×4 (16), 4×8 (32), 8×8 (64), 8×16 (128), to 16×16 (256). From the six figures, we can see that (i) the theoretical speedups has the same trend with the real speedups; (ii) for uniform traffic model, the speedup usually increases when the network size is scaled up; and (iii) for hotspot traffic model, the speedup reaches its maximum when the network size is scaled up to a certain size and becomes decreasing when the network size goes on increasing.

5.5.2. Effect of Traffic Models.
Figure 12 shows the effect of traffic models on the real and theoretical speedups of both integer and floating-point Wavefront Computation, Figure 13 shows the effect of traffic models on the real and theoretical speedups of both integer and floating-point Vector Norm, and Figure 14 shows the effect of traffic models on the real and theoretical speedups of Block Matching Algorithm in Motion Estimation.

(i) For uniform traffic model, consistent with the theoretical speedup performance model, the real speedup increases as the network size is scaled up, no matter the data type is integer or floating-point. This is because the contention latency induced by uniform

traffic is not enough to kill the performance improvement introduced by the parallelization. However, it can slow down the performance improvement.

(ii) Because a hotspot traffic model incurs heavy contention latency, the speedup increases when the network size is small but begins decreasing when the network size is scaled up to a certain finite value. Using (17), we can calculate the value of network size (N) for the maximal speedup. (i) For Wavefront Computation with integer data type, $N_{\mathrm{opt}} \approx 50$, so Figure 12(a) shows that both the theoretical and the real speedups go up from 1×1 (1) to 4×8 (32), the speedups on 8×8 (64) are approximately equal to the speedups on 4×8 (32), and the speedups turn to fall as the network size goes on increasing to 16×16 (256). (ii) For Wavefront Computation with floating-point data type, $N_{\mathrm{opt}} \approx 255$, so Figure 12(b) shows both the theoretical and the real speedups ascend when the network size is from 1×1 (1) to 16×16 (256). (iii) For Vector Norm with integer data type, $N_{\mathrm{opt}} \approx 44$, so Figure 13(a) shows that both the theoretical and the real speedups go up from 1×1 (1) to 4×8 (32), the speedups on 8×8 (64) are approximately equal to the speedups on 4×8 (32), and the speedups fall when the network size goes on increasing to 16×16 (256). (iv) For Vector Norm with floating-point data type, $N_{\mathrm{opt}} \approx 226$, so Figure 13(b) shows both the theoretical and the real speedups ascend when the network size is from 1×1 (1) to 16×16 (256). (v) For Block Matching Algorithm, $N_{\mathrm{opt}} \approx 15$, so Figure 14 shows that both the theoretical and the real speedups go up from

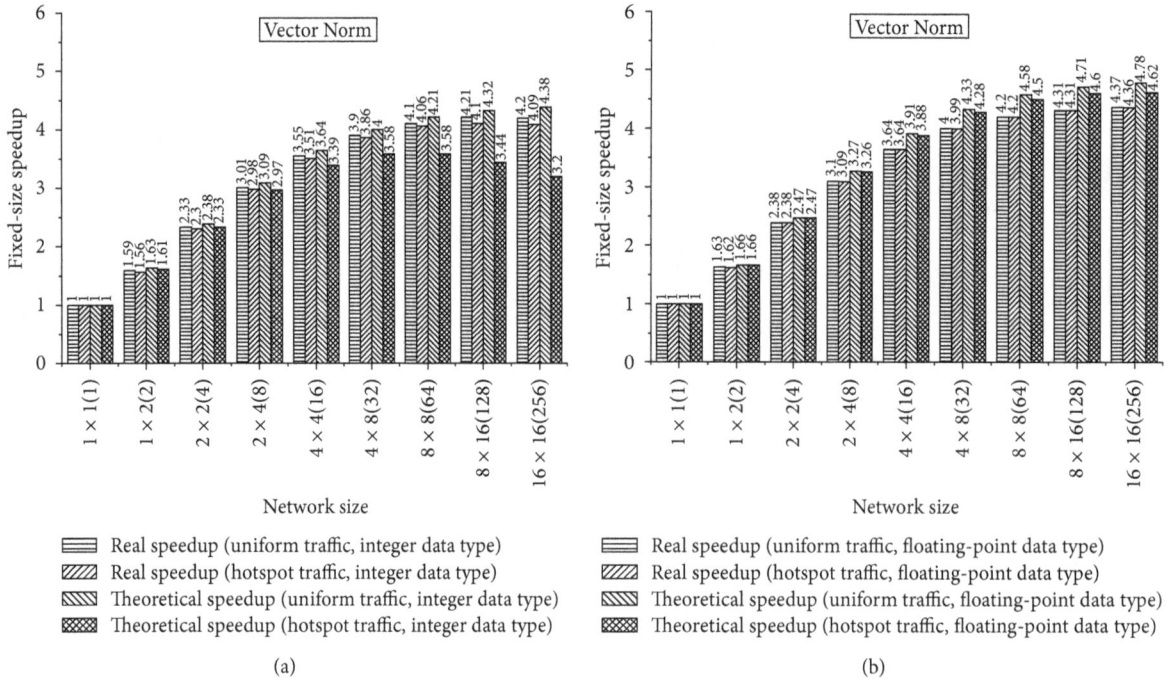

FIGURE 13: Effect of traffic models: Vector Norm with (a) integer data type and (b) floating-point data type.

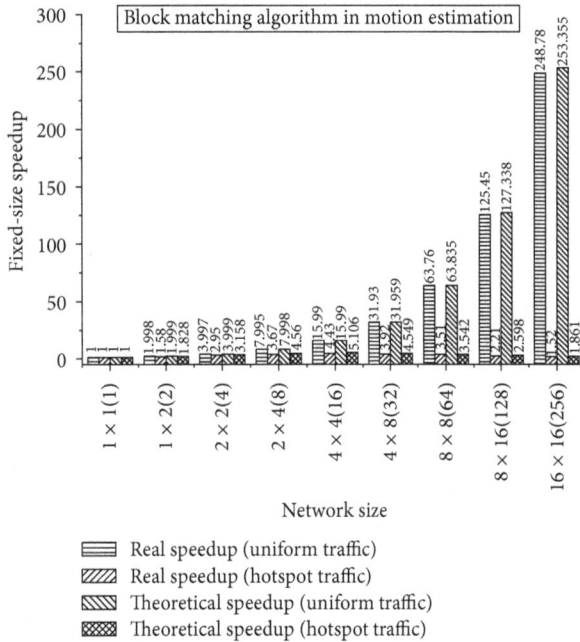

FIGURE 14: Effect of traffic models: Block Matching Algorithm in Motion Estimation.

accesses flow towards the central PM node and hence results in huge network contention.

(iii) Because hotspot traffic model consumes much more network contention latency than uniform traffic model, the speedup with hotspot traffic model is smaller than that with uniform traffic model for the same network size. The difference becomes larger when the network size is increasing.

5.5.3. Effect of Noncommunication/Communication Ratio. Figure 15 shows the effect of noncommunication/ communication ratio on the real and theoretical speedups of Wavefront Computation under both uniform and hotspot traffic models, and Figure 16 shows the effect of noncommunication/communication ratio on the real and theoretical speedups of Vector Norm under both uniform and hotspot traffic models.

(i) For the same network factors, the theoretical and real speedups for the floating point data type is higher than those for the integer data type. This is as expected because when the noncommunication time increases, the portion of communication latency becomes less significant, thus achieving higher performance.

(ii) For hotspot traffic model, the increase of noncommunication/communication ratio shifts the optimal network size (N_{opt}) to a larger value. For integer data type, $N_{\mathrm{opt}} \approx 50$ for Wavefront Computation and $N_{\mathrm{opt}} \approx 44$ for Vector Norm. For floating-point data type, $N_{\mathrm{opt}} \approx 255$ for Wavefront Computation and $N_{\mathrm{opt}} \approx 226$ for Vector Norm.

1×1 (1) to 4×4 (16) but fall when the network size is from 4×8 (32) to 16×16 (256). From Figure 14, we can see that the speedup under hotspot traffic model is very small, because the Block Matching Algorithm in Motion Estimation makes a large number of memory

Wavefront computation (a)

Fixed-size performance vs. Network size

Real speedup (uniform traffic, integer data type)
Real speedup (uniform traffic, floating-point data type)
Theoretical speedup (uniform traffic, integer data type)
Theoretical speedup (uniform traffic, floating-point data type)

Wavefront computation (b)

Fixed-size speedup vs. Network size

Real speedup (hotspot traffic, integer data type)
Real speedup (hotspot traffic, floating-point data type)
Theoretical speedup (hotspot traffic, integer data type)
Theoretical speedup (hotspot traffic, floating-point data type)

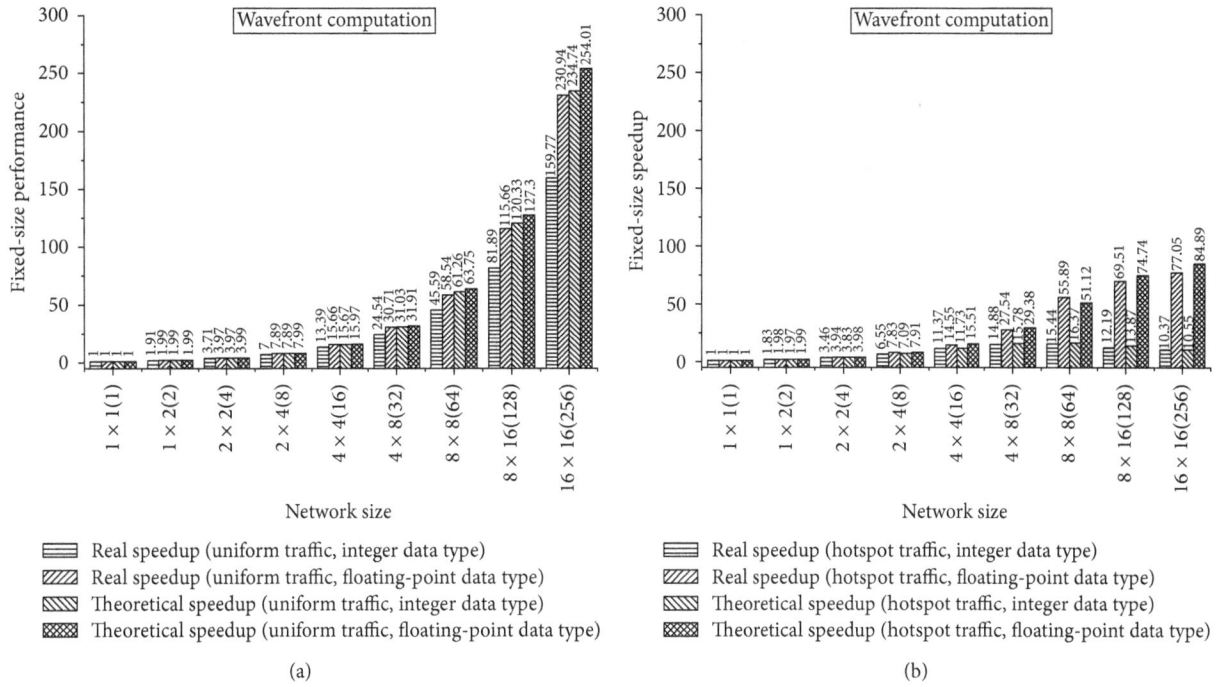

FIGURE 15: Effect of noncommunication/communication ratio: Wavefront Computation with (a) uniform traffic model and (b) hotspot traffic model.

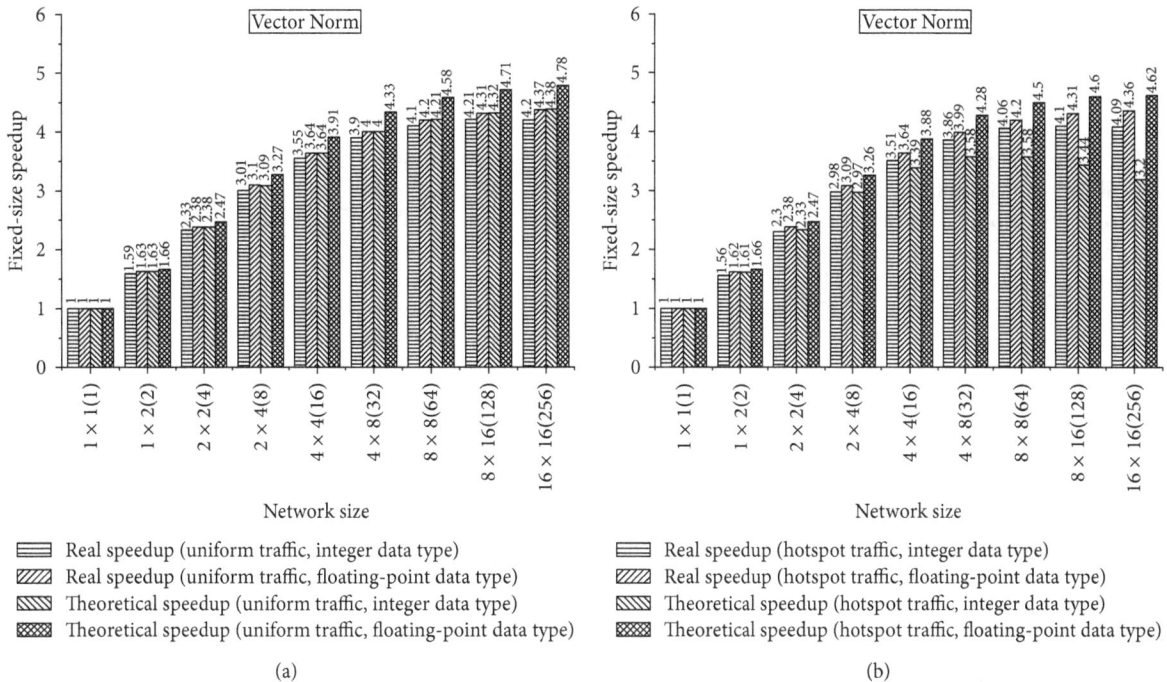

Vector Norm (a)

Fixed-size speedup vs. Network size

Real speedup (uniform traffic, integer data type)
Real speedup (uniform traffic, floating-point data type)
Theoretical speedup (uniform traffic, integer data type)
Theoretical speedup (uniform traffic, floating-point data type)

Vector Norm (b)

Fixed-size speedup vs. Network size

Real speedup (hotspot traffic, integer data type)
Real speedup (hotspot traffic, floating-point data type)
Theoretical speedup (hotspot traffic, integer data type)
Theoretical speedup (hotspot traffic, floating-point data type)

FIGURE 16: Effect of noncommunication/communication ratio: Vector Norm with (a) uniform traffic model and (b) hotspot traffic model.

6. Applicability and Limitation

6.1. Applicability. The target architectures and applications of our study are homogenous OLPCs and data-parallel applications, respectively. Homogenous OLPCs are such an on-chip computing platform that have a number of computing cores that run in and an on-chip network that provides concurrent pipelined communication, and data-parallel applications represent a wide range of applications whose data sets can be partitioned in parallel into data subsets handled individually by the same program running in different processor cores. Scalability is the common characteristic of both. Considering that homogenous OLPCs and data-parallel applications match each other very well in nature and hence data-parallel applications can obtain good speedup in homogenous OLPCs, the performance model proposed by the paper is applicable to homogenous OLPCs for data-parallel applications.

The performance model is general for homogenous OLPCs and a variety of data-parallel applications. For any particular application, a customized many-core platform such as application-specific architecture and hardware accelerator will be superior, but a NoC-based homogenous OLPC would be better than such a custom-designed hardware architecture when a variety of data-parallel applications share the same OLPC. The custom-designed many-core platform is specific so as not to be in the range of the general homogenous OLPCs. GPU (Graphic Processing Unit) is such kind of hardware accelerator for graphic processing as the name suggests. Although GPGPU (General-Purpose GPU) exhibits generality to some extent by providing programmability in its GPUs, it is still specific because the programmable GPU adopts a special structure for accelerating the graphic processing applications and the interconnections in GPGPU is special for such as stream processing and data shuffling that are common in graphic processing. Therefore, GPGPU is not in the range of homogenous OLPCs. Besides data-parallel applications, there exist other applications that do not have the scalability feature, so the proposed model is not applicable to those applications' performance analysis.

6.2. Limitation. The proposed performance model is not suitable for many-core platforms in specific application areas and the applications without the characteristic of scalability. The purpose of the model is to offer a general but workable way to estimate and evaluate the performance of homogenous OLPCs for data-parallel applications. Because the network communication latency and the ways of storing data of data-parallel applications are two of the most significant factors affecting the performance of homogenous OLPCs, when we establish the speedup performance model, the network communication latency and the data storing ways are stressed out and modeled in detail. Therefore, the processor behavior such as cache hierarchy and cache miss is not considered. We assume that all of the data are moved from the external memory to the appointed on-chip memory regions in different nodes before the system handles the data and the performance is measured from the time when the system begins handling the data, even if the data is continuously fed from the external memory, part of the latency can be hidden in the process of data handling, and we emphasize analyzing the effect of the data storing ways, so the model does not describe the situation that data are moved from the external memory.

7. Conclusion

Understanding the speedup potential that homogenous OLPC computing platforms can offer is a fundamental question to continually pursuing higher performance. This paper has focused on analyzing the performance of homogeneous OLPCs for data-parallel applications. Because the enhancement of application performance in OLPCs may be restricted by the increasing network communication latency even though the number of cores increases, one main issue for the analysis is to properly capture the network communication. We first detailed a network communication latency model by proposing two abstract concepts (*equivalent serial packet* and *equivalent serial communication*). Then, based on the network communication latency model, we have proposed the performance model. By considering the uniform and hotspot traffic models, the performance model has two detailed forms to reflect the distributed way and the centralized way of storing data of data-parallel applications. Essentially, the performance model revisits Amdahl's Law under the context of homogenous OLPCs. Theoretic analysis and real application experiments demonstrate that our model provides a feasible way to estimate and evaluate the performance of data-parallel applications onto homogenous OLPCs.

In the future, we plan to extend the performance model by considering the cache hierarchy and cache miss and the external memory access. Another direction is to emphasize studying the effect of topologies and communication protocols on the performance models of homogenous OLPCs.

Notations

k: Number of nodes in each dimension

N: The number of processor nodes, $N = k^2$

s: The number of subtasks in the serial part of a program

p: The number of subtasks or communications in the parallel part of a program

α: The ratio between the serial part and the parallel part in a program, $\alpha = s/p$

τ_c: The time of a communication

τ_{nc}:   The execution time of a subtask, that is, noncommunication time

H: Average hop count of transmitting a packet

τ_{1hop}:   The time of transmitting a packet in one hop

τ_t: Average time of transmitting a packet in the network

M: The number of packets in a communication

γ: The number of equivalent serial packets in a communication

ω: The number of equivalent serial communications in a program

T_T: The communication overhead of a program on OLPCs

S: Speedup

S_{max}:   Maximal speedup

S_{min}:   Minimal speedup.

Conflict of Interests

The authors declare that there is no conflict of interests regarding the publication of this paper.

Acknowledgments

The research is partially supported by the Hunan Natural Science Foundation of China (no. 2015JJ3017), the Doctoral Program of the Ministry of Education in China (no. 20134307120034), and the National Natural Science Foundation of China (no. 61402500).

References

[1] S. Borkar, "Thousand core chips—a technology perspective," in *Proceedings of the 44th ACM/IEEE Design Automation Conference (DAC '07)*, pp. 746–749, June 2007.

[2] M. Horowitz and W. Dally, "How scaling will change processor architecture," in *Proceedings of the IEEE International Solid-State Circuits Conference, Digest of Technical Papers (ISSCC '04)*, vol. 1, pp. 132–133, February 2004.

[3] A. Jantsch and H. Tenhunen, *Networks on Chip*, Kluwer Academic Publishers, New York, NY, USA, 2003.

[4] T. Bjerregaard and S. Mahadevan, "A survey of research and practices of network-on-chip," *ACM Computing Surveys*, vol. 38, no. 1, pp. 1–51, 2006.

[5] J. D. Owens, W. J. Dally, R. Ho, D. N. Jayashima, S. W. Keckler, and L.-S. Peh, "Research challenges for on-chip interconnection networks," *IEEE Micro*, vol. 27, no. 5, pp. 96–108, 2007.

[6] G. Amdahl, "Validity of the single processor approach to achieving large scale computing capabilities," in *Proceedings of the American Federation of Information Processing Societies Conference*, pp. 383–385, Atlantic City, NJ, USA, April 1967.

[7] X. Li and M. Malek, "Analysis of speedup and communication/computation ratio in multiprocessor systems," in *Proceedings of the Real-Time Systems Symposium*, pp. 282–288, 1988.

[8] J. Paul, "Amdahl's law revisited for single chip systems," *International Journal of Parallel Programming*, vol. 35, no. 2, pp. 101–123, 2007.

[9] S. Cho and R. G. Melhem, "Corollaries to Amdahl's law for energy," *Computer Architecture Letters*, vol. 7, no. 1, pp. 25–28, 2008.

[10] M. D. Hill and M. R. Marty, "Amdahl's law in the multicore era," *Computer*, vol. 41, no. 7, pp. 33–38, 2008.

[11] G. Loh, "The cost of uncore in throughput-oriented manycore processors," in *Proceedings of the Workshop on Architectures and Languages for Throughput Applications (ALTA '08)*, Beijing, China, June 2008.

[12] P. P. Pande, C. Grecu, M. Jones, A. Ivanov, and R. Saleh, "Performance evaluation and design trade-offs for network-on-chip interconnect architectures," *IEEE Transactions on Computers*, vol. 54, no. 8, pp. 1025–1040, 2005.

[13] W. Dally and B. Towles, *Principles and Practices of Interconnection Networks*, Morgan Kaufmann, 2004.

[14] Gaisler, "Leon3 processor," http://www.gaisler.com/doc/Leon3%20Grlib%20folder.pdf.

[15] M. Millberg, E. Nilsson, R. Thid, S. Kumar, and A. Jantsch, "The Nostrum backbone—a communication protocol stack for networks on chip," in *Proceedings of the 17th International Conference on VLSI Design*, pp. 693–696, January 2004.

[16] W. Zhu, V. C. Sreedhar, Z. Hu, and G. R. Gao, "Synchronization state buffer: supporting efficient fine-grain synchronization on many-core architectures," in *Proceedings of the 34th Annual International Symposium on Computer Architecture (ISCA '07)*, pp. 35–45, June 2007.

[17] ITU, "Draft ITU-T recommendation and final draft international standard of joint video specification (ITU-T Rec. H.264 — ISO/IEC 14496-10 AVC)," 2003.

5

Theoretical and Experimental Investigation of Direct Detection Optical OFDM Systems with Clipping and Normalization

Jiang Wu[1] and Zhongpeng Wang ⓘ[2,3]

[1]*School of Information, Zhejiang Sci-Tech University, Hangzhou 310023, China*
[2]*School of Information and Electronic Engineering, Zhejiang University of Science and Technology, Hangzhou 310023, China*
[3]*State Key Laboratory of Millimeter Waves, Southeast University, Nanjing 210096, China*

Correspondence should be addressed to Zhongpeng Wang; wzp1966@sohu.com

Academic Editor: Jit S. Mandeep

A data clipping and normalization technique is employed to improve the performance of the overall direct detection optical orthogonal frequency division multiplexing (DCO-OFDM) system. A detailed analysis of clipping distortion introduced by digital clipping and normalization is provided. The normalization operation amplifies the clipped data signal to the maximum input amplitude of a digital-to-analog converter (DAC). Based on the analysis, a BER formula of the proposed scheme is derived over the AWGN channel and single fiber channel. Performance of an optical clipped OFDM with normalization is assessed through numerical simulations and Monte Claro simulation over the AWGN channel. Theoretical analysis and simulation results both show that the clipping and normalization scheme can greatly improve the BER of an optical OFDM. In particular, BER performance of the proposed transmission scheme was measured in a practical OFDM transmission platform. The measured experimental results show that the clipped and amplified OFDM signal exhibits superior performance in comparison with the conventional OFDM signal. The received sensitivity at a BER of 10^{-3} for a 4 Gsamples/s (2.6667 Gbits/s) clipped and normalized OFDM signal with clipping ratio of 4 after 100 km standard single-mode fiber (SMF) transmission was improved by 4.3 dB when compared with the conventional OFDM system. The measured results also showed that the clipped OFDM signal exhibits superior performance in comparison with the conventional OFDM signal. Therefore, a clipping and normalization at the transmitter is most effective, and a substantial performance improvement can be obtained by a simple normalization after clipping.

1. Introduction

In recent years, orthogonal frequency division multiplexing (OFDM) modulation has been proposed in optical communications due to its advantage of having robustness to fiber transmission impairments such as chromatic dispersion [1–3]. In general, optical OFDM communication can be categorized as coherent optical OFDM (CO-OFDM) and direct detection optical OFDM (DDO-OFDM). Due to the low cost and complexity of DDO-OFDM, DDO-OFDM is mainly considered in short-range optical communication systems [4]. However, the optical OFDM signal in both types of transmission systems has a disadvantage of high PAPR. The high peak-to-average power ratio (PAPR) can lead to larger nonlinear effects, which can cause optical signal

intensity fluctuation and degrade the BER performance of systems [5, 6]. For single-channel DDO-OFDM systems, high PAPR gives rise to fiber nonlinearity such as self-phase modulation (SPM). This results in the nonlinear phase shift induced by SPM. On the contrary, high PAPR requires large dynamic range of nonlinear devices such as digital-to-analog converters (DACs), power amplifiers, and external modulators. These add the cost of systems. Therefore, mitigating nonlinearity distortion in optical OFDM by reducing PAPR becomes the important issue. In order to reduce PAPR, many methods have been proposed for PAPR reduction, such as companding [7], clipping [8], DCT precoding [9], DFT precoding [10, 11], and other methods [12, 13]. Recently, a fiber nonlinearity equalizer, which is based on support vector classification for optical OFDM, has been proposed [14]. Among these methods, the precoding technique draws great attention for its effective PAPR reduction. The precoding method utilizes additional DCT (or DFT) and IDCT (or IDFT) at the transmitter and receiver, respectively. So this method increases the computational complexity of the system.

Among the PAPR reduction methods, the clipping is the simplest. The idea of the clipping technique is clipping the signal components that exceed some clipping threshold. In general, the clipping technique can cause impairment to a communication system. In practical system, clipping may be employed before or after DACs. Clipping, which is employed before DACs, is called soft clipping, whereas clipping used after DACs is called hard clipping. A number of papers have studied the clipping effect on the radio frequency (RF) OFDM signals [15–17]. However, there is an important difference between DDO-OFDM and RF-OFDM: the RF baseband signal is complex-valued, whereas time-domain signal in DDO-OFDM is real-valued and non-negative [18]. A real-valued time signal can be obtained when the Hermitian symmetry condition must be satisfied in OFDM subcarriers. After that, a DC bias is added to the real-valued signal resulting in a nonnegative valued signal for DDO-OFDM. In addition, for optical OFDM system, high PAPR can give rise to fiber nonlinearity which in turn results in performance degradation of systems. Thus, some theory and analyses in RF-OFDM are not directly applied to optical OFDM. A number of papers have analyzed the clipping effects in optical OFDM systems [8,19–21]. In [22, 23], the performance of IM/DD optical OFDM with the digital baseband distortion has been analyzed and evaluated by simulations. However, these researches concern on how to decrease the negative effect of clipping on the performance.

The study results of the literature [24] showed that the degradation in the error rates is very small for clipping with the clipping ratio of 6 dB in frequency selective fading channels. The researched results of the paper [25] showed that the performance of the overall communication system can be improved by a baseband clipping method. So, the benefits of the clipping on the OFDM system can be obtained by fixing the appropriate value of clipping ratio. In fact, the performance degradation of systems caused by clipping is negligible when the proper clipping lever is used, whereas obtaining PAPR reduction by clipping can help to mitigate nonlinearities effects of fiber and Mach–Zehnder modulator. However, the advantage of the nonlinearity mitigation by clipping is often be neglected in optical OFDM systems. The clipped optical OFDM signal with low PAPR may obtain two benefits compared to the conventional optical OFDM signal: one is that clipped OFDM results in nonlinear interference reduction caused by the Kerr effect in a fiber-optic channel [26–28] and the second is that clipped OFDM may reduce to the impact of the MZM nonlinear on the system [5,29]. Thus, proper clipping in optical OFDM may improve the BER of optical OFDM.

In this paper, we mainly investigate the effect of clipping DDO-OFDM systems. As we know, there is less attention on the benefits of the clipping technique for PAPR of the baseband OFDM signal in a direct detection OFDM communication system. Inspired by the idea in paper [24, 25], we proposed a data clipping and normalization scheme, which limits the high-power spikes signal and amplifies the clipped low amplitude signal. We drive an analytical expression for the BER of the DDO-OFDM system based on data clipping and normalization. The BER expression quantitatively shows how the data clipping and normalization scheme improves the BER performance. We compare our analytical results to simulation results in an additive white Gaussian noise (AWGN) channel. The simulation results agree well with the analytical results. In particular, BER performance of the proposed transmission scheme is measured in a practical OFDM transmission platform. The experimental results show that the BER performance of the proposed system can be greatly improved compared with the conventional optical OFDM system and conventional clipped optical OFDM system. Thus, the proposed joint data clipping and normalization technique can be used for the optical communication system design.

The paper is organized as follows. In Section 2, the principle of a direct detection optical OFDM system with data clipping and normalization is presented. In Section 3, the theoretical analysis is given and the BER formulas of the clipped and normalized optical OFDM are derived. Section 4 presents the simulation results. Section 5 provides the experimental configuration, and Section 6 gives measured results. Finally, Section 7 concludes this paper.

2. System Principle

In this work, we employ a joint data clipping and normalization technique in the transmitter for an optical DDO-OFDM system to improve the BER performance of a system. Figure 1 illustrates the block diagram of the proposed DDO-OFDM transmitter, in which N subcarriers is used. At the transmitter, an incoming bit stream is first mapped into an M-ary quadrature amplitude modulation (M-QAM) symbol data stream. Then, a data symbol vector with N size is formed as $X = [0 \ X_1 \ ... \ X_{N/2-1} \ 0 \ X^*_{N/2-1} \ ... \ X^*_1]$. In order to produce a real-valued time-domain OFDM signal x_n, the input of the IFFT X_k is Hermitian symmetric, that is,

$$X_k = X^*_{N-k}, \quad 1 \le k \le N - 1,$$
$$X_k = 0, k = 0, \quad N/2, \tag{1}$$

where $*$ denotes complex conjugate. Thus, performing an N-point IFFT on X produces real-valued signal x, that is,

$$x_n = \frac{1}{\sqrt{N}} \sum_{k=0}^{N-1} X_k \exp\left(\frac{j2\pi kn}{N}\right), \quad 0 \le n \le N - 1. \tag{2}$$

After IFFT, the OFDM signal is fed into the data clipping blocks, where the signal is clipped by the data clipping block. We consider that the symmetric clipping is used. The operation of the data clipping (or baseband clipping) is represented as follows [31]:

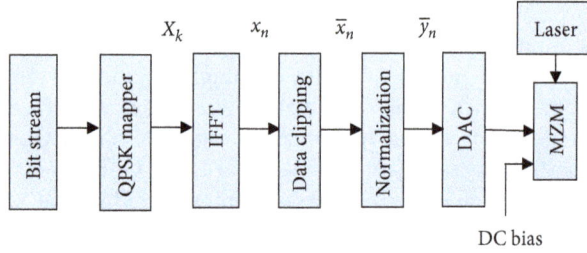

FIGURE 1: Block diagram of the proposed OFDM transmitter.

$$\bar{x} = g(x_n) = \begin{cases} x_n & \text{if } |x_n| \leq A_{\text{clip}}, \\ A_{\text{clip}} \dfrac{x_n}{|x_n|} & \text{if } |x_n| \geq A_{\text{clip}}. \end{cases} \quad (3)$$

The clipping ratio (CR) is defined by the following equation:

$$\text{CR} = \frac{A_{\text{clip}}^2}{P_x}, \quad (4)$$

where P_x is the average power of the transmitted signal and A_{clip} is the maximum allowed signal amplitude of the clipped OFDM signal, and it can be written as

$$A_{\text{clip}} = \sqrt{\text{CR} \cdot P_x}. \quad (5)$$

Higher clipping level values can be applied to the signal for reducing the clipping noise at the expense of increasing the electrical power of the signal. Hence, there is a trade-off between power efficiency and noise in the selection of the clipping level or bias. When A_{clip} is smaller than the maximum allowable amplitude of a DAC device, A_{max}, the system performance can be improved by normalizing the clipped OFDM signal. The clipped and normalized signal, which is the output of the normalization, is written as

$$\bar{y}_n = \frac{A_{\text{max}}}{A_{\text{clip}}} \bar{x} = \begin{cases} G x_n & \text{if } |x_n| \leq A_{\text{clip}}, \\ G\left(A_{\text{clip}} \dfrac{x_n}{|x_n|}\right) & \text{if } |x_n| \geq A_{\text{clip}}, \end{cases} \quad (6)$$

where $1 \leq G \leq (A_{\text{max}}/A_{\text{clip}})$. G is the normalized factor. The maximum allowable normalized factor can be expressed as

$$G_{\text{max}} = \frac{A_{\text{max}}}{A_{\text{clip}}} = \frac{A_{\text{max}}}{\sqrt{\text{CR} \cdot P_x}} = \frac{A_{\text{max}}}{\sqrt{P_x}} \cdot \text{CR}^{-1/2}. \quad (7)$$

Obviously, the smaller the value of clipping ratio is, the higher the normalization factor is. From Equation (6), we can see that the small signal, whose amplitude is below threshold A, is amplified by G times. When $G = 1$, the output of the normalization is the conventional clipped OFDM signal. In order to utilize fully the dynamic range of a DAC device, the clipped signal $\bar{x}(n)$ is normalized to $\bar{y}(n)$ with a peak-to-peak of $2A_{\text{max}}$, which is the maximum input amplitude of a DAC device. The signal $\bar{y}(n)$ is converted into the continuous OFDM signal $y(t)$ by a DAC device in order to drive a Mach–Zehnder modulator (MZM).

According to Bussgang's theorem [30], any nonlinear function of $x(n)$ can be decomposed into a scaled version of $x(n)$ plus a distortion term $d(n)$ that is uncorrelated with $x(n)$. For example, the clipped signal can be written as

$$\bar{x}(n) = \alpha x(n) + d(n). \quad (8)$$

Scaling factor α can be calculated as follows [30]:

$$\alpha = 1 - e^{-\text{CR}} + \frac{\sqrt{\pi}}{2} \sqrt{\text{CR}} \cdot \text{erfc}(\sqrt{\text{CR}}). \quad (9)$$

Then, the clipped signal $\bar{x}(n)$ is amplified by G times, and it can be written as

$$\bar{y}(n) = G \cdot \bar{x}(n) = G \cdot \alpha \cdot x(n) + G \cdot d(n). \quad (10)$$

In the optical link, the electrooptic conversion is performed by a standard Mach–Zehnder modulator (MZM) biased at the quadrature point. The signal $y(t)$ is biased with a biased voltage V_{DC} in order to ensure the waveform is nonnegative, and then, the signal is intensity modulated onto the optical carrier. The biased and clipped time-domain signal can be expressed as $\tilde{y}(t) = y(t) + V_{\text{DC}}$. This signal $\bar{y}(t)$ is then used to drive an idea optical modulator, where the output optical power is a replica of the corresponding electrical-drive signal.

The fiber nonlinearity remains sufficiently low when the optical launch power is not too high. In this case, the optical fiber link can be modeled as a linear channel [31]. Hence, the optical signal and noise can be assumed independent. Similar to the analysis in [31], polarization-mode dispersion (PMD) and fiber nonlinearity are neglected, and group-velocity dispersion (GVD) is the only fiber impairment considered. In the optical electric field domain, the fiber is modeled as a linear system with the transfer function given by Barros and Kahn [31]:

$$H(\omega) = e^{j\omega^2 (\beta_2/2)L}, \quad (11)$$

where β_2 is the fiber GVD parameter and can be expressed as $\beta_2 = -D\lambda^2/2\pi c$. Here, λ represents the carrier wavelength, and c is the speed of light. In addition, L is the fiber length, and ω is the angle frequency. Assume $h(t)$ is the impulse response of the fiber channel, and it equal to the inverse Fourier transformation of $H(\omega)$.

At the receiver, the PD detector converts the received optical signal to electrical signal. The sampled discrete signal can be expressed as

$$\begin{aligned} r(n) &= \bar{y}(n) \otimes h(n) + w(n) \\ &= G\bar{x} \otimes h(n) + w(n) \\ &= G\alpha x(n) \otimes h(n) + Gd(n) \otimes h(n) + w(n), \end{aligned} \quad (12)$$

where $w(n)$ is AWGN and $\otimes$ denotes circular convolution. By taking the DFT of Equation (12), the frequency domain data on the kth subarrier can be written as

$$R_k = G\alpha X_k H_k + G D_k H_k + W_k, \quad (13)$$

where H_k is the channel frequency response on the kth subcarrier and W_k is the additive white Gaussian noise in the frequency domain. When zeros forcing (ZF) equalization is employed, the coefficient of the kth subcarrier is given by

$$Q_{ZF,k} = \frac{1}{H_k}, \quad k = 0, 1, ..., N-1. \qquad (14)$$

After equalization, the received signal can be expressed as

$$\widehat{R}_k = G\alpha X_k + GD_k + H_k^{-1}W_k. \qquad (15)$$

3. Theoretical Analysis

3.1. BER Performance over AWGN Channel. In this subsection, we firstly analyze the BER performance of the proposed transmission scheme for the AWGN channel. At the receiver of the proposed system, the clipped OFDM signal with normalization after transmission over the AWGN channel can be expressed as

$$\begin{aligned} r(n) &= G \cdot \overline{x}(n) + w(n), \\ &= G \cdot \alpha \cdot x(n) + G \cdot d(n) + w(n). \end{aligned} \qquad (16)$$

The frequency signal at the kth subcarrier can be given as

$$R_k = G\alpha X_k + GD_k + W_k. \qquad (17)$$

Obviously, the clipping operation reduces the power of the signal. The loss power due to clipping can be estimated. It is can be given by the following equation [32]:

$$\sigma_d^2 = 2 \cdot \sqrt{\frac{2}{\pi}} \cdot \sigma_x^2 CR^{-3/2} e^{-CR/2}. \qquad (18)$$

Base on Equation (16), the SNR of the OFDM signals can be expressed as

$$\lambda = \frac{\alpha^2 G^2 \sigma_x^2}{G^2 \sigma_c^2 + \sigma_0^2} = \frac{\alpha^2 \sigma_x^2}{\sigma_d^2 + (\sigma_0^2/G^2)}. \qquad (19)$$

In order to take the useful dynamic range of a DAC, the allowable maximum normalization gain G_{max} in Equation (7) is adopted. By substituting Equations (7) and (18) into Equation (19), the SNR under the maximum normalization gain can be written as

$$\begin{aligned} \lambda^{am} &= \frac{\alpha^2 \sigma_x^2}{\sigma_d^2 + \sigma_0^2/G^2}, \\ &= \frac{\left[1 - e^{-CR^2} + (\sqrt{\pi}/2)\sqrt{CR} \cdot \operatorname{erfc}(\sqrt{CR})\right]^2 \cdot \sigma_x^2}{2 \cdot \sqrt{2/\pi} \cdot \sigma_x^2 CR^{-3/2} e^{-CR/2} + \sigma_0^2 \cdot \sigma_x^2 \cdot \left(\sqrt{CR}/A_{max}^2\right)}. \end{aligned} \qquad (20)$$

Thus, the theoretical BER formula of the clipped and normalized M-QAM OFDM can be expressed from [33]:

$$P_{b,MQAM}^{am} = \left(\frac{4 - 2^{(2-m/2)}}{m}\right) Q\left(\sqrt{\frac{3}{(M-1)}\lambda^{am}}\right), \qquad (21)$$

where $m = \log_2 M$ is the number of bits per constellation point. The Q function is defined as

$$Q(x) = \frac{1}{\sqrt{2\pi}} \int_x^\infty e^{-t^2/2} \, dt. \qquad (22)$$

When $G = 1$, Equation (20) becomes the SNR of the conventional clipped OFDM, and it can be expressed as

$$\lambda^c = \frac{\left[1 - e^{-\gamma^2} + \sqrt{\pi}/2\gamma \cdot \operatorname{erfc}(\gamma)\right]^2 \cdot \sigma_x^2}{2 \cdot \sqrt{2/\pi} \cdot \sigma_x^2 \gamma^{-3} e^{-\gamma^2/2} + \sigma_0^2}. \qquad (23)$$

Therefore, the theoretical BER expression of conventional OFDM over the AWGN channel is given from [33]:

$$P_{b,MAQM} = \left(\frac{4 - 2^{(2-m/2)}}{m}\right) Q\left(\sqrt{\frac{3}{(M-1)}\lambda^c}\right). \qquad (24)$$

3.2. BER Performance for Single-Mode Fiber Channel. Based on Equation (15), the signal-to-noise-and-distortion ratio (SNDR) of the kth subcarrier for the clipped and normalized OFDM system is given by

$$\begin{aligned} \lambda_{k,G}^{ZF} &= \frac{E\left[|G\alpha X_k|^2\right]}{E\left[|GD_k + H_k^{-1}W_k|^2\right]} = \frac{G^2\alpha^2 \cdot \sigma_X^2}{G^2\sigma_D^2 + H_k^{-2}\sigma_W^2} \\ &= \frac{\alpha^2 \cdot \sigma_X^2}{\sigma_D^2 + (1/G^2) \cdot (\sigma_W^2/H_k^2)}, \end{aligned} \qquad (25)$$

where σ_X^2 is the variance of the signal X_k, σ_D^2 is the variance of the signal D_k, and σ_W^2 is the variance of the AWGN noise W_k.

Thus, the BER performance of every subcarrier can be expressed as

$$P_{b,k}^{am} = \left(\frac{4 - 2^{(2-m/2)}}{m}\right) Q\left(\sqrt{\frac{3}{(M-1)}\lambda_{k,G}^{ZF}}\right). \qquad (26)$$

The overall performance of the clipped and normalized system can be expressed as

$$P_b^{am} = \frac{1}{N} \sum_{n=0}^{N-1} P_{b,k}^{am}. \qquad (27)$$

When $G = 1$ in Equation (25), the clipped and normalized OFDM system is converted into the conventional clipped OFDM system. Then, the signal-to-noise-and-distortion ratio (SNR) of the kth subcarrier at the receiver for the conventional clipped OFDM system is given by

$$\lambda_k^{ZF} = \frac{E\left[|\alpha X_k|^2\right]}{E\left[|D_k + H_k^{-1}W_k|^2\right]} = \frac{\alpha^2 \cdot \sigma_X^2}{\sigma_D^2 + (\sigma_W^2/H_k^2)}. \qquad (28)$$

Therefore, the BER of every subcarrier of the clipped system without normalization can be expressed as

$$P_{b,k} = \left(\frac{4 - 2^{(2-m/2)}}{m}\right) Q\left(\sqrt{\frac{3}{(M-1)}\lambda_k^{ZF}}\right). \qquad (29)$$

The overall performance of the clipped OFDM system can be expressed as

$$P_b = \frac{1}{N} \sum_{n=0}^{N-1} P_{b,k}. \tag{30}$$

Compared with Equations (25) and (28), when $G > 1$, we can conclude that

$$\lambda_{k,G}^{ZF} \geq \lambda_k^{ZF}. \tag{31}$$

Furthermore, there is $P_b^{am} < P_b$. Thus, a joint clipping and normalization scheme can be employed in optical OFDM in order to improve BER performance. The BER of the simulation and measured results in the following section show that the BER performance of the clipped and normalized OFDM signal can be greatly improved compared with the conventional clipped OFDM and conventional OFDM signals.

4. Simulation Results

In the following simulation, there are two criterions: equal average power and equal maximum amplitude criterions. The equal average power criterion is employed in many PAPR reduction techniques. In this case, the clipping obtains PAPR reduction at cost of degradation in BER performance. Obviously, the clipping decreases the average power of the OFDM signal, and the BER performance can be slightly improved if the loss average power of the signal can be compensated. In most reported literatures, the negative effect of clipping on BER performance has been widely concerned. Many techniques such as channel coding and iterated signal detector are employed to reduce the effect of clipping on systems. However, the effect of clipping on BER is not severe when the clipping ratio is greater than 6 dB.

4.1. Q Factor of the Clipped OFDM by Simulation. Q factor usually serves as another metric to evaluate the quality of signals in terms of its SNR. The higher the value of Q factor, the lower the bit error rate (BER). We first research the effect of the clipping with and without normalization on the Q factor of signals. The measured or estimated BER is also expressed as a Q factor using the following equation [34]:

$$Q(\text{dB}) = 20 \, \text{Log}\left(\sqrt{2} \times \text{erfc}^{-1}(2 \times \text{BER})\right), \tag{32}$$

where erfc^{-1} is the inverse complementary error function. Based on Equation (32), the obtained BER can be converted to the corresponding Q factor in dB.

Figure 2 shows the relationship between Q factor and clipping ratio (CR) over the AWGN channel at the received SNR of 6 dB. The clipping ratio is varied from 2 to 12, and thus, the Q factor simulation results are taken. From the graph, it is analyzed that, for the clipping and normalization scheme, as the clipping ratio increases, the Q factor increases when the value of clipping ratio is smaller than 4, but it decreases when the value of clipping ratio is higher than 4. The maximum Q factor was found to be 12 dB at the optimum clipping ratio of 4. However, for the conventional clipping scheme, the Q factor increases, when the value of clipping ratio is below 6, but the Q factor was found to be

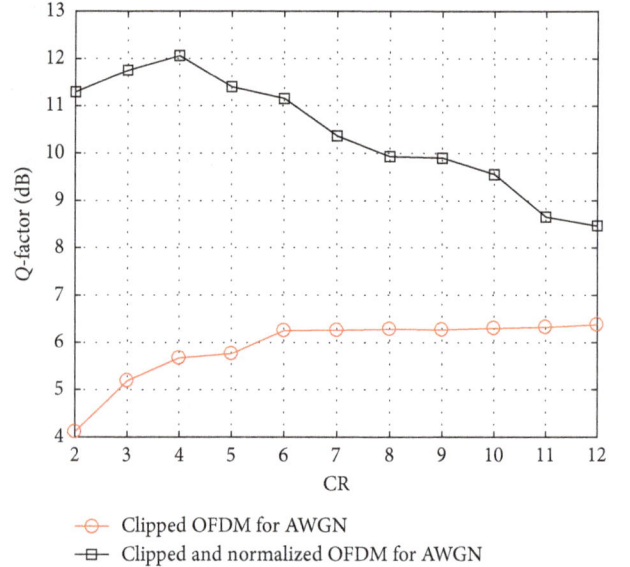

FIGURE 2: Q factor values as a function of the clipping ratio in clipped QPSK OFDM with and without normalization over the AWGN channel.

about 6.2 dB at clipping ratio of 6, but it does not almost increase with the increasing of clipping ratio. Therefore, the clipped and normalized OFDM system can obtain the best BER performance over the AWGN channel at clipping ratio (CR) of 4.

For a given received SNR of 8 dB, Figure 3 shows the relationship of the Q factor between clipping ratio in clipped QPSK OFDM with and without normalization after 100 km fiber transmission. The results which are similar to that of Figure 2 can also be obtained. The maximum Q factor value of the clipped OFDM signal with normalization was found to be 12 dB at clipping ratio of 4. The Q factor decreases with increase of the clipping ratio when the value of clipping ratio is higher than 4. For the conventional clipped OFDM, the Q factor increases when the value of clipping ratio varies from 2 to 6. However, the improvement Q factor is very little when the value of the clipping ratio varies from 6 to 12. Therefore, similar to Figure 2, the OFDM system can obtain the best BER performance over 100 km fiber channel at clipping ratio (CR) of 4.

From the experiment results in Figures 2 and 3, it can be seen that data normalization can improve the quality of the clipped OFDM signals over AWGN and fiber channel. The benefit of clipping can be achieved by data normalization. Therefore, clipping based on normalization can offer better BER as compared to the conventional clipping with the same level clipping ratio. This is also be verified by following experiments.

4.2. BER Performance over AWGN Channel. In the following subsection, we first evaluated the BER performance over the AWGN channel by theoretical numerical and simulation methods under without normalization case. Figure 4 shows the BER comparison for different clipping ratios for clipped OFDM without the normalization case. In this case, the

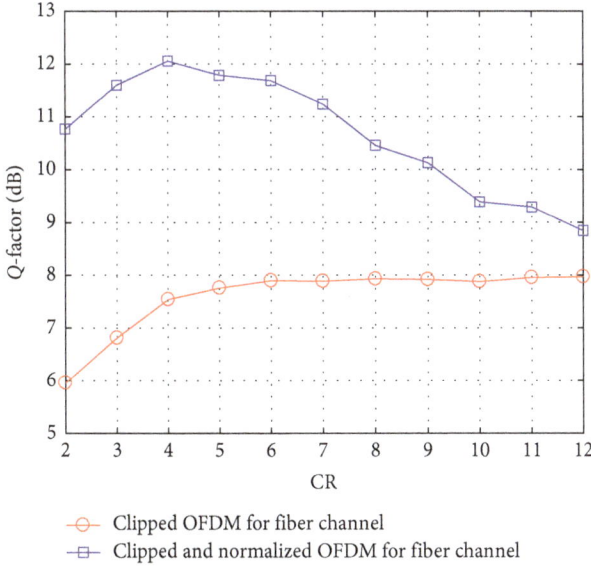

FIGURE 3: Q factor values as a function of the clipping ratio in clipped QPSK OFDM with and without normalization over the fiber channel, and the transmission is 100 km.

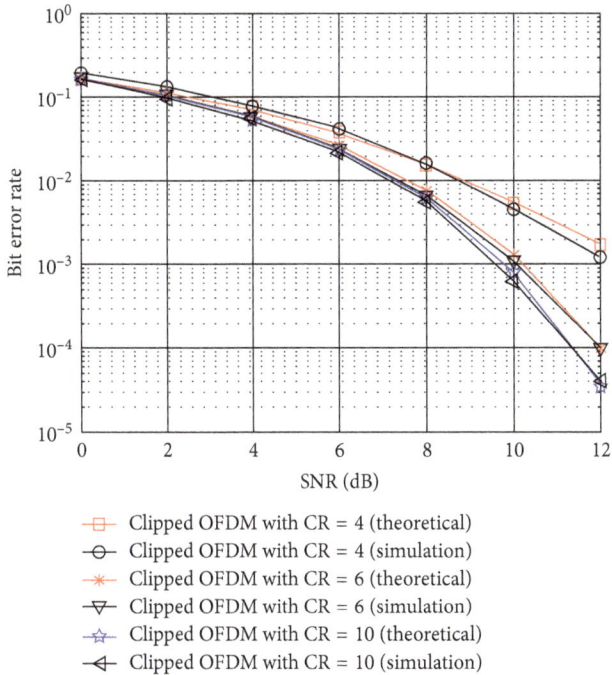

FIGURE 4: BER performance comparison for different clipping ratios by simulation and theoretical methods over the AWGN channel without the normalization case.

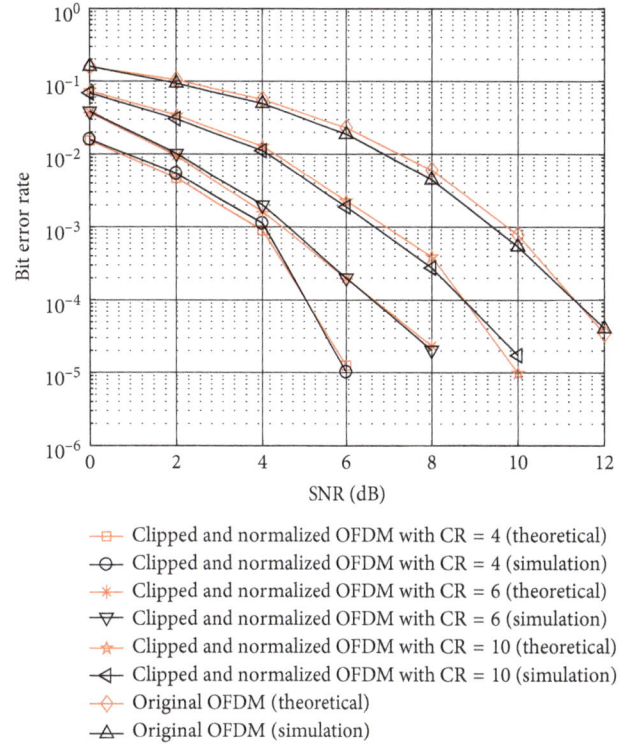

FIGURE 5: BER performance comparison for different clipped and normalized OFDM by simulation and theoretical methods over the AWGN channel.

BER performance of systems. The simulation results agree with the theoretical numerical results. These also verify the effectiveness of Equation (21).

4.3. BER Performance over Fiber Channel. For a short distance fiber-optic system with low data rate, fiber dispersion is the main impairment of degrading system performance. The optical signals affected by fiber nonlinearity are minimal. Therefore, in our simulation experiment, we neglect the effect of fiber nonlinearity so that we focus on studying how to improve the system performance by combating dispersion impairment. We evaluated the BER performance of clipped OFDM over the fiber channel with and without normalization. The main simulation parameters are shown in Table 1. Some of the parameters of the OFDM system are given according to the 802.16a OFDM physical layer (PHY) specification [35]. The number of points of FFT/IFFT, number of data subcarriers, and pilot subcarriers are fixed at 256, 192, and 8. In addition, 56 null subcarriers serve as the guard band in order to enable the signal to naturally decay and create the FFT brick wall shaping [35]. Cyclic prefixing (CP) with 32 samples symbols is inserted to combat intersymbol interference (ISI) due to channel dispersion. We use D and λ to state the dispersion coefficient and wavelength of continuous lightwave (CW) laser, respectively.

Figure 6 shows the simulation results for clipped optical OFDM without normalization over the fiber channel. From Figure 6, we can see that the clipping can degrade the BER

clipping degraded the BER of systems. The theoretical numerical results also are shown in Figure 4. It is clear that simulation results agree well the theoretical results. The experiment results verify the validity of Equation (24). The performances of the clipped OFDM with normalization are compared in Figure 5 for the AWGN channel by theoretical numerical and simulation. The experiment results show that the joint data clipping and normalization can improve the

TABLE 1: Simulation parameters.

Dispersion coefficient, D	17 ps/(nm km)
Optical carrier wavelength (λ)	1550 nm
Modulation	QPSK
Sample rate	4 Gs/s
FFT size	256
Number of data subcarriers	192
Number of pilot subcarriers	8
Null subcarriers	56
Length of CP	32
L (length of the fiber)	100 km

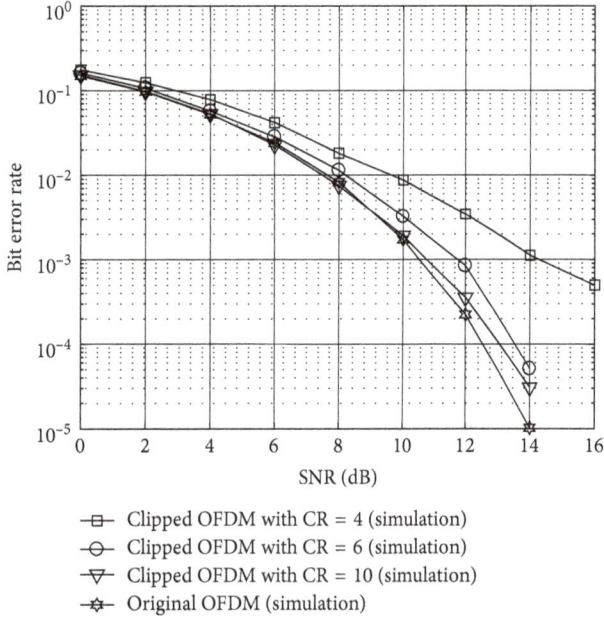

-□- Clipped OFDM with CR = 4 (simulation)
-○- Clipped OFDM with CR = 6 (simulation)
-▽- Clipped OFDM with CR = 10 (simulation)
-✳- Original OFDM (simulation)

FIGURE 6: BER performance comparison for different clipped and normalized OFDM without normalization over the fiber channel.

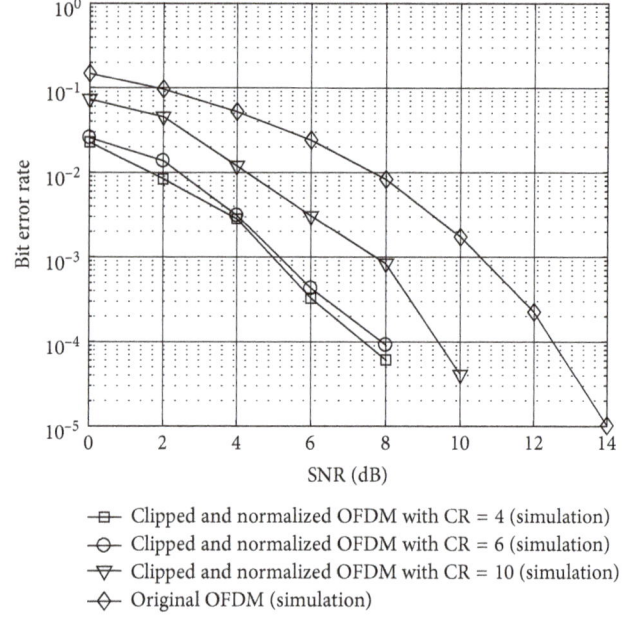

-□- Clipped and normalized OFDM with CR = 4 (simulation)
-○- Clipped and normalized OFDM with CR = 6 (simulation)
-▽- Clipped and normalized OFDM with CR = 10 (simulation)
-◇- Original OFDM (simulation)

FIGURE 7: BER performance comparison for different clipped and normalized OFDM over the fiber channel by simulation.

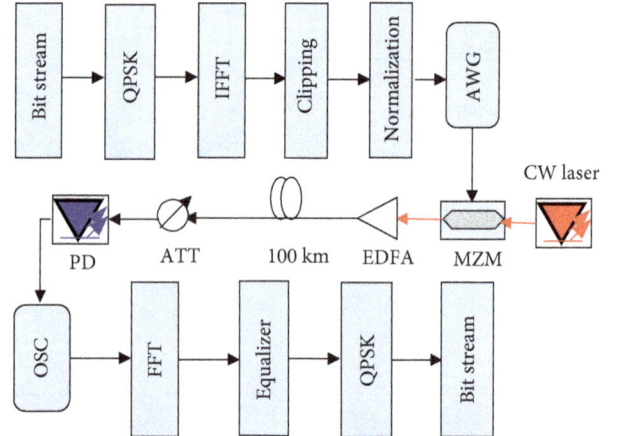

FIGURE 8: Experimental setup (EDFA: erbium-doped fiber amplifier; ATT: attenuator; PD: photodiode; OSC: oscilloscope; AWG: arbitrary waveform generator; ATT: attenuator).

performance under without normalization. But the effect of clipping on BER performance is very small when clipping ratio is bigger than 10 dB. At the bit error rate (BER) 10^{-3}, the BER performance can be degraded about 3.5, 1.3, and 0.2 dB for clipped OFDM with clipping ratio of 4, 6, and 10 compared to that of the original OFDM.

Figure 7 shows simulation results of the clipped and normalized OFDM over the fiber channel. It can be seen that the normalization can greatly improve the BER performance of systems. At BER 10^{-3}, the BER performance of the clipped and normalized OFDM can improve about 5.2, 5.1, and 2.4 dB SNR gain than original OFDM for CR of 4, 6, and 10, respectively.

However, the obtained performance improvement is at the cost of average power increasing. All different clipped and amplified OFDM signals are amplified to the same maximum amplitude, which is the maximum allowable amplitude of DAC in the transmitter end. The more the normalization factor is, the lower the clipping ratio is. Meantime, the higher the average power is, the lower the clipping ratio is.

5. Experimental Setup

Figure 8 depicts the experimental setup employed to evaluate the performance improvement of an optical direct detection OFDM system provided by the data clipping and normalization technique. In our experiment, 4 Gs/s OFDM signals are employed. An OFDM frame based on QPSK modulation has 256 subcarriers, among which 192 (96 ∗ 2) subcarriers are used for data transmission, 8 pilot subcarriers are used for channel estimation and synchronization, and 56 subcarriers are set to zero as the guard interval. In addition, 32 samples symbols are used as cyclic prefixing to avoid the interblock interference.

The time-domain QPSK OFDM waveforms are first generated in a Matlab program and uploaded onto an arbitrary waveform generator (AWG) operated at 4 GS/s to generate the corresponding OFDM analog signal with the peak-peak value of 1 V. The net bit rate was 4 Gs/s ∗ 192/2/256 ∗ 256/(256 + 32) ∗ 2 (bits/symbol for QPSK) = 2.6667 Gbits/s. Then, the analog OFDM signal is modulated via an external Mach–Zehnder modulator (MZM). The MZM is biased at the quadrate point with a biased voltage of 2.2 V. In the experiment, the value of V_π is setup at 3.5 V. The optical signal is obtained by using a continuous lightwave (CW) laser with nominal wavelength of 1549.261 nm. The generate signal is injected into erbium-doped fiber amplifier (EDFA) with a noise figure of 5 dB to adjust the proper launched power into fiber. Then, the resulting optical signal is launched into a 100 km standard single-mode fiber (SSMF). The attenuation and dispersion coefficients of the fiber are 0.19 dB/km and 17 ps/(nm km), respectively. At the receiver side, the signal is sampled using a digital storage oscilloscope (DSO) with 10 Gsamples/s and applied to the OFDM demodulator. After demodulation, the received signal is equalized and applied to demapper.

6. Experimental Results

In this section, we will measure the performance of the clipped QPSK OFDM signal of 4 Gs/s with and without normalization after 100 km standard SMF transmission in our transmission experiment platform. The measured experiment results will verify that a joint clipping and normalization scheme can improve the BER performance of the system.

6.1. Q Factor of the Clipped OFDM with Normalization. Figure 9 shows Q factor versus launch optical power after 100 km transmission for various values of CR. For the clipped and normalized OFDM system, the optimum Q factor is about 12.7 dB with launch power 6 dBm for clipped OFDM with CR of 4. However, the optimum Q factor is 10.2 dB with launch power 6.0 dBm for the 4 Gs/s original OFDM signal. At low fiber launch power, the transmission performance is mainly by amplified spontaneous (ASE) noise. Hence, the Q factor is improved with an increase in the fiber launch power, as shown in Figure 9. On the contrary, when the fiber launch power is over 6 dBm for the proposed system, the Q factor is decreased with an increase in the fiber launch power due to the impact of fiber nonlinearity.

High launch optical power can cause the fiber nonlinearity impairments, which includes self-phase modulation (XPM), cross-phase modulation (XPM), and four shift mixing (FWM). The nonlinear distortion can be approximated as a phase shift of the transmission signal. And the resulting phase shift causes the degradation of the performance of an optical system. The introduced shift phase shift can be expressed as follows [36]:

$$\phi_{NL}(t) = \gamma L_{eff} P(t), \qquad (33)$$

where γ is the nonlinear coefficient of the fiber, $P(t)$ is the instantaneous optical power, and L_{eff} is the nonlinear effective length and is expressed as follows [36]:

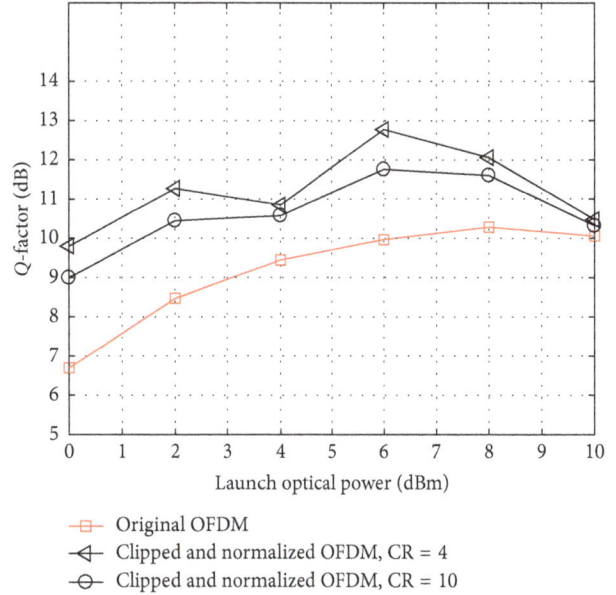

FIGURE 9: Q factor values as a function of the launch power in 4 Gs/s QPSK DCO-OFDM after 100 km of transmission.

$$L_{eff} = \frac{1 - e^{-\alpha L}}{\alpha}, \qquad (34)$$

where L is the length of fiber and α is the attenuation coefficient.

Based on Equation (33), the greater the launch power, the greater the phase shift, and therefore, the lower the Q factor. When the fiber nonlinearity is not the main impairment of degrading system performance, the effect of fiber nonlinearity can be neglected. From Figure 9, it can be seen that the clipped and amplified OFDM with clipping ratio of 4 can obtain the Q factor improvement of 2.9 dB when compared to that of the original OFDM under launch optical power of 6 dB case. When the launch optical power is higher than the optimum value of 6 dB, the Q factor value begins to decline due to the fiber nonlinearity.

From Figure 9, the measured experiment results confirm that the joint data clipping and normalization scheme can greatly improve the BER performance of systems. The approach only is done in the digital baseband and does not need any hardware equipment.

6.2. BER of the Clipped OFDM without Normalization. Figure 10 shows the measured BER performance results at the launched optical power of the 6 dBm case for the conventional clipped OFDM systems. As shown in Figure 3, the clipped system has excellent transmission performance because of the fiber nonlinearity mitigation by clipping when compared with the conventional OFDM system. Compared with the conventional OFDM system at BER = 10^{-3}, the data clipping can improve the BER performance approximately 1.3, 2.2, 0.8, and 0.5 dB with CR at 4, 6, 8, and 10, respectively. From Figure 10, it can be seen that the optimum power clipping ratio is 6 under without normalization. In the optimum clipping case, the performance of the system can

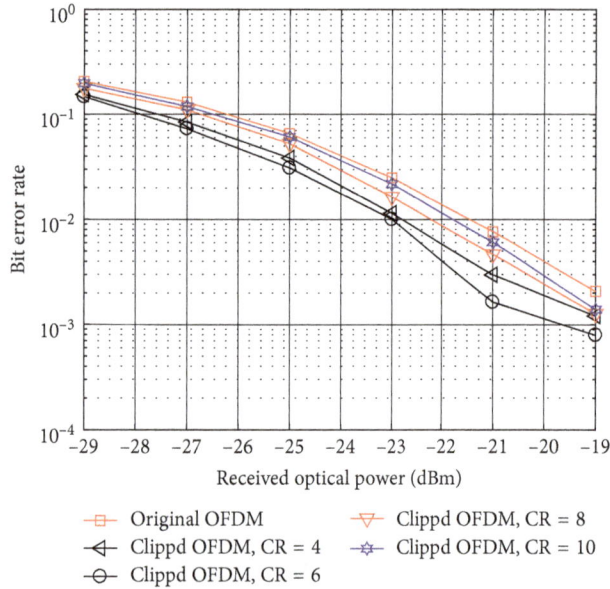

FIGURE 10: Comparison of the BER performance of the clipped QPSK OFDM signals with the launch power of 6 dBm.

obtain the greatest improvement. This is because that the decreasing of PAPR of clipped OFDM is helpful to reduce influence of nonlinearities of the fiber and MZM modulator on the BER performance of the optical OFDM systems [26, 27]. The results of Figure 10 show that proper clipping can improve the BER performance of optical OFDM transmission systems.

6.3. BER of the Clipped OFDM with Normalization. Figure 11 shows the measured BER performance comparisons of the clipped and normalized OFDM signals with the launch power of 6 dBm at CR of 4, 6, 8, and 10 and the normalization gain is $G_{max} = (A_{max}/\sqrt{P_x}) \cdot CR^{-1/2}$. In fact, the lower the value of CR, the bigger the gain G is.

From Figure 11, we can see that the sensitivity of the received signals of the clipped and normalized OFDM system can be improved about 4.3, 4, 3.5, and 3 dB with CR of 4, 6, 8, and 10 compared with conventional OFDM, respectively. The performance of the proposed system with CR of 4 is the most best.

Comparing Figure 10 with Figure 11, we can see that the sensitivity of the clipped and normalized OFDM system is higher than that of the conventional clipped OFDM system at the difference value of CR. These measured experiment and the former simulation experiment results both verify that the joint data clipping and normalization scheme can greatly improve the BER of systems.

7. Conclusions

We proposed a digital clipped and normalized DDO-OFDM system and theoretically described its principles. The BER formula of the proposed scheme is derived for the AWGN channel and fiber channel. Analytical results show that there is an optimum clipping ratio in clipped and normalized

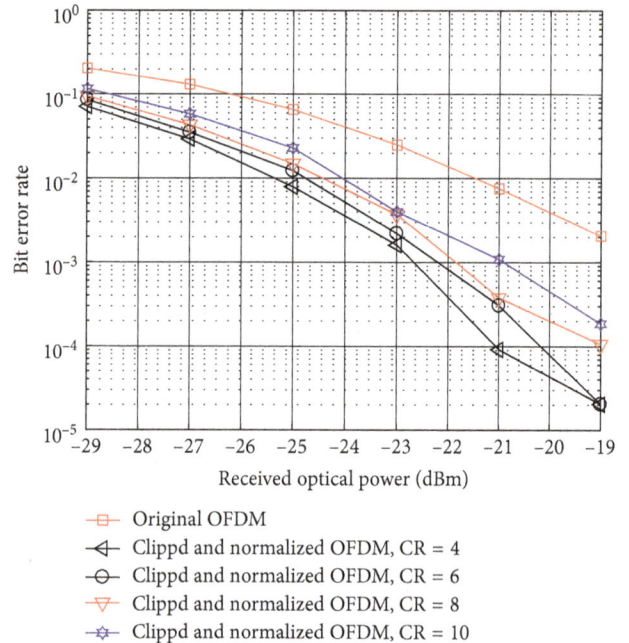

FIGURE 11: Comparison of the BER performance of the clipped and normalized QPSK OFDM signals with the launch power of 6 dBm.

OFDM under maximum allowable amplitude constraint. The improvement of a direct detection optical OFDM system achieved by the data clipping and normalization has been assessed experimentally and analyzed theoretically. The measured experimental, simulation, and theoretical analysis results all show that the joint clipping and normalization method can greatly improve the BER of systems. The proposed scheme is an efficient scheme without adding any hardware equipment. For the 4 Gs/s clipped and normalized QPSK OFDM signal with CR = 4, the received sensitivity of the systems can be substantially improved by approximately 4.3 dB as compared with the conventional OFDM systems. The experimental results also show that the conventional clipped OFDM scheme outperforms the conventional OFDM scheme. Additionally, the main advantage of the proposed scheme is easy implementation in the optical system.

Conflicts of Interest

The authors declare that they have no conflicts of interests.

Acknowledgments

This work was supported in part by the Zhejiang Provincial Natural Science Foundation of China under LY17F050005, Open Fund of the State Key Laboratory of Millimeter Waves (Southeast University, Ministry of Education, China) under K201214, and Key Laboratory of University Wireless

Communications (BUPT), Ministry of Education, PR China under KFKT-2018101.

References

[1] J. Amstrong, "OFDM for optical communication," *Journal of Lightwave Technology*, vol. 27, no. 3, pp. 189–204, 2009.

[2] J. Armstrong and A. J. Lowery, "Power efficient optical OFDM," *Electronics Letters*, vol. 42, no. 6, pp. 370–372, 2006.

[3] C. Sánchez, B. Ortega, J. L. Wei, J. Tang, and J. Capmany, "Analytical formulation of directly modulated OOFDM signals transmitted over an IM/DD dispersive link," *Optics Express*, vol. 21, no. 6, pp. 7651–66, 2013.

[4] W. Shieh, H. Bao, and Y. Tang, "Coherent optical OFDM: theory and design," *Optics Express*, vol. 16, no. 2, pp. 841–859, 2008.

[5] Y. London and D. Sadot, "Nonlinear effects mitigation in coherent optical OFDM system in presence of high peak power," *Journal of Lightwave Technology*, vol. 29, no. 21, pp. 3275–3281, 2011.

[6] Y. Gao, J. Yu, J. Xiao, Z. Cao, F. Li, and L. Chen, "Direct-detection optical OFDM transmission system with pre-emphasis technique," *Journal of Lightwave Technology*, vol. 29, no. 14, pp. 2138–2145, 2011.

[7] S. Azou, Ş. Bejan, P. Morel, and A. Sharaiha, "Performance improvement of a SOA-based coherent optical-OFDM transmission system via nonlinear companding transforms," *Optics Communications*, vol. 336, no. 4, pp. 177–183, 2015.

[8] C. R. Berger, Y. Benlachtar, R. I. Killey, and P. A. Milder, "Theoretical and experimental evaluation of clipping and quantization noise for optical OFDM," *Optics Express*, vol. 19, no. 18, pp. 17713–17728, 2011.

[9] M. Sung, J. Lee, and J. Jeong, "DCT-precoding technique in optical fast OFDM for mitigating for fiber nonlinearity," *IEEE Photonics Technology Letters*, vol. 25, no. 22, pp. 2209–2212, 2013.

[10] L. Cai, Y. Qi, T. Jiang et al., "Investigation of coherent optical multi-band DFT-S OFDM in long haul transmission," *IEEE Photonics Technology Letters*, vol. 24, no. 19, pp. 1704–1707, 2012.

[11] T. A. Truong, M. Arzel, H. Lin, B. Jahan, and M. Jézéque, "DFT precoded OFDM—an alternative candidate for next generation PONs," *Journal of Lightwave Technology*, vol. 32, no. 6, pp. 1228–1238, 2014.

[12] J. Xiao, J. Yu, X. Li et al., "Hadamard transform combined with companding transform technique for PAPR reduction in an optical direct-detection OFDM system," *Journal of Optical Communications and Networking*, vol. 4, no. 10, pp. 709–714, 2012.

[13] H. Chen, J. Yun, J. Xiao, Z. Cao, L. Fan, and L. Chen, "Nonlinear effect mitigation based on PAPR reduction using electronic pre-distortion technique in direct-detection optical OFDM system," *Optical Fiber Technology*, vol. 19, no. 5, pp. 387–391, 2013.

[14] N. Tu, S. Mhatli, E. Giacoumidis, L. Van Compernolle, M. Wuilpart, and P. Megret, "Fiber nonlinearity equalizer based on support vector classification for coherent optical OFDM," *IEEE Photonics Journal*, vol. 8, no. 2, pp. 1–9, 2016.

[15] H. Chen and A. M. Haimovich, "Iterative estimation and cancellation of clipping noise for OFDM signals," *IEEE Communications Letters*, vol. 7, no. 7, pp. 305–307, 2003.

[16] L. Wang and C. Tellambura, "Analysis of clipping noise and tone-reservation algorithms for peak reduction in OFDM systems," *IEEE Transactions on Vehicular Technology*, vol. 57, no. 3, pp. 1675–1694, 2008.

[17] L. Wang and C. Tellambura, "A simplified clipping and filtering technique for PAR reduction in OFDM systems," *IEEE Signal Processing Letters*, vol. 12, no. 6, pp. 453–456, 2005.

[18] Z. Yu, R. J. Baxley, and G. T. Zhou, "EVM and achievable data rate analysis of clipped OFDM signals in visible light communication," *EURASIP Journal on Wireless Communications and Networking*, vol. 2012, no. 1, pp. 1–16, 2012.

[19] L. Nadal, M. S. Moreolo, J. M. Fàbrega, and G. Junyent, "Low complexity PAPR reduction techniques for clipping and quantization noise mitigation in direct-detection O-OFDM systems," *Optical Fiber Technology*, vol. 20, no. 3, pp. 208–216, 2014.

[20] S. Dimitrov, S. Sinanovic, and H. Haas, "Clipping noise in OFDM-based optical wireless communication systems," *IEEE Transactions on Communications*, vol. 60, no. 4, pp. 1072–1081, 2012.

[21] J. Armstrong and B. J. C. Schmidt, "Comparison of asymmetrically clipped optical OFDM and DC-biased optical OFDM in AWGN," *IEEE Communications Letters*, vol. 12, no. 5, pp. 343–345, 2008.

[22] E. Vanin, "Performance evaluation of intensity modulated optical OFDM system with digital baseband distortion," *Optics Express*, vol. 19, no. 5, pp. 4280–4293, 2011.

[23] L. Chen, B. Krongold, and J. Evans, "Theoretical characterization of nonlinear clipping effects in IM/DD optical OFDM systems," *IEEE Transactions on Communications*, vol. 60, no. 8, pp. 2304–2312, 2012.

[24] K. R. Panta and J. Armstrong, "Effect of clipping on the error performance of OFDM in frequency selective fading channels," *IEEE Transactions on Wireless Communications*, vol. 3, no. 2, pp. 668–671, 2004.

[25] H. Qian, R. Raich, and G. T. Zhou, "On the benefits of deliberately introduced baseband nonlinearities in communication systems," in *Proceedings of IEEE International Conference on Acoustics, Speech, and Signal Processing*, pp. 905–908, Atlanta, GA, USA, September 2004.

[26] G. Bosco, R. Cigliutti, A. Nespola et al., "Experimental investigation of nonlinear interference accumulation in uncompensated links," *IEEE Photonics Technology Letters*, vol. 24, no. 14, pp. 1230–1232, 2012.

[27] P. Poggiolini, A. Carena, V. Curri, G. Bosco, and F. Forghieri, "Analytical modeling of nonlinear propagation in uncompensated optical transmission links," *IEEE Photonics Technology Letters*, vol. 23, no. 11, pp. 742–744, 2011.

[28] A. Demir, "Nonlinear phase noise in optical-fiber-communication systems," *Journal of Lightwave Technology*, vol. 25, no. 8, pp. 2002–2032, 2007.

[29] H. Chen, J. He, J. Tang et al., "Nonlinear distortion evaluation of MZM with equivalent mathematical model calculation in IM/DD OOFDM transmission system," *Optics Communications*, vol. 316, pp. 31–36, 2014.

[30] H. Ochiai and H. Imai, "Performance analysis of deliberately clipped OFDM signals," *IEEE Transactions on Communication*, vol. 50, no. 1, pp. 89–101, 2002.

[31] D. J. F. Barros and J. M. Kahn, "Comparison of orthogonal frequency-division multiplexing an on-off keying in amplified direct-detection single-mode fiber systems," *Journal of lightwave technology*, vol. 28, no. 12, pp. 1811–1820, 2010.

[32] D. Mestdagh, P. Spruyt, and B. Biran, "Effect of amplitude clipping in DMT-ADSL transceivers," *Electronics Letters*, vol. 29, no. 15, pp. 1354-1355, 1993.

[33] J. G. Proakis, *Digital Communications*, McGraw-Hill, New York, NY, USA, 1995.

[34] A. V. T. Cartaxo and T. M. F. Alves, "Theoretical and experimental performance evaluation methods for DD-OFDM systems with optical amplification," *Journal of Microwaves Optoelectronics and Electromagnetic Applications*, vol. 10, no. 1, pp. 82–94, 2011.

[35] IEEE Std 802.16a, *IEEE Standard for Local and Metropolitan Area Networks-Part 16: Air Interface for Fixed Broadband Wireless Access Systems-Amendment 2: Medium Access Control Modifications and Additional Physical Layer Specifications for 2-11 GHz*, IEEE, Piscataway, NJ, USA, 2003.

[36] B. Foo, B. Corcoran, C. Zhu, and A. J. Lowery, "Distributed nonlinearity compensation of dual-polarization signals using optoelectronics," *IEEE Photonics Technology Letters*, vol. 28, no. 20, pp. 2141–2144, 2016.

6

Applying Pulse Width Modulation in Body Coupled Communication

Miltiadis Moralis-Pegios, Pelagia Alexandridou, and Christos Koukourlis

Telecommunications Systems Laboratory, Electrical and Computer Engineering Department, Democritus University of Thrace, 67100 Xanthi, Greece

Correspondence should be addressed to Christos Koukourlis; c.koukourlis@gmail.com

Academic Editor: Zhenjiang Zhang

We study the application of Pulse Width Modulation (PWM) technique in Body Coupled Communication. The term Body Coupled Communication is used in order to specify that the human body or a part of it, such as an arm, is used as a path that transfers digital information. The digital information is either generated due to coupling of the body, for example, in medical equipment as a result of a measurement, or generated by external circuitry attached somewhere onto the body and is transferred to a terminal by touching it. In this paper, the latter case will be described, where, for illustration purposes, the touch of the human hand on a doorknob triggers the unlocking mechanism.

1. Introduction

As more mobile devices such as cellular phones, PDAs, mp3 players, tablets, and notebooks become necessary and almost indispensable nowadays, there is an increasing necessity for an easier and more comfortable method to connect each other. Body Area Networks (BANs) propose new ways of wireless communications that allow an interaction between the user and devices placed in close proximity or coupled directly with the user's body. There are several proposed technologies appropriate for a BAN. One of the candidate technologies is Body Coupled Communication (BCC) or Intrabody Communication (IBC) which was firstly proposed by Zimmerman [1]. Body Coupled Communication, in which the human body is used as a signal transmission medium, is becoming a promising solution for BANs including a large variety of interesting potential applications.

As signals pass through the human body, electromagnetic noise and interference have reduced influence on transmissions. This technology has several advantages over the conventional RF approaches. Since its operation is based on near-field coupling (galvanic or capacitive), most of the signal from the transmitter is restricted to the body area without the possibility to be collected or interfere with other external

RF devices. Moreover, since the communication frequency can be lowered without affecting any antenna size, the power consumption of the transmitter and the receiver is also much reduced compared to Bluetooth or Zigbee applications. According to BCC approach, a low frequency carrier (less than 1 MHz) is used; so no energy is propagated, minimizing remote eavesdropping and interference with other BANs. All the above characteristics make this technology a very promising solution for BANs, health monitoring applications, and identification systems like entering a security PIN in a bank automated teller machine terminal.

In this paper, a wearable identification system that uses BCC technology is proposed but mainly the application of a specific digital modulation technique in Body Coupled Communications is studied. This technique is a digital version of the well-known Pulse Width Modulation (PWM), which provides special spectral characteristics and simplicity of implementation. Generally speaking, PWM is an "analog" modulation technique which employs a square wave carrier. This modulation carries the information in the duty cycle of the modulated square wave carrier. In our case, we employ this digital version of PWM, named B-PWM, by using only two distinct duty cycles, as the two different symbols that represent the two logic states, one and zero. It will be explained later that the two distinct duty cycles are preferable to be complementary; that is, they must sum up to 100%. For the case described later, we selected 25% and 75%, representing "logic zero" and "logic one," respectively.

In order to have a very compact and small sized transmitter, we developed our modulator around a Microchip general purpose microcontroller, almost without any other component. It includes the clock oscillator and can be programmed in order to accomplish the modulation and also store the required identification data. It can be powered by a single 3 to 5.5 V power supply, which can be accommodated by one or two Lithium battery cells; or it can be charged by inductive coupling when in close proximity to the receiver, just before the transmission.

It is well known from the literature [1, 2] that a signal can be transferred via the human body by utilizing either of two methods: the galvanic method and the capacitive method. The specific digital modulation proposed in this paper is suitable for either case. Also, from the literature it comes out that most of the systems consider as appropriate for signal transmission through the human body a center frequency in the range of hundreds of kHz up to a few MHz, for example, 500 kHz [3]. Due to the body behavior, there is no guarantee that the receiver's frequency characteristic is flat. Instead, it depends on the electronics and the behavior of the human body itself. In an attempt to explain this behavior, some approaches [1, 4] propose an electrical model of the human body for BCC.

2. Body Coupled Communication

2.1. Propagation Methods. As referred above, two coupling methods have been proposed in the literature: the capacitive and the galvanic coupling. In both approaches, a pair of electrodes is used at both transmitter and receiver ends in order to carry out communication. Capacitive coupling makes use of a single "signal" electrode, while the second "ground" electrode remains on air. On the other hand, galvanic coupling requires direct contact between the pair of electrodes and the human body. A significant difference between these two approaches is that in the capacitive coupling the environment can affect the behavior and the quality of the transmitted signal, while in galvanic coupling approach communication is affected more by the body tissue dielectric parameters [5].

2.2. Modulation Methods. In previous studies, several modulation schemes have been tested. OOK [6] and FSK [7, 8] are the most commonly used modulation methods regarding Body Coupled Communication systems. In other studies [3, 9], MSK and BPSK are proposed. In our experiments, OOK, FSK, and PWM have been tested. Although OOK has low energy and spectrum requirements and the implementation is simple, it shows low tolerance to noise. On the other hand, although FSK overcomes noise effects to some extent, comparing to OOK and B-PWM, the implementation is complex and less energy efficient. The digital Pulse Width Modulation

scheme proposed in this paper proves to be a suitable and appealing modulation technique for BCC technology, since it meets the basic requirements, that is, low energy consumption, simple implementation, and noise tolerance, as well as other practical advantages that will be described in Section 3.

2.3. Characterization. Before designing and implementing a WBAN (Wireless Body Area Network), one must consider the key design requirements of it. The present study is examined with the following criteria.

Suitability. The signal that passes through the human body must be suitable and not harmful for the human's health. In our implementation, carrier frequency ranges between 500 kHz and 1 MHz and the current flow is below 20 mA; thus, it meets this requirement according to the international regulations [10, 11].

Reliability. Reliability is a crucial factor for every communication system. For that reason, we take into account the influence of the human body and external noise on the signal. The proposed method's spectral behavior is identical for both logic stages of the transmitted signal.

Energy Efficiency. There is no doubt that power consumption is of key importance in wearable technology. Thus, we have chosen just a low-power microcontroller for the transmitter's circuit.

Size. The device size is also a crucial issue for every wearable device. In the present implementation, the transmitter consists of a single low-power 8-pin microcontroller (that includes an internal oscillator) and one pair of electrodes.

Security. In this paper, due to the design of an identification system, security is a significant factor, although in the literature more sophisticated methods have been proposed [12]. The communication system presented here provides communication between two devices that both have direct contact to the body. The signal is transmitted only through the human body; that is, it is not radiated; so there is no possibility of eavesdropping or interference with other WBANs without the user's notice.

Data Rate. A high data rate is a challenge for every communication system. In the literature, data rates up to 10 Mbps [13] have been mentioned. The nature of the application that is developed here does not require high data rate; so we merely set it to 10 kbps.

3. Binary Pulse Width Modulation (B-PWM)

3.1. Spectral Properties. The adoption of B-PWM is based on its inherent characteristic, as shown later, of occupying the same spectrum for both cases of symbol transmission (logic zero and logic one) due to the complementary values of selected duty cycles. This property is crucial for the front end stage of Body Coupled or Intrabody Communication designs

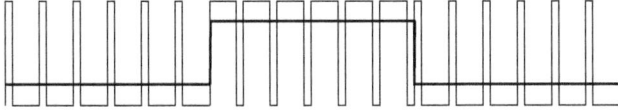

FIGURE 1: Pulse stream with two complementary duty cycles (25% and 75%) corresponding to logic "0" and logic "1." The bold line on top is the associated average value with voltage levels corresponding to the logic states.

[14, 15], because the behavior of the body in association with the front end electronics usually exhibits strange or not easily predictable behavior.

The chopping action of B-PWM can move the spectrum of the modulated signal anywhere by changing the repetition rate of the carrier pulse stream. In Figure 1, two discrete pulse streams (symbols) of complementary duty cycles (25% and 75%) are shown. We will show that the spectral content of all the segments (symbols) of the waveform is the same, under the assumption that the waveforms are periodic and they can be analyzed in Fourier series. The amplitudes of the corresponding harmonics prove to be equal but of 180° phase difference.

It is well known that average value of a periodic waveform is as follows:

$$A_0 = A \cdot \frac{\tau}{T_0}, \tag{1}$$

where A is the amplitude of the pulses, τ is the "on" time, and T_0 is the period of the waveform while the ratio τ/T_0 is the duty cycle. If we consider the two distinct symbols, s_0 and s_1, of complementary duty cycles, it is obvious that s_1 can be produced from s_0 with a simple linear process:

$$s_1(t) = -s_0(t) + A; \tag{2}$$

that is, s_1 is derived from s_0 when inverted and raised by A volts plus a time shift. Since the time shift does not generate any new spectral components and because of this linear relation, the spectral lines of both waveforms are identical, one by one, despite being with phase shift of 180° (i.e., inverted), with the only exception among the spectra of the two waveforms to be the spectral line at zero Hertz, that is, at DC. This similarity in spectral content of the two symbols employed ensures that the behavior of the body/receiver combination will be identical for both cases of signaling, that is, logic "one" and logic "zero."

As mentioned earlier, in our case, the B-PWM modulated carrier is generated at the output of a microcontroller powered by a single supply. This means that the transmitted waveform has a certain DC content during the duration of one bit which is equal to the mean value of the waveform. This DC content (which differs for the two logic levels) is where the digital information is stored. The whole system behaves as high pass stage not only due to the coupling with the body but, in some cases, due to the front end of the receiver as well. So, in any case, someone can consider that the information, which lies in the DC component of each symbol,

would be lost due to this high pass behavior. Actually, as it will be shown later, the DC component is indeed lost, but not the information, since the information lies in the zero-crossings of the modulated waveform and a restoration of the received waveform in rail-to-rail voltage levels takes place at the receiver. Since the duty cycle percentages stay at the prescribed values, there is no loss of information, as long as the demodulator is a simple Low Pass Filter.

3.2. Optimal Duty Cycle Selection. According to Figure 1, we can initially assume that the difference between the two complementary values of duty cycle should be the maximum in order to have the most distinct levels of the two mean values (bold line in Figure 1). This drives to an extreme selection of 0% and 100% duty cycle pair, which is not acceptable because there is no chopping of the signal as referred in Section 3.1. On the other hand, the maximization of the voltage difference between the two levels at the demodulation stage, as described above, is not the only criterion. This will be clarified later.

In Figure 2, a part of a modulated waveform, as it should appear after the front end of the receiver, is shown. Due to the coupling and the electronics of the receiver, the waveform is weighted above and below a horizontal straight line in order to result in zero average value. As described earlier, due to the capacitive coupling of the receiver/detector and the high pass behavior of the front end, the low frequency content of the transmitted signal (the mean value included) will be lost. In Figure 3, the output of the transmitter is shown while in Figure 4 the block diagram of the receiver is given. The actual views of the signals (oscilloscope captures) at the receiver are the ones shown in Figure 5 after the 1st and the 2nd stages. In the upper trace of Figure 5, an oscilloscope capture at the output of the first (current to voltage converter) stage is shown. It is clear that the low frequency content is lost and the waveform is weighted above and below the zero mean value (zero volts line). In the lower trace of the same figure, the output of the second (inverting voltage comparator) stage is given and it is shown that the data carrying waveform is restored in to its two original levels. Also in Figure 6 (upper trace), the same restored waveform at the voltage comparator's output is given in different time scale, followed (lower trace of Figure 6) by the waveform produced at the output of the demodulating Low Pass Filter. In the lower trace of Figure 6, it is shown that the two different duty cycles give two distinct voltage levels despite being rounded, which correspond to the two logic states of the data. The same waveform (output of the LPF) is also shown at the upper trace of Figure 7 in order to be compared to the output of the Pulse Shaping Stage of the receiver.

As stated earlier, at first glance, in Figure 5 (upper trace), it seems that the information lying in the voltage levels of Figure 1 (bold line) is lost. Fortunately, this is not the case, because the information actually lies in the zero-crossings of the modulated signal, according to the duty cycle: the signal, as shown in Figure 2 or Figure 5 (upper trace), will be restored at the original levels at the receiver after a voltage comparator stage. Since the signal can be distorted and noise is added, it is

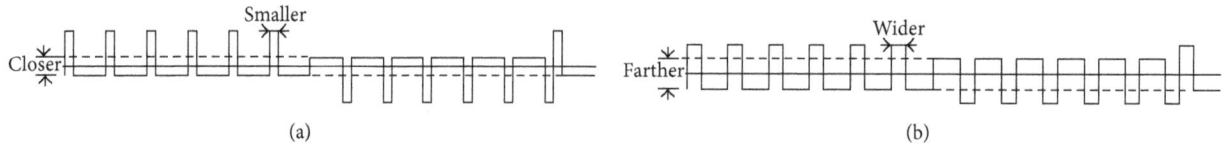

(a)   (b)

FIGURE 2: Two different cases of complementary duty cycles are shown. The left trace (a) has more distinct duty cycles. The right trace (b) has duty cycles closer to 50%. Both traces are weighted above and below the zero mean value (horizontal) line.

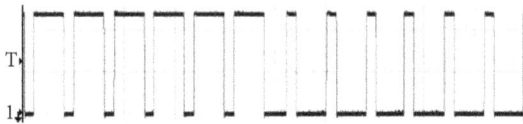

FIGURE 3: Oscilloscope capture of the B-PWM waveform at the output of the transmitter, similar to the one of Figure 1.

now evident that the two levels (corresponding mean values or envelopes) of the signal close to the zero line, Figure 2, must be as distinct as possible. This requirement drives at duty cycles closer to 50%. This is opposite to our former aspect of the most distinct duty cycles. In Figure 2, the two cases of modulated waveforms of the same initial amplitude but of different pairs of complementary duty cycles are shown. The left trace (a) has more distinct duty cycles, more apart from 50%, while the right trace (b) is closer to 50%. It is evident that the right trace (b) has an advantage since the two envelopes around the straight (zero mean value) line are in larger distance (more distinct) than the corresponding ones of the left trace (a). As the 3rd stage, the demodulating one, is a Low Pass Filter which accepts the restored but still modulated B-PWM waveform and has to give the demodulated data at its output, the most distinct the duty cycles are, the best the voltage level separation appears. So, a compromise at the duty cycle selection is necessary. Since the lower value of the duty cycle has to be somewhere among 0% and 50% while the upper value of the duty cycle has to be in the range among 100% and 50%, the average values of the two ranges are selected, that is, 25% and 75%, in order to maintain the complementarity of the pair (summing 100%), because in this way we obtain the most distinct voltage levels at the output of the demodulation filter while preserving the chopping action of the transmitted waveform.

4. Hardware Implementation and Experimental Results

In order to evaluate the proposed modulation, a prototype transmitter-receiver pair has been built. As referred above, the whole transmitter, including the B-PWM modulation, is implemented around a Microchip general purpose microcontroller which also includes the clock oscillator. The microcontroller generates a 64-bit key at an output pin. The PWM signal is applied to the human body via a coupler. In the present implementation, the coupler is composed of two copper electrodes separated by a 5×2.5 cm dielectric

(Figure 8). The one electrode is connected to the ground, while the second one is connected to the output pin.

The coupler's position on the human body influences the communication too [16, 17]. In our experiments, the pair of the transmitter's electrodes was placed along the hand. It turned out that the signal's strength is higher when applied on the forearm and lower when applied over the bones of the wrist. Also, the signal's strength varies in relation to the carrier frequency too, as it is shown in Figure 9.

In Figure 10, the experimental setup is shown. The battery-powered transmitter (small board) is placed on a human hand. The electrode is held in place, touching the skin, by the assistance of two rubber bands.

On the receiver's side, there is a single electrode that receives the B-PWM signal by simply touching it. Various types of electrodes were tested for the receiver. It turned out that a single small sized electrode was sufficient. The detected B-PWM signal, after being restored, passes through a simple Low Pass Filter with a cutoff frequency above the data rate in order to get the average value and thus demodulate it. The demodulated signal is restored in the appropriate form for the USART peripheral and enters the microcontroller. The microcontroller compares the input data with the valid key pool that has already been in memory. If the input data matches to a valid key, an electrically operated lock opens. If they do not match or there is no signal, the lock remains closed and a relevant message is shown at the LCD screen. A similar mechanism could be adopted and applied to bank ATM systems, providing additional security to the identification of the user.

5. Conclusion

In this paper, we presented B-PWM as a modulation scheme suitable for Body Coupled Communication. B-PWM proved to be a promising modulation technique for such an application. An identification system that uses BCC was developed and tested. The transmitter and the receiver are isolated and communicate to each other only through the human body.

The major advantage of the proposed method is its simplicity of implementation, especially at the modulator, comparing to [18], for example. This is most critical when the Body Coupled Communication is to be applied in general purpose devices including wearable electronic devices applications like wrist-type computers such as smartwatches. The proposed modulation can be easily implemented almost only in software of any microcontroller based system without the need of any other component, except a contact surface to the skin. Also, the method can be extended in higher

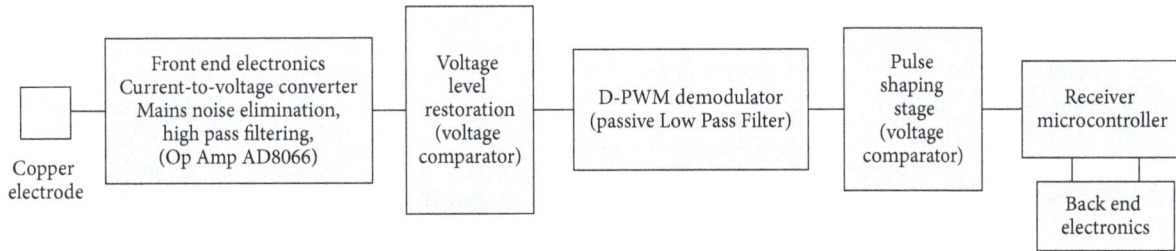

FIGURE 4: Stages (block diagram) of the receiver.

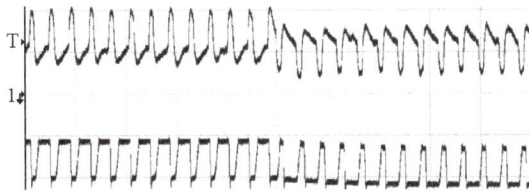

FIGURE 5: Upper trace: oscilloscope capture at the output of the first (current to voltage converter) stage. It is shown that the low frequency content is lost and the waveform is weighted above and below the zero mean value. Lower trace: output of the second (inverting voltage comparator) stage. The waveform is restored in to its original (two) levels.

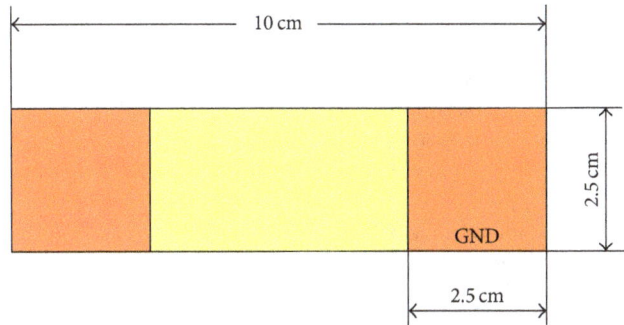

FIGURE 8: Transmitter's coupler electrode. The actual electrode is made from copper covered printed circuit board. The copper has been removed from the middle area.

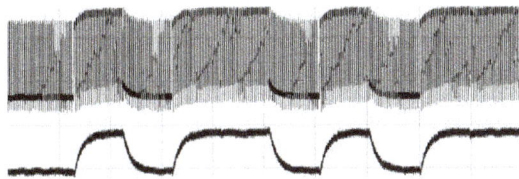

FIGURE 6: Oscilloscope capture at the outputs of the second (voltage comparator, different time scale) as the above and the third (demodulation Low Pass Filter) stages of the receiver. The two different duty cycles give two distinct voltage levels (despite being rounded) corresponding to the logic states.

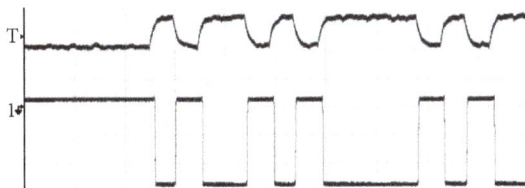

FIGURE 9: Relative signal strength versus frequency.

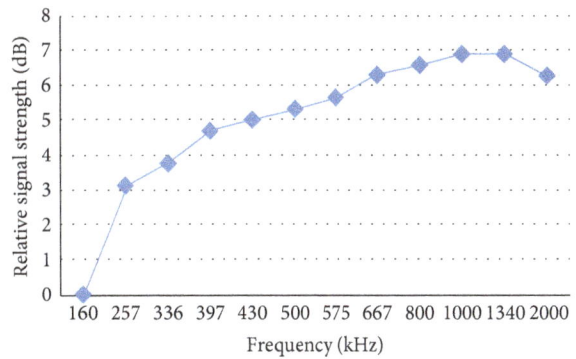

FIGURE 7: Oscilloscope capture at the outputs of the third stage (demodulation Low Pass Filter) as in lower trace of Figure 6 and Pulse Shaping Stage (data output) of the receiver. The signal is finally restored to the voltage levels of the demodulated data waveform.

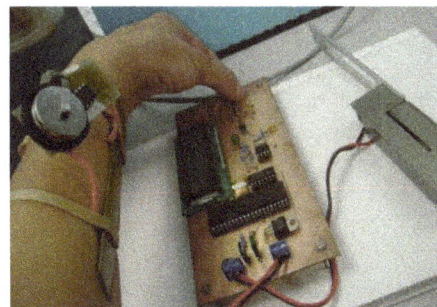

modulation levels, that is, using more than two symbols (duty cycles) giving the opportunity to pack more bits per symbol in cases where the demand of higher bit rate is necessary. The demodulator, at the receiver side, is also easily implemented as a Low Pass Filter. In most of the cases, there is no need of clock extraction.

FIGURE 10: Human hand wearing the transmitter and touching the receiver board. At the right side, for demonstration purposes, is shown an electrically operated door lock.

Conflict of Interests

The digital modulation method described in this paper is a part of a proposal submitted for possible financing. The decision regarding this financing is pending. In the case of favorable decision, the relevant royalties will be divided among the authors, the finance authority, and the Democritus University of Thrace.

References

[1] T. G. Zimmerman, "Personal area networks: near-field intrabody communication," *IBM Systems Journal*, vol. 35, no. 3-4, pp. 609–617, 1996.

[2] N. Seyed Mazloum, *Body-coupled communications—experimental characterization, channel modeling and physical layer design [Ph.D. thesis]*, Chalmers University of Technology, 2008.

[3] M. Wegmueller, A. Lehner, J. Froehlicr et al., "Measurement system for the characterization of the human body as a communication channel at low frequency," in *Proceedings of the 27th Annual International Conference of the Engineering in Medicine and Biology Society (IEEE-EMBS '05)*, pp. 3502–3505, IEEE, September 2005.

[4] G. S. Anderson and C. G. Sodini, "Body coupled communication: the channel and implantable sensors," in *Proceedings of the IEEE International Conference on Body Sensor Networks (BSN '13)*, pp. 1–5, IEEE, Cambridge, Mass, USA, May 2013.

[5] B. Kibret, M. Seyedi, D. T. H. Lai, and M. Faulkner, "The effect of tissues in galvanic coupling intrabody communication," in *Proceedings of the IEEE 8th International Conference on Intelligent Sensors, Sensor Networks and Information Processing: Sensing the Future (ISSNIP '13)*, pp. 318–323, April 2013.

[6] T. Leng, Z. Nie, W. Wang, F. Guan, and L. Wang, "A human body communication transceiver based on on-off keying modulation," in *Proceedings of the 2nd International Symposium on Bioelectronics and Bioinformatics (ISBB '11)*, pp. 61–64, November 2011.

[7] K. Hachisuka, A. Nakata, T. Takeda et al., "Development of wearable intra-body communication devices," *Sensors and Actuators A: Physical*, vol. 105, no. 1, pp. 109–115, 2003.

[8] K. Partridge, B. Dahlquist, A. Veiseh et al., "Empirical measurements of intrabody communication performance under varied physical configurations," in *Proceedings of the 14th Annual ACM Symposium on User Interface Software and Technology (UIST '01)*, pp. 183–190, ACM, November 2001.

[9] J. A. Ruiz, J. Xu, and S. Shimamoto, "Propagation characteristics of intra-body communications for body area networks," in *Proceedings of the 3rd IEEE Consumer Communications and Networking Conference (CCNC '06)*, vol. 1, pp. 509–513, January 2006.

[10] B. H. E. A. Camelia Gabriel, "European Council recommendation, human exposure to electromagnetic fields high frequency (10 khz to 300 ghz)," in *Electricity and Magnetism in Biology and Medicine*, pp. 73–76, 1999.

[11] J. Cleveland and F. Robert, "Federal communications commission (FCC), guidelines for human exposure to radiofrequency radiation,radiation, federal register 38653 100 (1985)," in *Mobile Communications Safety*, Telecommunications Technology & Applications Series, pp. 79–144, Springer US, 1997.

[12] S. D. Kim, S. M. Lee, and S. E. Lee, "Secure communication system for wearable devices wireless intra body communication," in *Proceedings of the IEEE International Conference on Consumer Electronics (ICCE '15)*, pp. 381–382, January 2015.

[13] M. Shinagawa, M. Fukumoto, K. Ochiai, and H. Kyuragi, "A near-field-sensing transceiver for intrabody communication based on the electrooptic effect," *IEEE Transactions on Instrumentation and Measurement*, vol. 53, no. 6, pp. 1533–1538, 2004.

[14] P. Harikumar, M. I. Kazim, and J. J. Wikner, "An analog receiver front-end for capacitive body-coupled communication," in *Proceedings of the NORCHIP Conference*, pp. 1–4, IEEE, Cpenhagen, Denmark, November 2012.

[15] M. H. Maruf, A. Korishe, and S. Roy, "Analysis of analog receiver front end sections for body-coupled communication," in *Proceedings of the International Conference on Informatics, Electronics and Vision (ICIEV '14)*, pp. 1–5, IEEE, May 2014.

[16] F. Koshiji, K. Sasaki, D. Muramatsu, and K. Koshiji, "Input impedance characteristics of wearable transmitter electrodes for intra-body communication," in *Proceedings of the 1st IEEE Global Conference on Consumer Electronics (GCCE '12)*, pp. 362–363, October 2012.

[17] J. Bae and H.-J. Yoo, "The effects of electrode configuration on body channel communication based on analysis of vertical and horizontal electric dipoles," *IEEE Transactions on Microwave Theory and Techniques*, vol. 63, no. 4, pp. 1409–1420, 2015.

[18] M. Wang, Z. Wang, J. Li, and F. Wan, "Architectural hardware design of modulator and demodulator for galvanic coupling intra-body communication," in *Proceedings of the 7th Biomedical Engineering International Conference (BMEiCON '14)*, pp. 1–4, November 2014.

7

Adaptive Complex-Valued Independent Component Analysis based on Second-Order Statistics

Yanfei Jia[1,2] and Xiaodong Yang[1,3]

[1]*College of Information and Communication Engineering, Harbin Engineering University, Heilongjiang 150001, China*
[2]*College of Eletrical and Information Engineering, Beihua University, Jilin 132012, China*
[3]*Collaborative Research Center, Meisi University, Tokyo 1918506, Japan*

Correspondence should be addressed to Yanfei Jia; jia_yanfei@163.com

Academic Editor: Panajotis Agathoklis

This paper proposes a two-stage fast convergence adaptive complex-valued independent component analysis based on second-order statistics of complex-valued source signals. The first stage constructs a cost function by extending the real-valued whiten cost function to a complex-valued domain and optimizes the cost function using a complex-valued gradient. The second stage uses the restriction that the pseudocovariance matrix of the separated signal is a diagonal matrix to construct the cost function and the geodesic method is used to optimize the cost function. Compared with other adaptive complex-valued independent component analysis, the proposed method shows a faster convergence rate and smaller error. Computer simulations were performed on synthesized signals and communications signals. The simulation results demonstrate the validity of the proposed algorithm.

1. Introduction

Blind source separation (BSS) is the separating of a set of source signals from a set of mixed signals without the aid of information (or with very little information) about either the source signals or the mixing process. Independent component analysis (ICA) is an attractive approach for solving blind source separation problems. ICA can be divided into real-valued ICA and complex-valued ICA according to the mixed signals. Complex-valued ICA is widely used to estimate the mixing matrix or to separate complex-valued mixed signals, such as frequency domain signals [1, 2], digital communication signals [3, 4], functional magnetic resonance imaging signals [5], and power system signals [6].

Studies of complex-valued ICA can be divided into three categories. The first category includes methods based on a nonlinear function, such as complex-valued fastICA (C-fastICA) [7], noncircular complex fastICA (NC-fastICA) [8], complex maximization of non-Gaussiantiy (CMN) [9], complex-valued ICA by entropy bound minimization (CEBM) [10], complex-valued ICA by entropy rate bound minimization (CERBM) [11], and others [3, 12]. The second category includes methods that are based on kurtosis or higher-order cumulants, such as joint approximative diagonalization of eigenmatrix (JADE) [4], kurtosis maximization (KM) [13], pseudo-Euclidean gradient iteration ICA (GEGI-ICA) [14], and others [15–17]. The third category includes methods based on second-order statistics, such as strong-uncorrelating transform (SUT) [18, 19] and its adaptive algorithms [2, 20–22] and pseudo-uncorrelating transform (PUT) [23]. Recently, the performance and separability of complex-valued Gaussian mixtures of SUT method have also been studied [24, 25]. Every complex-valued ICA category has its own merits and appropriate application conditions. The methods based on second-order statistics have a simple structure and low computation complexity and are suitable for complex Gaussian and non-Gaussian noncircular signals. In contrast, ICA methods in the first and second categories are not suitable for use with complex Gaussian noncircular signals.

The major advantage of SUT is that *"whenever applicable, remains perhaps the simplest and most accessible approach"* [24]. SUT is a batch algorithm and cannot be used to process signals in real time, so some adaptive complex-valued ICA algorithms have been proposed based on second-order statistics [2, 20–22]. Compared with other complex-valued ICA strategies, adaptive complex-valued ICA algorithms based on second-order statistics are simpler in structure and do not require the probability density of the real and imaginary parts of a complex-valued source signal to be non-Gaussian. The Scott method [20] proposes an updating formula of the separating matrix for adaptive complex-valued ICA without mathematical speculation. The Cong method [2] simultaneously uses diagonal covariance and pseudocovariance noncircular signals as the cost function to deduce the adaptive complex ICA. The convergence condition of the Scott and Cong methods requires that the covariance and pseudocovariance of the separated signal are simultaneously diagonal. For example, if only the covariance of the separated signal is diagonal, the method is unable to reach convergence until the pseudocovariance is also diagonal. This requirement could affect convergence speed. The Yang method [22] uses a two-step serial updating method to make the separated signals satisfy the above convergence condition. In the second step, Yang uses the orthogonal method to force the separating matrix to be a unitary matrix. This changes the updating direction of the separating matrix and leads to slow convergence speed.

To increase the rate of convergence, a fast complex-valued ICA method is proposed in this work. The proposed method first extends the real-valued whitening process to a complex-valued domain to provide unit variance for the processed signal. Second, this work uses the restriction that the pseudocovariance matrix of the separated signals is a diagonal matrix to construct cost function and optimize the cost function using the geodesic method. This avoids computing the square root and inverse of the separating matrix and also keeps the separating matrix to be an orthogonal matrix, without any forcing operation. This improves the convergence speed of the proposed method compared to the other adaptive methods.

2. Complex-Valued ICA and Second-Order Statistics

2.1. Complex-Valued Linear ICA Model. Generally, a linear complex-valued ICA model that is noise-free can be expressed as follows:

$$x = As, \tag{1}$$

where $s = [s_1, s_2, \ldots, s_n]^T$ is the unknown column vector of source signals, n is the number of source signals, A is the unknown complex-valued mixing matrix, $x = [x_1, x_2, \ldots, x_m]^T$ is the column vector of observed complex-valued mixed signals, and m is the number of observed signals. The components of the source signals are mutually independent. Most complex-valued ICA algorithms assume that the number of observed signals is not less than the source signals, and only one Gaussian source signal is allowed. The aim of complex-valued ICA is to search the separating matrix and estimate the source signals and mixing matrix. Given that complex-valued ICA does not utilize any information about the source signals or mixing matrix, it has some indeterminacy in amplitude, sequence, and phase. This indeterminacy

does not affect the shape of the estimated source signal waveform, which contains most information about source signals.

2.2. Second-Order Statistics of Complex-Valued Signals. Assume a complex-valued random column vector $s = s_R + js_I$, where s_R and s_I are the real and imaginary part of s, respectively, and $j = \sqrt{-1}$. The expectation $E[\cdot]$ of the random vector s is defined as follows:

$$E[s] = E[s_R] + jE[s_I]. \tag{2}$$

Its covariance matrix cov(s) is defined as follows:

$$\text{cov}(s) = E\left[(s - E(s))(s - E(s))^H\right], \tag{3}$$

where $(\cdot)^H$ denotes the Hermitian transpose. Its corresponding pseudocovariance matrix is defined as follows:

$$p\,\text{cov}(s) = E\left[(s - E(s))(s - E(s))^T\right], \tag{4}$$

where $(\cdot)^T$ denotes the matrix transpose. The covariance matrix together with the pseudocovariance matrix is the full expression of second-order statistics [19]. If the pseudocovariance matrix equals zero, the random vector is considered circular or proper. If both the covariance matrix and pseudocovariance matrix of the random vector are diagonal with nonzero diagonal elements, the random vector is noncircular or improper, and components of the random vector are called strong uncorrelated components.

2.3. Complex-Valued ICA Based on SUT. For any complex random vector x, if the vector can be transformed into a random vector s by use of a nonsingular square matrix w, where $s = [s_1, s_2, \ldots, s_n]^T = wx$ has covariance that is a unit matrix and pseudocovariance that is a diagonal matrix with diagonal elements between zero and one, then the matrix w is called SUT. If the observed signal is the complex random vector x and the source signal is s, then the SUT is the separating matrix in complex-valued ICA. The procedure for complex-valued ICA based on SUT is as follows [18].

(1) Whitening the complex-valued observed signals x: the whitening procedure is given by

$$z = Bx = \text{cov}(x)^{-1/2} x, \tag{5}$$

where the whitening matrix B is the inverse of the matrix square root of the covariance matrix and z is the whitened signal with a unit covariance matrix.

(2) Determining the separating matrix of the whitened signal by use of Takagi's factorization: this is done according to

$$p\,\text{cov}(z) = U\Lambda U^T. \tag{6}$$

From (5) and (6) we obtain the separating matrix $w = U^H B$.

3. Proposed Adaptive Complex-Valued ICA

In this section, we describe an adaptive fast convergence complex-valued ICA algorithm based on second-order statistics, used in the SUT method. This is unlike other adaptive complex-valued ICA methods that simultaneously force separated signals to comply with second-order statistics. Instead, this method uses an adaptive serial updating method to realize the SUT. First, we use an adaptive method to whiten the observed signals. The cost function used in real-value whitening is directly extended to the complex-valued signal. The cost function is given as follows:

$$J(w) = -\frac{1}{2}\left[\log\left(\det\left(ww^H \right) \right) - \sum_{i=1}^{n} E\left\{ |y_i|^2 \right\} \right], \quad (7)$$

where w is the whitening matrix and y_i is the ith whitening signal. In complex-valued signal processing, the steepest descent direction of cost function (7) is

$$\frac{\partial J(w)}{\partial w^*}$$
$$= -\frac{1}{2}\left(\frac{\partial \log\left(\det\left(ww^H \right) \right)}{\partial w^*} - \frac{\partial \sum_{i=1}^{n} E\left\{ |y_i|^2 \right\}}{\partial w^*} \right) \quad (8)$$
$$= -\frac{1}{2}\left[2\left(ww^H \right)^{-1} w - 2E\left(yx^H \right) \right]$$
$$= -\left(ww^H \right)^{-1} w + E\left(yx^H \right),$$

where x is the observed signal, $y = wx$, $\partial w^H/\partial w^* = I$, and $\partial w^T/\partial w^* = 0$. To avoid computing the matrix inverse, a complex-valued natural gradient is used to simplify (8):

$$\Delta w = \frac{\partial J(w)}{\partial w^*}\left(w^H w \right)$$
$$= \left[-\left(ww^H \right)^{-1} w + E\left(yx^H \right) \right]\left(w^H w \right) \quad (9)$$
$$= E\left(yy^H \right) w - w = \left[E\left(yy^H \right) - I \right] w.$$

So, adaptive whitening can be expressed as follows:

$$w(k+1) = w(k) + \mu\left[I - E\left(yy^H \right) \right] w(k). \quad (10)$$

If we use the instantaneous value instead of the expected value in (10), we obtain the adaptive real-time whitening method:

$$w(k+1) = w(k) + \mu\left[I - yy^H \right] w(k). \quad (11)$$

Second, we must modify the separated signals to satisfy a diagonal pseudocovariance matrix while keeping the covariance matrix as a unit matrix. We use the cost function in [22], which can be expressed as follows:

$$J_2(v) = \frac{1}{2}\left\| \Lambda - E\left[zz^T \right] \right\|^2, \quad (12)$$

where $z = vy$, v is the separating matrix of the whitened signals and Λ is the diagonal matrix of $E[zz^T]$. The ordinary gradient with v^* is as follows:

$$\frac{\partial J_2(v)}{\partial v^*} = \frac{1}{2}\frac{\partial \left\| \Lambda - E\left[zz^T \right] \right\|^2}{\partial v^*}$$
$$= \frac{1}{2}\frac{\partial \mathrm{tr}\left(\left(\Lambda - E\left[zz^T \right] \right)\left(\Lambda - E\left[zz^T \right] \right)^H \right)}{\partial v^*} \quad (13)$$
$$= \frac{1}{2}\frac{\partial \mathrm{tr}\left(\left(\Lambda - vE\left[yy^T \right] v^T \right)\left(\Lambda - vE\left[yy^T \right] \right)^H \right)}{\partial v^*}$$
$$= -\left(\Lambda - vE\left[yy^T \right] v^T \right) E\left[v^* y^* y^H \right].$$

The update of v can be written as follows:

$$v(k+1) = v(k)$$
$$+ \mu\left(\Lambda - v(k) Pv(k)^T \right) E\left[v(k)^* P^* \right], \quad (14)$$

where $P = E[yy^T]$ is the correlated matrix of the whitened signal. At the convergence point, the pseudocovariance matrix of the separated signal is diagonal. To keep the covariance matrix of the separated signal as a unit matrix, the separating matrix v must be a unitary matrix. In [22], they directly used the method of fixed-point fastICA to force the separating matrix to be a unitary matrix:

$$v(k+1) = v(k+1)\left(v(k+1)^H v(k+1) \right)^{-1/2}. \quad (15)$$

This approach has two major drawbacks. One is that (16) changes the steepest gradient direction in every iteration, which slows the convergence speed. The second is that (16) must compute the square root and the inverse of the separating matrix in every iteration, which increases the algorithm computation complexity, slowing the time of convergence.

To overcome this problem, we use a geodesic method to search the optimized separating matrix v. The geodesic method causes the separating matrix to move on the surface of the orthogonal matrix to converge to a local minimum without a forcing operation. The geodesic method is given by

$$v(k+1) = Pv(k), \quad (16)$$

where

$$P = \exp\left[u\left(\Delta v(k+1) v(k)^H - v(k) \Delta v(k+1)^H \right) \right]. \quad (17)$$

If $v(k)$ is a unitary matrix, then $v(k+1)$ is also a unitary matrix. By using the geodesic method, we do not need additional operations to make the separating matrix be an orthogonal matrix and change its search direction.

Using the geodesic method with self-tuning [26] to optimize the cost function (12), we can describe a fast convergence complex-valued ICA method. The implementation process of the proposed adaptive ICA method is as follows:

(1) Initialize the whitening matrix and separating matrix using unit matrix, learning rate μ_1 and μ_2, and iterative number for optimizing (7) and (12), respectively.

(2) Use (10) to whiten the observed signal x and obtain the whitening signal $y = wx$ and whitening matrix w.

(3) Compute the gradient of the cost function in Riemannian space, which can be expressed as follows:

$$G(k) = (\Lambda - R) E\left[F^H\right] - E[F]\left(\Lambda - R^H\right), \qquad (18)$$

where Λ is a diagonal matrix with diagonal elements R, $R = v(k)Pv(k)^T$, $P = E[yy^T]$, and $F = v(k)^* P^H$.

(4) Compute the rotation matrix $D(k) = \exp(-\mu_2 G(k))$ and $Q(k) = D(k)D(k)$.

(5) If $J \geq (\mu_2/2)\text{real}\{\text{trace}\{G(k)G(k)^H\}\}$, $\mu_2 = 2\mu_2$, where

$$J = \|\Lambda - R\|^2 - \|\Lambda_2 - Q(k)RQ(k)^T\|^2, \qquad (19)$$

where Λ and Λ_2 are diagonal matrices corresponding to R and $Q(k)RQ(k)^T$, respectively.

(6) If $J < (\mu_2/2)\text{real}\{\text{trace}\{G(k)G(k)^H\}\}$, $\mu_2 = 0.5\mu_2$.

(7) Update the separating matrix

$$v(k+1) = \exp\left(-\mu_2 G(k)\right) v(k). \qquad (20)$$

(8) If $\text{real}\{\text{trace}\{G(k)G(k)^H\}\}$ is sufficiently small, then STOP; else return to step (3).

4. Experimental Results and Analysis

In order to test the algorithm, we used five synthesized signals with different spectral coefficients, three digital communication signals with different spectral coefficients, and three synthesized signals of which two signals have same spectral coefficients as the source signals. For simplicity, we directly used the expectation of the signal instead of the instantaneous value. Quality of separation was assessed using the performance index (PI), a widely used index in ICA. PI can be expressed as [27]

$$\text{PI}(H) = \frac{1}{n(n-1)} \left\{ \sum_{i=1}^{n} \left(\sum_{j=1}^{n} \frac{|h_{ij}|}{\max_l |h_{il}|} - 1 \right) \right.$$

$$\left. + \sum_{j=1}^{n} \left(\sum_{i=1}^{n} \frac{|h_{ij}|}{\max_l |h_{lj}|} - 1 \right) \right\}, \qquad (21)$$

where h_{ij} is the (i, j) element of the global system matrix $H = wBA$, wB is the separating matrix of mixed signal, A is the mixing matrix, and $\max_l |h_{il}|$ and $\max_l |h_{lj}|$ are the maximum absolute value of the elements in the i row and j column vector H, respectively. When perfect separation is achieved, the performance index is zero. "In practice, the value of performance index 10^{-2} gives quite a good performance" [27]. The smaller the value of PI, the better the performance.

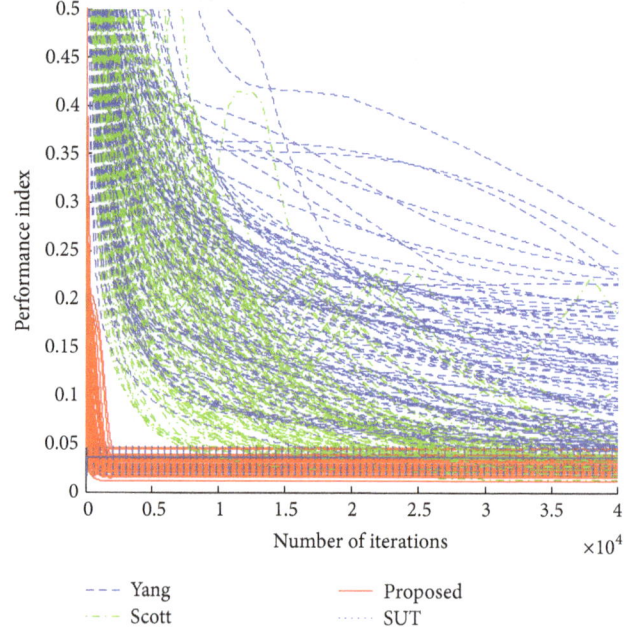

FIGURE 1: Convergence curves of four methods with synthesized signals.

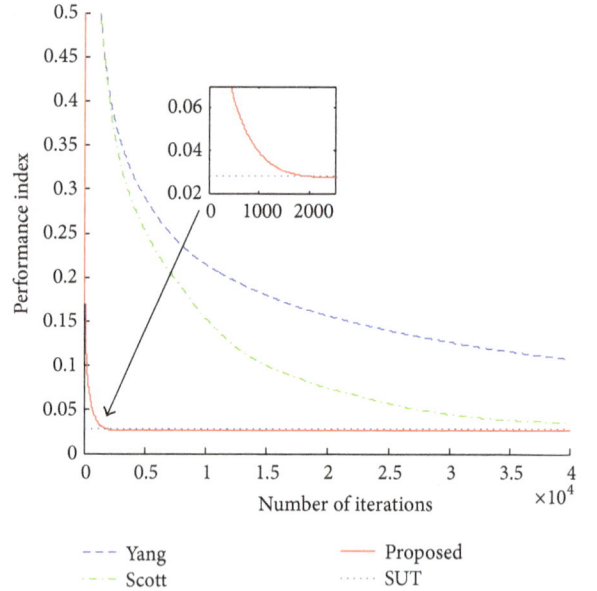

FIGURE 2: Average convergence curves of four methods with synthesized signals.

In the first experiment, five complex-valued synthesized source signals with 10000 samples were used, constructed as follows:

$$s_k(t) = \left[N_{(0,1)}(t) + \sin\left(\frac{\pi}{100k}t\right) \right]$$

$$+ j\left[N_{(0,k)}(t) + \cos\left(\frac{\pi}{100k}t\right) \right], \qquad (22)$$

where $k = 1, 2, 3, 4, 5$, $N_{(0,k)}(t)$ is a sample drawn from a normal random distribution within $(0, k)$, and $j = \sqrt{-1}$.

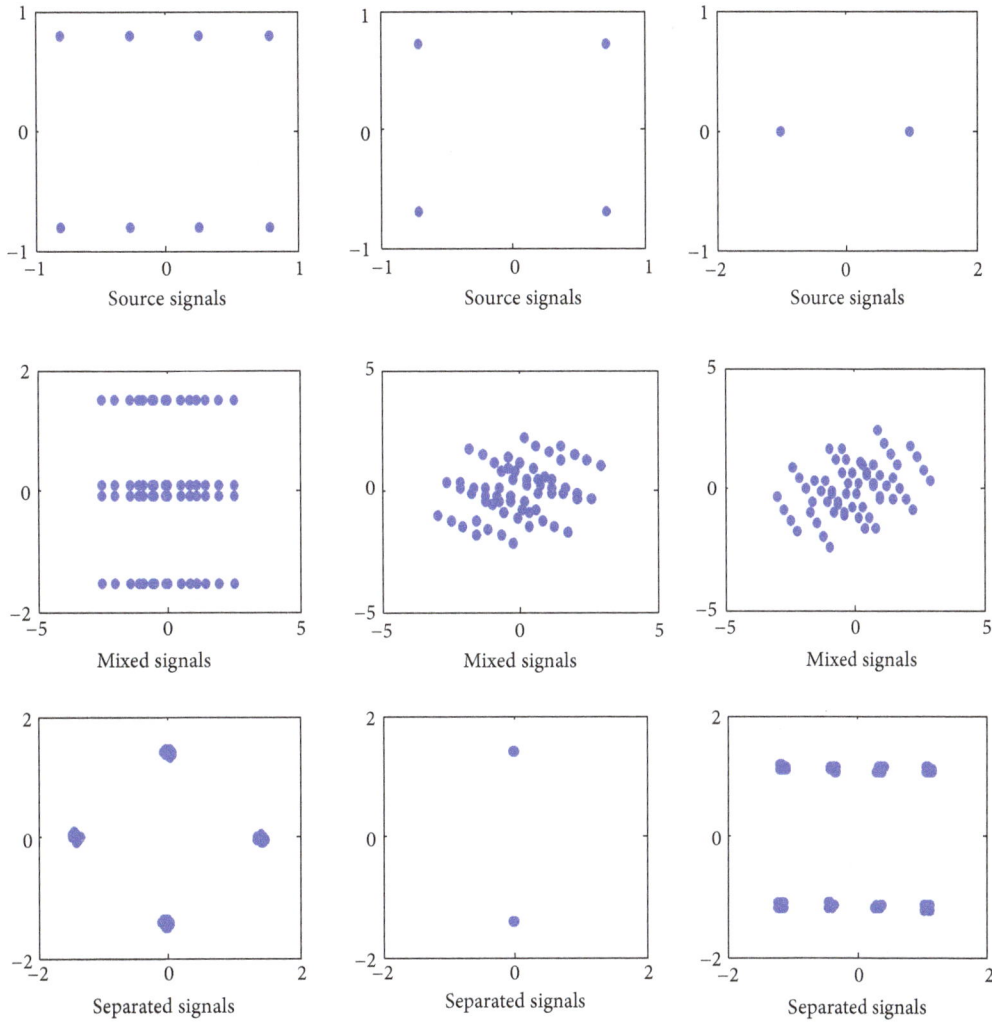

FIGURE 3: Original signals, mixed signals, and separated signals in the digital communication system.

The mixing matrix is a complex-valued random matrix with real and imaginary parts generated from a random uniform distribution between 0 and 1. All algorithms have the same learning rate of 0.01 and were run 100 times. Each time, the source signal and mixing matrix was independently generated.

In contrast, convergence curves are shown in Figure 1 that correspond to the four methods: Yang method [22], Scott method [20], SUT method [18], and our proposed method. Every method has 100 convergence curves, and every convergence curve corresponds to results from one run. The SUT method is a batch method without iterative computations. Therefore, the convergence curves are straight lines. From Figure 1, we can see that all the convergence curves of the proposed method are more closer than the other adaptive methods except for the SUT method. This suggests that the proposed method shows improved, stable performance for different mixed sources that is better than the other adaptive methods. The SUT method shows the smallest fluctuation range, followed by the proposed method, Scott method, and

then the Yang method. This indicates that the proposed method is more suitable for processing different mixed signals than the other adaptive methods, except for the SUT method. Although the performance of SUT is more stable than the other methods for separating different mixed signals, its realization involves Takagi's factorization that is difficult to implement and is not suitable for real-time separation of mixed signals. The adaptive complex-valued BSS method is easy to perform and is more appropriate for real-time separation of mixed signals.

Average convergence curves for the four methods are shown in Figure 2. From Figure 2, we see that the Yang method does not converge to a stationary point until 40000 iterations; the Scott method starts to converge close to 35000 iterations; the proposed method starts to converge after about 2000 iterations. Thus, the proposed method has a faster convergence speed. The performance index is larger than the proposed method when the Scott method converges to a stationary point. This indicates that the proposed method has a smaller error than the Scott method. The performance

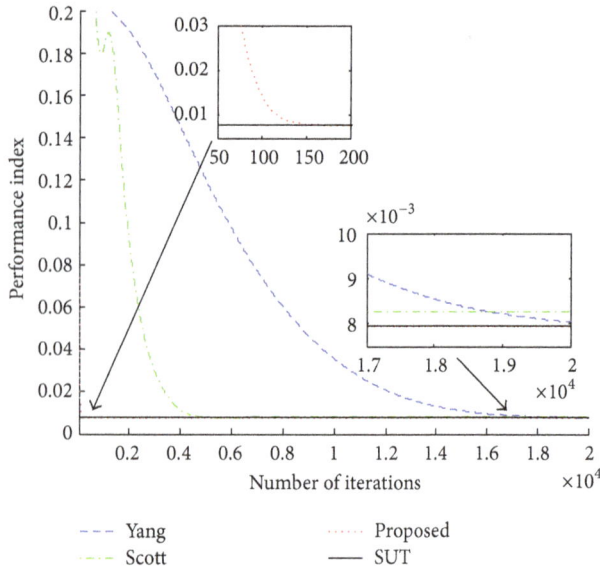

FIGURE 4: Average convergence curves of the four methods for digital communication signals.

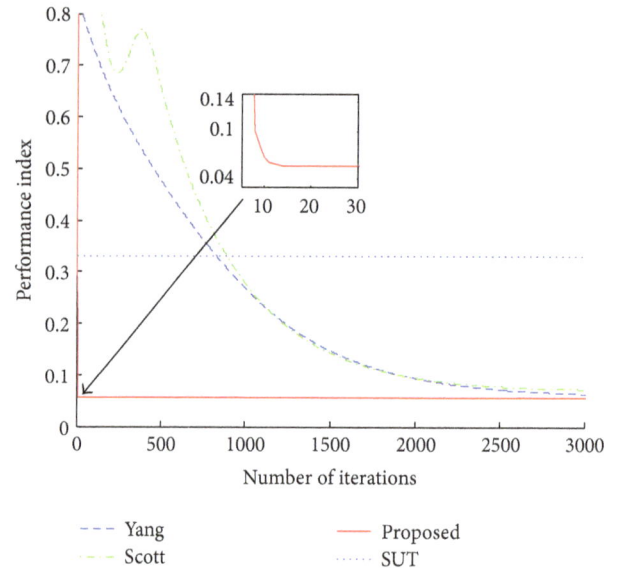

FIGURE 5: Average convergence curves of all methods for source signals; two of these signals have the same spectral coefficients.

indices of the proposed method and the SUT method are very similar, indicating that the two methods have almost the same amount of error.

In the second experiment, we supposed that three digital communication signals (8QAM, 4QAM, and BPSK) impinge on a uniform linear antenna array with three elements from directions of 10°, 25°, and 70°. In Figure 3, the first row gives the original source signals, the second row gives the three mixed signals that are separately received by the three elements of antenna, and the third row provides the separated signals obtained using the proposed method. Comparing the source signals with the separated signals, we see that the constellation of separated signals is almost the same as the source signals, except the sequence, amplitude, and phase, which are inherently indeterminate. This shows that the proposed method is valid for the supposed communication signals.

The average convergence curves for the four methods are shown in Figure 4 from an average of 100 different simulation runs with a learning rate of 0.01. From Figure 4, we see that the proposed method starts to converge after 150 iterations, the Scott method starts to converge after 4500 iterations, and the slowest to converge is the Yang method, which starts to converge after 17000 iterations. Thus, the proposed method has faster convergence than the other adaptive methods. When the proposed method convergences to the stationary point, the performance index curve of proposed method and SUT method are the same. This means that the two methods have the same error for the communicating signals.

In the third experiment, three random complex-valued signals were used as source signals, with spectral coefficients of 0, 0.6, and 0.6. Their imaginary and real parts were generated by a random uniform distribution function. Average convergence curves from an average of 100 different simulation runs with a learning rate of 0.01 are shown for

the four methods in Figure 5. From Figure 5, we see that the performance indexes of Yang method, the Scott method, and the proposed method are less than 0.1 at the stationary point. The average performance index of SUT is about 0.33, which is far greater than 0.1. According to [27], this means that the three adaptive methods successfully separated the mixed signals but the SUT method failed for the mixed signals. The SUT method includes Takagi's factorization to factorize the pseudocovariance matrix. Therefore, it is not suitable for noncircular signals with the same spectral coefficients.

The proposed method has two stages. The convergence curves shown in all figures are the convergence curves only for the second stage. For the first stage, the whitening signal converges to the unit matrix in first experiment after about 600 iterations and after about 100 iterations in the second and third experiments. Compared with other methods, the total iterations required for the proposed method are far less than other methods.

5. Conclusions

This paper proposes an adaptive complex-valued ICA method for noncircular signals based on second-order statistics and the geodesic method. The proposed method has faster convergence and smaller error than the other adaptive methods. For different mixing source signals, the proposed method has better performance and faster convergence than the Scott method. For source signals with different spectral coefficients, the proposed method and the SUT method have almost the same error. However, the SUT method is not suitable for source signals that some of source signals have the same spectral coefficients.

Competing Interests

The authors declare that they have no competing interests.

Acknowledgments

This work was supported by the National Natural Science Foundation of China (61271115) and the Foundation of Jilin Educational Committee (2015235).

References

[1] S. Nagarajaiah and Y. Yang, "Blind modal identification of output-only non-proportionally-damped structures by time-frequency complex independent component analysis," *Smart Structures and Systems*, vol. 15, no. 1, pp. 81–97, 2015.

[2] F. Cong, Q.-H. Lin, P. Jia, X. Shi, and T. Ristaniemi, "Second order impropriety based complex-valued algorithm for frequency-domain blind separation of convolutive speech mixtures," in *Proceedings of the 21st IEEE International Workshop on Machine Learning for Signal Processing (MLSP '11)*, pp. 1–6, Beijing, China, September 2011.

[3] G. Qian, P. Wei, and H. Liao, "Efficient variant of noncircular complex fastica algorithm for the blind source separation of digital communication signals," *Circuits, Systems, and Signal Processing*, vol. 35, no. 2, pp. 705–717, 2016.

[4] J. F. Cardoso and A. Souloumiac, "Blind beamforming for non-gaussian signals," *IEE Proceedings, Part F: Radar and Signal Processing*, vol. 140, no. 6, pp. 362–370, 1993.

[5] W. Du, G.-S. Fu, V. D. Calhoun, and T. Adalı, "Performance of complex-valued ICA algorithms for fMRI analysis: importance of taking full diversity into account," in *Proceedings of the IEEE International Conference on Image Processing (ICIP '14)*, pp. 3612–3616, Paris, France, October 2014.

[6] F. Karimzadeh, S. Esmaeili, and S. H. Hosseinian, "Method for determining utility and consumer harmonic contributions based on complex independent component analysis," *IET Generation, Transmission and Distribution*, vol. 10, no. 2, pp. 526–534, 2016.

[7] E. Bingham and A. Hyvärinen, "A fast fixed-point algorithm for independent component analysis of complex valued signals," *International Journal of Neural Systems*, vol. 10, no. 1, pp. 1–8, 2000.

[8] M. Novey and T. Adali, "On extending the complex FastICA algorithm to noncircular sources," *IEEE Transactions on Signal Processing*, vol. 56, no. 5, pp. 2148–2154, 2008.

[9] M. Novey and T. Adali, "Complex ICA by negentropy maximization," *IEEE Transactions on Neural Networks*, vol. 19, no. 4, pp. 596–609, 2008.

[10] X.-L. Li and T. Adali, "Complex independent component analysis by entropy bound minimization," *IEEE Transactions on Circuits and Systems. I. Regular Papers*, vol. 57, no. 7, pp. 1417–1430, 2010.

[11] G.-S. Fu, R. Phlypo, M. Anderson, and T. Adalı, "Complex independent component analysis using three types of diversity: non-Gaussianity, nonwhiteness, and noncircularity," *IEEE Transactions on Signal Processing*, vol. 63, no. 3, pp. 794–805, 2015.

[12] W. Zhao, Y. Shen, Z. Yuan et al., "A novel method for complex-valued signals in independent component analysis framework," *Circuits, Systems, and Signal Processing*, vol. 34, no. 6, pp. 1893–1913, 2015.

[13] H. Li and T. Adali, "A class of complex ICA algorithms based on the kurtosis cost function," *IEEE Transactions on Neural Networks*, vol. 19, no. 3, pp. 408–420, 2008.

[14] V. James, B. Mikhail, and R. Luis, *A Pseudo-Euclidean Iteration for Optimal Recovery in Noisy ICA*, vol. 28 of *Advances in Neural Information Processing Systems*, Montreal Canadiens, 2015.

[15] C. Ji, Y. R. Wang, and X. Y. Wang, "A new complex blind source separation algorithm based on standard kurtosis," *Journal of Northeastern University(Natural Science)*, vol. 36, no. 5, pp. 614–617, 2015.

[16] R.-J. Wang, H.-F. Zhou, Y.-J. Zhan, and M.-Q. Chen, "An algorithm for adaptive complex blind source separation based on Newton update," *Acta Electronica Sinica*, vol. 42, no. 6, pp. 1125–1131, 2014.

[17] W. Zhao, Y. Wei, Y. Shen et al., "An efficient algorithm by kurtosis maximization in reference-based framework," *Radioengineering*, vol. 24, no. 2, pp. 544–551, 2015.

[18] J. Eriksson and V. Koivunen, "Complex-valued ICA using second order statistics," in *Proceedings of the IEEE Workshop on Machine Learning for Signal Processing*, pp. 183–192, São Luís, Brazil, October 2004.

[19] J. Eriksson and V. Koivunen, "Complex random vectors and ICA models: identifiability, uniqueness, and separability," *IEEE Transactions on Information Theory*, vol. 52, no. 3, pp. 1017–1029, 2006.

[20] C. D. Scott, J. Eriksson, and V. Koivunen, "Equivariant algorithm for estimating the strong-uncorrelating transform in complex independent component analyses," in *Independent Component Analysis and Blind Signal Separation: 6th International Conference, ICA 2006, Charleston, SC, USA, March 5–8, 2006. Proceedings*, vol. 3889 of *Lecture Notes in Computer Science*, pp. 57–65, Springer, Berlin, Germany, 2006.

[21] S. C. Douglas, J. Eriksson, and V. Koivunen, "Adaptive estimation of the strong uncorrelating transform with applications to subspace tracking," in *Proceedings of the IEEE International Conference on Acoustics Speech and Signal Processing Proceedings*, vol. 4, p. 4, Toulouse, France, May 2006.

[22] S.-Y. Yang, L.-Q. Zhao, C.-Z. Zhang, and Y.-F. Jia, "Research on adaptive real-time blind separation algorithm of complex mixed signals," *Systems Engineering and Electronics*, vol. 31, no. 5, pp. 1018–1021, 2009.

[23] H. Shen and M. Kleinsteuber, "Algebraic solutions to complex blind source separation," in *Proceedings of the 10th International Conference on Latent Variable Analysis and Source Separation*, pp. 74–81, Tel Aviv, Israel, March 2012.

[24] A. Yeredor, "Performance analysis of the strong uncorrelating transformation in blind separation of complex-valued sources," *IEEE Transactions on Signal Processing*, vol. 60, no. 1, pp. 478–483, 2012.

[25] D. Ramírez, P. J. Schreier, J. Vía, and I. Santamaría, "Testing blind separability of complex Gaussian mixtures," *Signal Processing*, vol. 95, pp. 49–57, 2014.

[26] T. E. Abrudan, J. Eriksson, and V. Koivunen, "Steepest descent algorithms for optimization under unitary matrix constraint," *IEEE Transactions on Signal Processing*, vol. 56, no. 3, pp. 1134–1147, 2008.

[27] A. Cichocki and S. I. Amari, *Adaptive Blind Signals and Image Processing*, John Wiley & Sons, Chichester, UK, 1st edition, 2002.

Improved RSSI Positioning Algorithm for Coal Mine Underground Locomotive

Bin Ge,[1] Kai Wang,[1] Jianghong Han,[2] and Bao Zhao[1]

[1]School of Computer Science and Engineering, Anhui University of Science & Technology, Huainan, Anhui 232001, China
[2]School of Computer and Information, Hefei University of Technology, Hefei 230009, China

Correspondence should be addressed to Bin Ge; bge@aust.edu.cn

Academic Editor: Peter Jung

Aiming at the large positioning errors of traditional coal mine underground locomotive, an improved received signal strength indication (RSSI) positioning algorithm for coal mine underground locomotive was proposed. The RSSI value fluctuates heavily due to the poor environment of coal mine underground. The nodes with larger RSSI value corrected by Gaussian-weighted model were selected as beacon nodes. In order to reduce the positioning error further, the estimated positions of the locomotives were corrected by the weighted distance correction method. The difference between actual position and estimated position of beacon node was regarded as the positioning error and was given a corresponding weight. The results of simulation show that the positioning accuracy of Gaussian-weighted model is better than statistical average model and Gaussian model and it has a high positioning accuracy after correcting positioning error correction. In the 10 m of communication range, positioning error can be maintained at 0.5 m.

1. Introduction

Due to the closed and harsh environment of coal mine and the narrow and intricate roadway, mine transportation accidents often occur. It always accounts for a large proportion in all kinds of coal mine accidents. Achieving coal mine underground locomotive positioning technology is significant for reducing economic losses and transportation accidents. It also not only can allocate resource reasonable, but also can repair locomotive timely when it breaks down.

At present, the underground positioning technologies are Bedford and Kennedy [1] proposed evaluation of ZigBee (IEEE 802.15.4) Time-of-Flight-based distance measurement for application in emergency underground navigation. Lei et al. [2] introduced a positioning technology of underground moving target which is based on Wi-Fi (wireless-fidelity) and Web GIS (geographic information system). It uses trilateration algorithm to calculate relative coordinate of the target. Hongpeng Chi et al. [3] introduced a laser-based positioning system. The laser light is reflected back to locomotive using time difference to calculate the distance and achieve positioning. Gao et al. [4] proposed a locomotive position system in coal mine which is based on the piezoelectric accelerometer, through integrating over the collected acceleration value of the electric locomotive and solving out the speed and position. Zhang et al. [5] used the inertial navigation system into the coal mine locomotive localization system. Han et al. [6] introduced a weighted centroid localization algorithm which is based on received signal strength indicator. It can access to the path of decline index dynamically, and then calculate its own location using weighted centroid algorithm.

These researches of the positioning technology provide some useful exploration for achieving the coal mine locomotive positioning, but they have defect in practical application. In literature [1, 2] they are simple but have low accuracy and do not suit high-precision mine operations. In literature [3] it is influenced by the environment and may get larger positioning errors in harsh environment. In literatures [4, 5] they have limits; for example, if locomotive gets rapidly speedy, positioning accuracy will become low and there is greater cumulative error and it takes longer time in positioning process. In literature [6] it is achieved easily and has better accuracy, but it is not stable and the error fluctuates greatly with the change of locomotive location.

According to the analysis, considering the low cost, complexity, and without any additional hardware device of RSSI location and arranging nodes sufficiently in the pit [7–9], this paper uses RSSI location and makes improvement. The positioning errors of RSSI positioning algorithm for coal mine underground locomotive have two primary factors. Firstly, the RSSI value is very unstable, because it is influenced by bad environment such as diffraction, multipath, and obstructions. Secondly, location algorithm is not precise enough and cannot deal with some inaccurate date effectively. The main methods of managing RSSI value are statistical mean model, weighted model, Gaussian model, and so forth. Literature [10] simulated several models and the results show that the highest ranging precision is Gaussian model. Literature [11] proposed a correction algorithm based on Gaussian anchor nodes; literature [12] proposed a D-Gaussian model, it can filtrate variance level interference. Literature [13] used hybrid filtering method to optimize RSSI value. This paper presents a Gaussian-weighted model to get more accurate distance, uses weighted least squares method to estimate location of the locomotive, and corrects positioning error using weighted distance correction method.

2. Wireless Communication Model and RSSI Ranging Model

2.1. Wireless Communication Model of Coal Mine. Because coal mine tunnel is narrow and has rails, the nodes cannot be arranged anywhere. In order to collect wireless signal of the nodes correctly, the beacon nodes are arranged on both sides of the track and spaced in a certain distance to reduce interference between the signals, as shown in Figure 1.

In Figure 1, Wireless AP (access point) is a node which can communicate with other wireless AP using wireless technology and A, B, C, and D are the beacon nodes which are arranged on both sides of the rail. When the locomotive is running to this location shown in Figure 1, it can communicate with beacon nodes and it is ensured that the number of the nodes which can communicate with locomotive is greater than or equal to 4. That ensures the calculation of exact location. In order to guarantee the number of communication nodes, the distance between nodes A and B should not exceed the range of the communication; the same goes for C and D, in Figure 1. To obtain more accurate RSSI values, the distance between A and B or C and D can be reduced. If the distance is short sufficiently, the number of communication nodes may be increased to five or more and the relative cost and complexity of the algorithm will increase.

2.2. RSSI Ranging Model. As the propagation distance of radio signal increase, the signal strength decay. It has different attenuation amplitude, in different environments. In WSN (Wireless Sensor Network), the commonly used wireless signal propagation models are: free space propagation model, log-distance path loss model, Hata model, log-distance distribution model, and so forth [14, 15].

Log-distance distribution model can be used to calculate the path loss when nodes received information. The model is shown as follows:

$$\text{PL}(d) = \text{PL}(d_0) + 10 \times n \times \lg\left(\frac{d}{d_0}\right) + X_\sigma, \quad (1)$$

where $\text{PL}(d)$ is the path loss after signal transmitting distance d, the unit of the path loss is dB, X_σ is a Gaussian distribution random variable whose mean value is 0 and standard deviation is σ (generally 4–10), n is path attenuation factor and always takes 2–5, d_0 is reference distance and usually equals 1 meter, and $\text{PL}(d_0)$ is a path loss after signal transmitting distance d_0 (generally 1 m).

The RSSI of nodes at distance d is

$$\text{RSSI} = P_{\text{send}} + P_{\text{amplify}} - \text{PL}(d), \quad (2)$$

where P_{send} is transmitting power and P_{amplify} is the gain of antenna.

Similarly

$$A = P_{\text{send}} + P_{\text{amplify}} - \text{PL}(d_0), \quad (3)$$

where A is the RSSI at the distance of d_0.

By (1), (2), and (3) the distance between unknown nodes and beacon nodes can be calculated:

$$d = 10^{((A-\text{RSSI}-X_\sigma)/10n)} d_0. \quad (4)$$

3. Improved RSSI Positioning Algorithm of Underground Locomotive

RSSI values show a great probability of randomness and volatility, when the wireless signal is propagating in coal mine. If only use formula (4) to calculate distance, x_σ is unknown and it is not a solution. Without x_σ or it takes 0 the fluctuation range of positioning error is large. Obviously this is not consistent with the positioning requirements of underground. To reduce the positioning error RSSI value correction is an essential part of the positioning algorithm. So a new correction model of RSSI value (Gaussian-weighted model) is proposed. Then estimate locomotive position using the least squares method. At last correct the estimation position using weighted distance correction.

3.1. Correct RSSI Value. Gaussian-weighted model combines the advantages of several models; concrete steps are as follows.

(1) High probability of RSSI values is selected by Gaussian model. When the unknown node receives several RSSI values from a same beacon node, the RSSI values with a probability density less than 0.6 (this value may be due to environment) are filtered by Gaussian model [13]. Arrange the remaining RSSI values from small to large: $\text{RSSI}_1 \leq \text{RSSI}_2 \leq \cdots \leq \text{RSSI}_m$, where m is number of remaining RSSI.

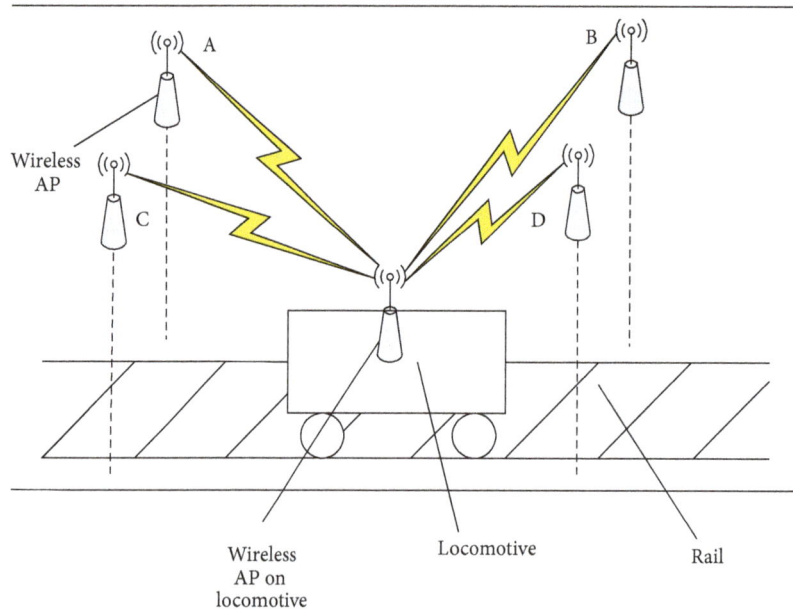

FIGURE 1: Wireless communication model.

(2) Calculate the weight of the RSSI values of m. Suppose that μ and σ^2 are the mean and variance of these RSSI values. Consider

$$\mu = \frac{1}{m}\sum_{i=1}^{m} \text{RSSI}_i,$$

$$\sigma^2 = \frac{1}{m}\sum_{i=1}^{m} \left(\text{RSSI}_i - \mu\right)^2. \tag{5}$$

Define the formula of weights:

$$w_i = \frac{1/\max\left\{\sigma^2, \left(\text{RSSI}_i - \mu\right)^2\right\}}{\sum_{i=1}^{m} 1/\max\left\{\sigma^2, \left(\text{RSSI}_i - \mu\right)^2\right\}}, \tag{6}$$

where $\max\{\sigma^2, (\text{RSSI}_i - \mu)^2\}$ is to make the weight value smoother. As long as the difference between RSSI value and mean is less than the overall variance, take σ^2 to determine the weight of size. On the contrary, it is $(\text{RSSI}_i - \mu)^2$. From (6) it can be drawn that the larger the difference, the smaller the weight.

(3) Calculate the RSSI value after correcting. Let the m RSSI value multiply with the weight values of themselves; then sum the multiplied values as the correction RSSI value. The formula is

$$\text{RSSI}_c = \sum_{i=1}^{m} w_i \times \text{RSSI}_i. \tag{7}$$

After correcting, formula (4) turns to

$$d = 10^{((A - \text{RSSI}_c)/10n)} d_0. \tag{8}$$

3.2. Weighted Least Square Method. Suppose that the coordinate of locomotive is $A(x, y)$ and the coordinates of beacon nodes A_i are (x_i, y_i), $i = 1, 2, \ldots, N$, where N is the number of beacon nodes and $N \geq 4$. There are N measurement equations, when each beacon node can communicate with the unknown node, which showed

$$D_i = d_i + v_i, \quad i = 1, 2, \ldots, N, \tag{9}$$

where $d_i = \sqrt{(x - x_i)^2 + (y - y_i)^2}$ and v_i is measurement error. Formula (9) is linearized and both sides are squared

$$-2x_i x - 2y_i y + x^2 + y^2 = (D_i - v_i)^2 - \left(x_i^2 + y_i^2\right). \tag{10}$$

Define R and R_i: $R = \sqrt{x^2 + y^2}$; $R_i = \sqrt{x_i^2 + y_i^2}$. Formula (10) is turned to

$$-2x_i x - 2y_i y + R^2 = D_i^2 - R_i^2 + v_i^2 - 2D_i v_i. \tag{11}$$

Simplifying formula (11) we can obtain the following:

$$\mathbf{h} = \mathbf{G}\boldsymbol{\theta} + \mathbf{v}, \tag{12}$$

where

$$\boldsymbol{\theta} = \begin{bmatrix} x & y & R^2 \end{bmatrix}^T, \qquad \mathbf{h} = \begin{bmatrix} D_1^2 - R_1^2 \\ D_2^2 - R_2^2 \\ \vdots \\ D_N^2 - R_N^2 \end{bmatrix},$$

$$\mathbf{G} = \begin{bmatrix} -2x_1 & -2y_1 & 1 \\ -2x_2 & -2y_2 & 1 \\ \vdots & \vdots & \vdots \\ -2x_N & -2y_N & 1 \end{bmatrix},$$

$$\mathbf{v} = \begin{bmatrix} v_1^2 - 2D_1v_1 & v_2^2 - 2D_2v_2 & \cdots & v_N^2 - 2D_Nv_N \end{bmatrix}^T.$$

$$(13)$$

The following is the least-square solution of formula (12):

$$\boldsymbol{\theta} = \left(\mathbf{G}^T\mathbf{W}\mathbf{G}\right)^{-1}\mathbf{G}^T\mathbf{W}\mathbf{h} = \begin{bmatrix} \theta_x & \theta_y & \theta_{R^2} \end{bmatrix}^T, \quad (14)$$

where $\mathbf{W}$ is the inverse matrix of covariance matrix $\mathbf{v}$ of Gaussian noise.

3.3. Correct Positioning Error. Because weighted least square method cannot correct some positioning error which is caused by some inaccurate date, estimated location must make further correction and calculate more precise location. The method is weighted distance correction. Estimate the position of beacon nodes as unknown nodes and then the difference of coordinates between estimated position and actual position as the positioning errors and give them appropriate weight. Finally calculate correction value. There are 3 steps.

(1) Every beacon node is used as unknown node to estimate its own estimated position and calculate the error between estimated position and actual position. Suppose that the estimated position is (x_{ai}, y_{ai}) which is calculated by the way of above. So the positioning error is

$$e_i = \begin{bmatrix} x_i - x_{ai} \\ y_i - y_{ai} \end{bmatrix}. \quad (15)$$

(2) Determine the weights of localization errors of every beacon node. Because in the coal mine the closer the distance between beacon node and unknown node, the more similar of environment and the closer positioning errors, its weight should be greater. The weights of N beacon nodes which are involved in positioning are defined as follows:

$$k_i = \frac{1/(1 + d_i)}{\sum_{i=1}^{N} 1/(1 + d_i)}, \quad (16)$$

where d_i is the distance from unknown node to N beacon nodes. They are calculated by formula (8).

(3) Calculate the final location coordinate. The correction values are obtained by multiplying formulae (15) and (16). Consider

$$e = \begin{bmatrix} e_x = \sum_{i=1}^{N} k_i \times (x_i - x_{ai}) \\ e_y = \sum_{i=1}^{N} k_i \times (y_i - y_{ai}) \end{bmatrix}. \quad (17)$$

So the final positioning coordinates of unknown node (locomotive) are

$$\begin{aligned} x &= \theta_x + e_x, \\ y &= \theta_y + e_y. \end{aligned} \quad (18)$$

3.4. Select Better Beacon Nodes. Selected beacon node also has a great influence on the positioning error. A feature of the wireless signal is path loss when it is spread; the closer the distance between the node and locomotive, the smaller the path loss and the smaller the effect of various positioning error [16]. Equation (4) shows that the RSSI value is greater when it has closer communication distance. So the beacon nodes of bigger RSSI value are selected. These beacon nodes have more decision power to the location of the locomotive that can improve the positioning precision of the algorithm. The specific process is as follows.

(1) When the number of beacon nodes is larger or equal to 4 which is communication with the locomotive $N_0 \geq 4$, arrange the RSSI values. Select ahead of N $(4 \leq N \leq N_0)$ RSSI values as beacon nodes and involve in the calculation of the next step.

(2) When $N_0 < 4$, space some time and continue to signal until $N_0 \geq 4$. Then execute step (1).

4. Process of Improved Algorithm

From the above analysis to improved algorithm for coal mine locomotive positioning, we can know that the process is divided into the following steps.

(1) Before receive positioning signal, beacon nodes calculate their own positioning errors using formula (15). Then they wait for positioning.

(2) After the positioning signal is emitted by locomotive, the nodes that can communicate with locomotive send the positioning errors which are calculated by step (1); then the locomotive records them. Do it repeatedly and record RSSI values. These exact RSSI values of each node can be calculated by formula (7).

(3) These RSSI values are arranged and select the number of N RSSI which is larger than others as beacon nodes.

(4) These RSSI values are translated into distance by formula (8). Then estimate the initial position of locomotive. At last determine the location of locomotive by formula (18).

5. Simulation and Experiment

Using MATLAB simulation platform for experiment, in order to make the simulation results more in line with the actual environment of locomotive, we selected a plane of $5\,\text{m} \times 100\,\text{m}$ and arranged nodes on both sides of the track. ZigBee technology is used as wireless communication technology. Because it shows low power consumption, low cost, short time delay, large network capacity, reliability, and security and

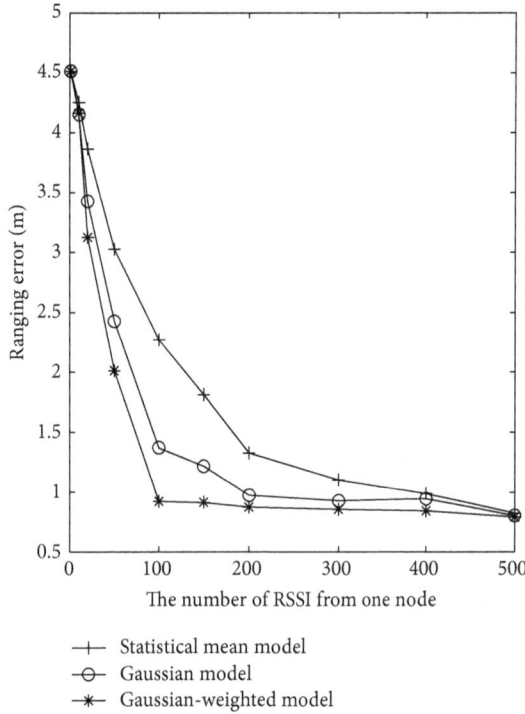

FIGURE 2: Relationship between the number of RSSI of a single node and ranging error.

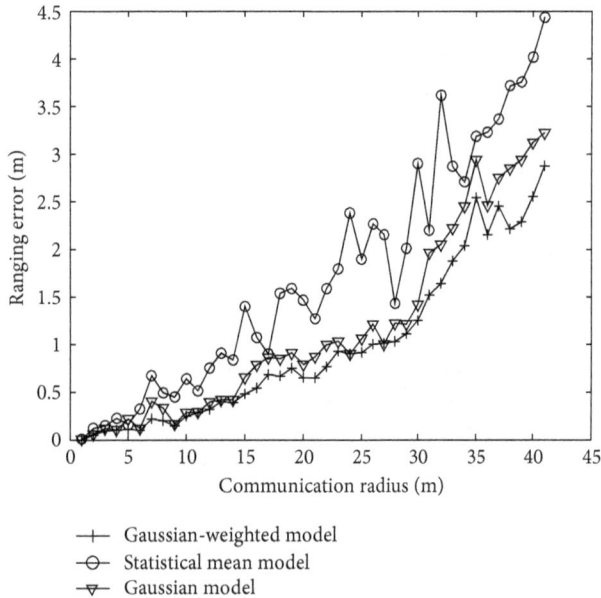

FIGURE 3: Comparison of the ranging error of RSSI.

TABLE 1: Comparison of the ranging error.

Error/m	Statistical mean model	Gaussian model	Gaussian-weighted model
Average error	2.2708	1.1689	0.9213
Maximum error	3.0328	1.8512	1.6789

time of locomotive. Secondly, it will cause more interference. The simulation is divided into the following three aspects.

(1) Compare correction models of RSSI value. Compare ranging accuracy among statistical mean model, Gaussian model, and Gaussian-weighted model. If it has higher ranging accuracy, its correction of RSSI values is more precise. The result is shown in Figures 2 and 3.

Figure 2 is a relationship between the number of RSSI of a single node and ranging error and communication radius is 20 m. The number of RSSI values of a single node has a great influence on ranging accuracy. When the number of RSSI value increases, the cost of the location is higher and the positioning accuracy increases. It can be seen from Figure 2 that the accuracy of the three models is increased, as the number of RSSI value increases, particularly within 100 the errors fall fastest. The Gaussian-weighted model has the highest ranging accuracy and the error decline is most obvious. When the number of RSSI values is more than 100, ranging error decreased slowly significantly and the positioning error of three models becomes closer. When the number of RSSI values is taken to 500, ranging error is very similar. From the above analysis, it is best to take 100 as the number of the RSSI values in this simulation. That costs less and the Gaussian-weighted model maintains high ranging accuracy.

Figure 3 shows the comparison of ranging error of the three models. The number of the RSSI value of each node is 100. As can be seen from Figure 3, the error of statistical mean model is maximum and the maximum ranging error is 4.5 m, approximating 10% of measurement distance. The error of Gaussian-weighted model is minimum and the maximum ranging error is 3 m, approximating 7% of measurement distance. With the expansion of communication radius, ranging errors of three models are increased. The error of statistical mean models fluctuates bigger and grows faster than the other two models. In order to obtain more accurate data, we took the communication radius 20 m and did experiment repeatedly to calculate the average error. The specific data are shown in Table 1.

The results can be taken from the experiment that the highest ranging precision is Gaussian-weighted model. Its average error is less than 1 m within communication radius of 20 m and maximum error does not exceed 2 m. The lowest ranging precision

can meet the need of mine. Suppose the initial coordinate of locomotive is $(0, 0)$, path attenuation factor n is 3.5, A takes -45 dB, X_σ is Gaussian distribution whose mean is 0, and standard deviation σ is 6. The speed of the locomotive is about 5 m/s. Six beacon nodes were in range of communication. Firstly, it will increase the communication and calculation

TABLE 2: Comparison of the positioning error.

Communication radius	5 m			10 m		
Ranging model	Statistical mean model	Gaussian model	Gaussian-weighted model	Statistical mean model	Gaussian model	Gaussian-weighted model
Positioning error (without correction)/m	0.4780	0.3526	0.3645	1.0838	1.0126	0.8213
Positioning error (correction)/m	0.4215	0.3215	0.3187	0.8838	0.6126	0.5213
Communication radius	20 m			40 m		
Ranging model	Statistical mean model	Gaussian model	Gaussian-weighted model	Statistical mean model	Gaussian model	Gaussian-weighted model
Positioning error (without correction)/m	1.9421	1.5765	1.3569	5.3751	3.4548	3.0521
Positioning error (correction)/m	1.6421	1.1765	0.9569	4.3751	2.7548	2.0521

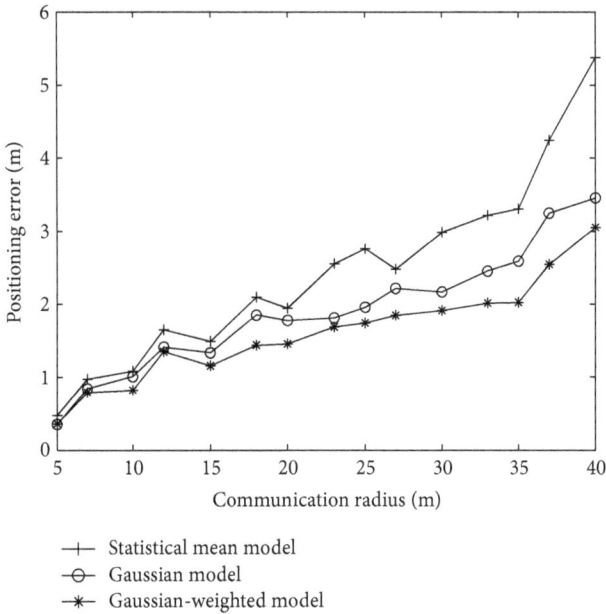

(a) Without positioning error correction

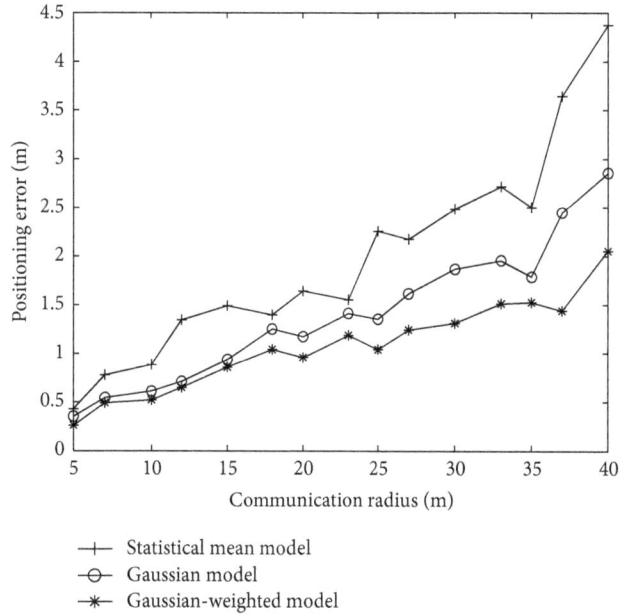

(b) With positioning error correction

FIGURE 4: Comparison of these positioning algorithms with or without positioning error correction.

is statistical mean model. Its average ranging error already exceeds 2 m within the communication radius of 20 m. Gaussian model is between them; compared to the statistical mean model the error is more stable. So the most accurate correction model of RSSI value is Gaussian-weighted model and its error is more stable.

(2) Comparison of positioning error. Firstly, compare positioning accuracy of statistical mean model, Gaussian model, and Gaussian-weighted models with or without weighted distance correction. The result is shown in Figure 4. Secondly, the positioning accuracies are compared with different numbers of the beacon nodes. The result is shown in Table 3.

From Figure 4(a) it can be seen that the minimum positioning error is Gaussian-weighted model and

with the increasing of communication radius it grows more stable. The maximum positioning error is statistical mean model and the error is not stable enough. Gaussian model is between the other two models and the error grows stable.

Comparing the positioning error of Figures 4(a) and 4(b), it can be seen that if algorithm is corrected by weighted distance correction, it is more accurate than without correcting. Gaussian-weighted model is the best. At a communication radius of 40 m, error of Gaussian-weighted model is reduced 33%, while the statistical mean model and the Gaussian model are reduced 19% and 22%. The specific comparison data is shown in Table 2.

TABLE 3: Comparison of the positioning error with different beacon nodes.

The size of N	4	6	8	10
Statistical mean model	1.7708	16589	1.5913	1.5237
Gaussian model	1.4028	1.2012	1.1189	1.0756
Gaussian-weighted model	1.2617	0.9636	0.9337	0.9165

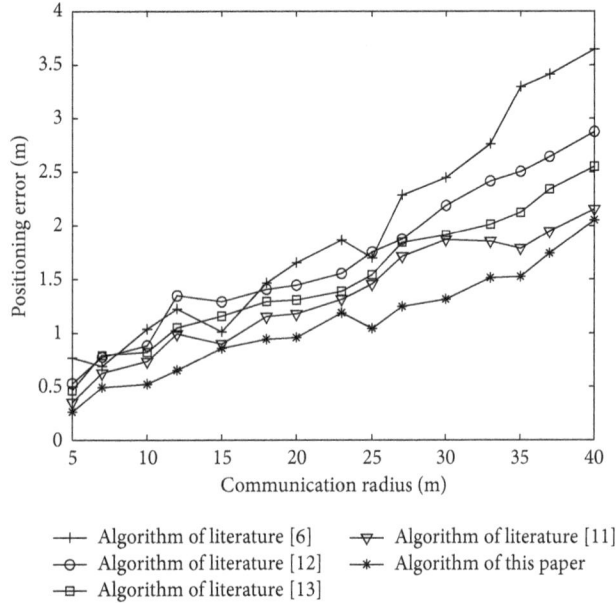

+— Algorithm of literature [6]   ▽— Algorithm of literature [11]
-○- Algorithm of literature [12]   -*- Algorithm of this paper
-□- Algorithm of literature [13]

FIGURE 5: Comparison of the positioning error among different algorithms.

In Table 3, communication radius is 20 m. From Table 3 it can be seen that positioning error is small when the number of beacon nodes increased. But with the increase of beacon nodes, the reduced amount of positioning error is decreased, particularly Gaussian-weighted model. This is because the amount of calculation of Gaussian-weighted model is larger than the Gaussian model and the statistical mean model. It will cost more time to perform the calculation when the beacon nodes are increased. So positioning error will become large when the locomotive is moving. From Table 3 and considering the cost, 6 is the most appropriate choice.

(3) For comparison of positioning algorithm, compare these algorithms of literatures [6, 11–13] and this paper in the same environment; the simulation result is shown in Figure 5. As can be seen from Figure 5, the positioning accuracy of this paper is highest. The algorithm of literature [6] has high positioning accuracy within 25 m. Out of 25 m, its positioning grows faster than others. The other three algorithms have higher positioning error than this paper, but they are more stable than in the literature [6].

The results of above experiment show that in the case of positioning error correction Gaussian-weighted model has highest positioning accuracy. The smaller the communication radius, the less the positioning error is. When the communication radius is 10 m, the positioning accuracy can be maintained at about 0.5 m.

6. Conclusions

(1) This paper analyzes a number of factors of locomotive positioning errors for coal mine and proposes a coal mine locomotive positioning algorithm based on RSSI. Because the RSSI value in coal mine is very unstable, it is corrected by weighted-Gaussian model. This model combines the advantages of Gaussian model and weighted model. It has high ranging precision in the coal mine of harsh environment.

(2) Just using weighted least square method to estimate the position of locomotive cannot reduce the impact of some inaccuracy dates effectively. Correcting the estimated position is necessary. This paper proposes a weighted distance correction method. The positioning errors of each beacon node are calculated using Gaussian-weighted model and weighted according to the distances from locomotive to the beacon node. The closer the distance is, the greater the weight is.

(3) The algorithm which is proposed is simulated by MATLAB. The results of experiment show that Gaussian-weighted model is the best model. In the case of positioning error correction, positioning error is descendant while communication radius is the same and the minimum positioning error is less than 0.5 m.

Conflict of Interests

The authors declare that there is no conflict of interests regarding the publication of this paper.

Acknowledgments

This work was financially supported by the National Natural Science Foundation of China (61070220 and 61170060), Anhui Provincial Natural Science Foundation of China (1408085ME110), Anhui Provincial Major Project of Colleges and Universities Natural Science (KJ2013ZD09), and Anhui Provincial Key Project of Colleagues and Universities Natural Science (KJ2012A096).

References

[1] M. D. Bedford and G. A. Kennedy, "Evaluation of ZigBee (IEEE 802.15.4) time-of-flight-based distance measurement for application in emergency underground navigation," *IEEE Transactions on Antennas and Propagation*, vol. 60, no. 5, pp. 2502–2510, 2012.

[2] M. Lie, D. Enjie, F. Qiyan et al., "Underground moving target positioning and historical trajectory extraction based on Wi-Fi

and WebGI," *Geography and Geo-Information Science*, vol. 28, no. 3, pp. 109–110, 2013.

[3] H. Chi, K. Zhan, and B. Shi, "Automatic guidance of underground mining vehicles using laser sensors," *Tunnelling and Underground Space Technology*, vol. 27, no. 1, pp. 142–148, 2012.

[4] Y. Gao, H. Sun, and Y. Yang, "The locomotive position system in coal pit based on the piezoelectric accelerometer," *Piezoelectrics and Acoustooptics*, vol. 34, no. 5, pp. 782–784, 2012.

[5] Q.-Z. Zhang, S.-B. Zhang, J. Wang, and H.-F. Bian, "Low-cost GPS/INS in-motion alignment model for open-pit mine transport truck monitoring and dispatch system," *Journal of the China Coal Society*, vol. 38, no. 8, pp. 1362–1367, 2013.

[6] D.-S. Han, W. Yang, Y. Liu, and Y. Zhang, "A weighted centroid localization algorithm based on received signal strength indicator for underground coal mine," *Journal of the China Coal Society*, vol. 38, no. 3, pp. 522–528, 2013.

[7] X. Wang, S. Yuan, R. Laur, and W. Lang, "Dynamic localization based on spatial reasoning with RSSI in wireless sensor networks for transport logistics," *Sensors and Actuators, A: Physical*, vol. 171, no. 2, pp. 421–428, 2011.

[8] R.-B. Zhang, J.-G. Guo, F.-H. Chu, and Y.-C. Zhang, "Environmental-adaptive indoor radio path loss model for wireless sensor networks localization," *International Journal of Electronics and Communications*, vol. 65, no. 12, pp. 1023–1031, 2011.

[9] Z. Wei, Z. Lv, C. Yang, J. Han, and L. Shi, "Wireless positioning method based on correctness judgment of received signal strength for mine locomotive," *Chinese Journal of Scientific Instrument*, vol. 35, no. 1, pp. 178–184, 2014.

[10] J.-W. Zhang, L. Zhang, Y. Ying, and F. Gao, "Research on distance measurement based on RSSI of ZigBee," *Chinese Journal of Sensors and Actuators*, vol. 22, no. 2, pp. 285–288, 2009.

[11] G. Wan, J. Zhong, and C. Yang, "Improved algorithm of ranging and locating based on RSSI," *Application Research of Computers*, vol. 29, no. 11, pp. 4156–4158, 2012.

[12] J.-Q. Xu, W. Liu, Y.-Y. Zhang, and C.-L. Wang, "RSSI-based anti-interference WSN positioning algorithm," *Journal of Northeastern University*, vol. 31, no. 5, pp. 647–650, 2010.

[13] W. Tao, Y. Zhu, and Z. Jia, "A distance measurement algorithm based on RSSI hybrid filter and least square estimation," *Chinese Journal of Sensors and Actuators*, vol. 25, no. 12, pp. 1748–1753, 2012.

[14] J. Zheng, C. Wu, H. Chu, and Y. Xu, "An improved RSSI measurement in wireless sensor networks," *Procedia Engineering*, vol. 15, pp. 876–880, 2011.

[15] G. Blumrosen, B. Hod, T. Anker, D. Dolev, and B. Rubinsky, "Enhanced calibration technique for RSSI-based ranging in body area networks," *Ad Hoc Networks*, vol. 11, no. 1, pp. 555–569, 2013.

[16] X. Luo, W. J. O'Brien, and C. L. Julien, "Comparative evaluation of received signal-strength index (RSSI) based indoor localization techniques for construction jobsites," *Advanced Engineering Informatics*, vol. 25, no. 2, pp. 355–363, 2011.

9

The Channel Compressive Sensing Estimation for Power Line based on OMP Algorithm

Yiying Zhang,[1] Kun Liang,[1] Yeshen He,[2] Yannian Wu,[2] Xin Hu,[2] and Lili Sun[2]

[1]College of Computer Science and Information Engineering, Tianjin University of Science & Technology, Tianjin, China
[2]China Gridcom Co., Ltd, Shenzhen, Guangdong, China

Correspondence should be addressed to Kun Liang; liangkun@tust.edu.cn

Academic Editor: Hui Cheng

Power line communication (PLC) can collect information by power line which increases the coverage and connectivity of the smart grid. In this paper, we analyze the transmission characteristics of the power line channel and model it with mathematics channel. The multipath effect of the power line channel is studied with a novel technology named compressive sensing herein. We also proposed a new method to the power line channel estimation based on compressive sensing. We can collect and extract the effective parameters of the power line channel to storage, which only take very little storage space. The simulation results show that the proposed approach can reduce the amount of processing data in the digital signal processing module and decrease the requirement for the hardware.

1. Introduction

Industry 4.0 employs CPS (Cyber-Physical System) to promote industrial production with interconnection and intelligence. And various communication technologies provide important support for the CPS. Power Line Communication, as a special communication technology, not only is designed to transmit electrical energy, but also is used to transmit multiple types of information [1]. However, due to the fact that the power line network is designed for the transmission of electrical energy and works in high electromagnetic phenomena environment, the characteristics of power line channel are different from other conventional communication channel greatly. Power line noise is very complex. It is not a single Gaussian white noise in the other usual communication environments and includes colored background noise and periodic impulse noise (asynchronous or synchronized), as shown in Figure 1.

According to the random in or out access of electrical equipment, PLC has a strong time-varying characteristic [2, 3]. The channel state information is essential for the relevant data detection, quantification, and interference suppression. Therefore, we need to further analyze and conduct research on the transmission characteristics and estimation methods of power line channel [4–6]. Generally, the traditional channel estimation method has three ways. Nonblind channel estimation is the most traditional channel estimation method, which mainly makes use of the pilot signal response channel's features on transmitting terminal. Blind channel estimation is a kind of channel estimation method which does not need to send pilot signal, Semiblind channel estimation is a compromise method between the above two methods [7]. The traditional channel estimation method requires a high speed analog to digital converter. In order to accurately estimate the channel characteristics on receiving terminal, we need to send long pilot signal and collect large sample data. There is no doubt that the hardware complexity and cost of the receiving terminal will increase.

Different from other areas of data compression for universal video, voice, image number, and so on, in order to be able to reliably analyze the power grid state, the data of power system stored after compression must be able to retain the perturbed feature quantities for each frequency band of the power quality [8]. In [8–10], authors adopt the wavelet transform to apply the power quality disturbance data compression and obtained a certain degree of compression. However,

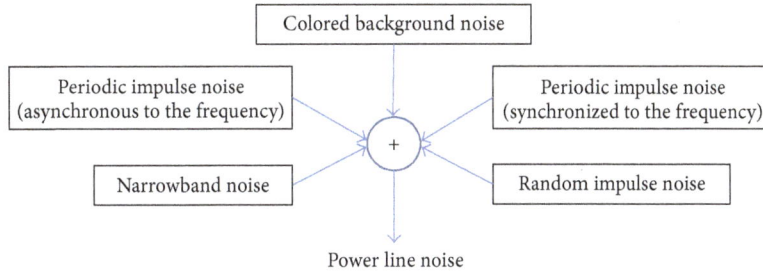

FIGURE 1: Power line noise.

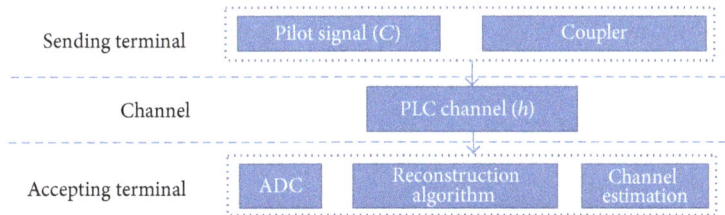

FIGURE 2: Block diagram of PLC estimation model.

the wavelet transform algorithm is complex, computationally intensive, real-time, and difficult to apply in real-time power quality monitoring system; meanwhile, the wavelet function is not unique, often using different wavelet analysis of the same signal analysis results which may vary greatly without adaptability.

Compression sensing includes two parts: the signal is measured on the measurement vector, and the signal is reconstructed by the measured value [11, 12]. Compression sensing theory shows that if the original signal is sparse on a certain base, the sampling frequency can be greatly reduced, and the original signal can be reconstructed exactly when the constrained equidistant condition between the observation matrix and the transform base is satisfied. The signal is sampled and compressed.

Therefore, researcher suggests that transmission characteristics of the power line channel are time-varying linear channel. We can estimate it based on OFDM pilot signal. However, this channel estimation mechanism ignores ADC device requirements on the receiving terminal. In the wireless communication scenario, a method of wireless channel estimation is based on compressive sensing. By orthogonal matching pursuit (OMP) algorithm on receiving terminal, we reconstruct the transmission characteristics of wireless channel [13–15]. This paper proposed an approach for the power line channel estimation based on compressive sensing. We analyzed the sparse characteristic of power line channel. Send the appropriate pilot signal by compressive sensing technology from the transmitting terminal. After the power line channel delay and attenuation, we extracted the effective features of power line channel to finish the power line channel estimation.

The rest of this paper is organized as follows. Section 2 presents PLC channel estimation model. Section 3 shows

compressive sensing estimation channel characteristics. Section 4 describes the simulation in detail and Section 5 evaluates our solution and gives the conclusion.

2. Power Line Communication Channel Estimation Model

Power line communication channel estimation model includes two parts: transmitting terminal and receiving terminals. Transmitting terminal is used to send appropriate pilot signal, amplified and coupled to the power line through the coupling circuit and then is influenced by the actual power line channel environment. Channel estimation at the receiver can get the transmission characteristics of power line channel.

As shown in Figure 2, the transmission signal includes the effective signal and pilot signal, amplified and coupled to the power line through the coupling circuit. The signal is attenuated by the power line channel and the interference of the noise. Through a coupling circuit, receiving terminal does the electrical isolation and receiving and then starts the digital signal processing after A/D (analog-to-digital signal conversion) by ADC [16].

Usually, those residential areas are generally used in the combination of radial and trunk distribution mode. There are a large number of nodes in the power line network, such as the branch structure and the impedance mismatch [17], which are shown in Figure 3.

These nodes cause the transmission signal on the power line to not be able to reach the receiving node directly from the sending node. There will be reflected and standing waves on different paths. The final receiving device received the superimposed signal via reflected and standing wave in the different paths. This makes the power line channel cause multipath effect, and the transmission characteristics show a certain frequency selective fading.

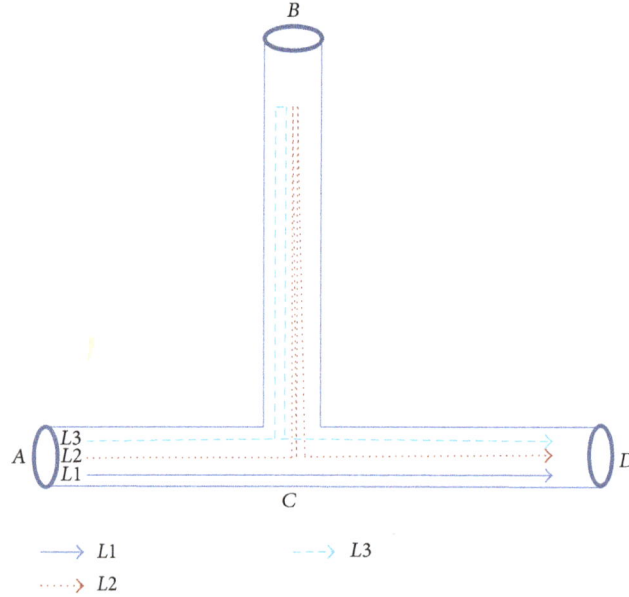

FIGURE 3: Branch circuit of power line communication.

Not only is the signal on power line directly transmitted from the sending node A to the receiving node D, but also it reflected many times to arrive at the receiving node D, forming the multipath effect. The signals in Figure 3 may have the transmission path as follows: (1) $L1$: $A \rightarrow C \rightarrow D$; (2) $L2$: $A \rightarrow C \rightarrow B \rightarrow C \rightarrow B \rightarrow C \rightarrow D$; (3) $L3$: $A \rightarrow C \rightarrow B \rightarrow C \rightarrow D$.

Since the multipath effect disperses the signal ability and arrives at the receiver with the different signal phase, it seriously influences the accepted effect. When there are n paths, frequency diversity can be achieved by changing the carrier frequency to improve the signal decision rate.

We suppose the channel transfer function of the ith path in multipaths is $H_i(f)$, so transfer function of the power line channel was formulated as follows:

$$H(f) = \sum_i H_i(f), \tag{1}$$

where $H(f)$ present the superposition of multipaths transmission characteristics. According to the reasons of multipath transmission, we analyzed the transmission characteristics on each path. There has been a path delay and the signal is attenuated with the increase of the transmission distance and frequency [18].

According to formula (1), we consider the main parameters affecting the characteristics of each path to establish the power line channel model. Meanwhile, for simulation, we simplify (1) as follows [19]:

$$H(f) = \sum_{i=1}^{N} g_i(f) e^{-(\alpha_0 + \alpha_1 f^k)d_i} e^{-j2\pi f\tau_i}, \tag{2}$$

in which parameters (a_0, a_1, and k) are obtained by field measurements on the reference channel. Coefficient $g_i(f)$ is the weight of each multipath channel generated randomly

TABLE 1: Parameters of transfer function model.

Model parameters	References
i	The number of paths; when the path is the shortest, $i = 1$
a_1, a_2	Attenuation parameter
k	Attenuation factor index; the typical value is 0.5~1
g_i	Weighted factor of path i; its absolute value is less than or equal to 1
d_i	The length of path i
τ_i	The delay of path i
f	Frequency

between $(-1, 1)$. The parameter d_i is the channel length of each path, with uniform distribution on $(0, L)$. L defined the upper limit of the path length on each scenario. In the power line channel model, main parameters in Table 1 can be used for modeling and analyzing the typical power line channel [8, 20]. Although we reduce parameters, they also can reflect the power line channel characteristics.

The mathematical model of power line channel shows that the power line channel is a multipath channel, which has frequency selective fading. The frequency response function reveals the sparse nature of the transmission characteristics of the power line channel.

3. Compressive Sensing Estimation Channel Characteristics

Due to the electromagnetic phenomena and other external causes, the environments usually affect the performance of

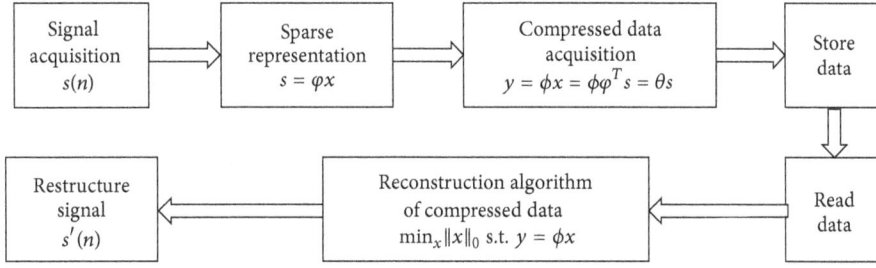

FIGURE 4: Process of CS.

device in power line communication. Therefore, it is very necessary to evaluate the channel accurately to acquire characteristic parameters of the channel impulse response.

3.1. Compressive Sensing Technology.
The process of compress sensing includes three steps: signal sparse representation, data compression, and reconstruction of compressed data as shown in Figure 4.

The compression sensing is a novel information acquisition theory, which mainly consists of two parts: the signal is projected on the measurement vector to obtain the measured value; the signal is reconstructed from the measured value. Usually, the power steady-state harmonic signal can be expressed as

$$f(t) = a_0 \cos(2\pi\gamma_0 t + \theta_0) + \sum_{m=1}^{L} a_m \cos(2\pi\gamma_m t + \theta_m), \quad (3)$$

where a_0, γ_0, and θ_0 are the amplitude, frequency, and phase of the fundamental component; L denotes the number of harmonics; a_m, γ_m, and θ_m denote the amplitude, frequency, and phase of the mth harmonic component. We assume the signal $f(t)$ to be s in time domain. And then, we can start the process of compress sensing as follows.

3.1.1. Signal Sparse Representation.
Assume the signal s has no sparsity in its time domain; it can be converted to another domain to obtain the sparsity projection x. Then, we can compress the sparsity x by the compress sensing technology. The transform domain projection process is the process of signal sparse representation, as shown in

$$s = \varphi x, \quad (4)$$

where s is the original signal without sparsity; φ denotes the projection matrix; x is the projection of s in the projection matrix φ, that is, the sparse signals.

3.1.2. Data Compression.
Let $x(n)$ be a digital signal sampled by the ADC; the dimension is N. If x is a sparse signal and the sparsity is K ($K \ll N$), that is, there are only K nonzero elements in the signal, then it can be compressed by the compressed sensing technology to reduce the dimension N

of the original signal $x(n)$ to the dimension M ($M \ll N$) and then we can get the compressed signal y as shown in

$$y = \phi x = \phi\varphi^T s = \theta s, \quad (5)$$

where y is the compressed signal; ϕ is an opposite observation matrix; x is the discrete signal; s is the original signal.

3.1.3. Reconstruction of Compressed Data.
Since the front-end hardware completed the data compression process, it reduces the storage requirements of analysis section. Then, we just focus on the back-end hardware compression algorithm for data recovery reconstruction work. Based on the above compression algorithm, the restore reconstructed of the compressed signal can be realized through 0-minimum-norm by

$$\min_x \quad \|x\|_0$$
$$\text{s.t.} \quad y = \phi x, \quad (6)$$

where x denotes the sparse signal with reconstruction; y is the restored signal after x's observations; 0-minimum-norm indicates the number of nonzero elements. The results of CS in sampling are as shown in Figure 5.

3.2. CS Channel Estimation.
The traditional least squares channel estimation response is equal length pilot blocks by transmitting the channel impact, so that the transmitted signal and channel impulse response of linear convolution are converted to circular convolution; received vector can be written as

$$y = p * h + n, \quad (7)$$

where $*$ denotes the circular convolution; P is the transmitted pilot signal.

$$y = Ch + n. \quad (8)$$

After constructing an appropriate pilot signal, it transmits the test pilot signal C and couples power line at the transmitting terminal through couplers. However, the pilot signal is affected due to channel transmission characteristics of the power line and the power line noise. Figure 6 shows the estimation model in the pilot point.

(a) The original signal

(b) The sampling signal by CS

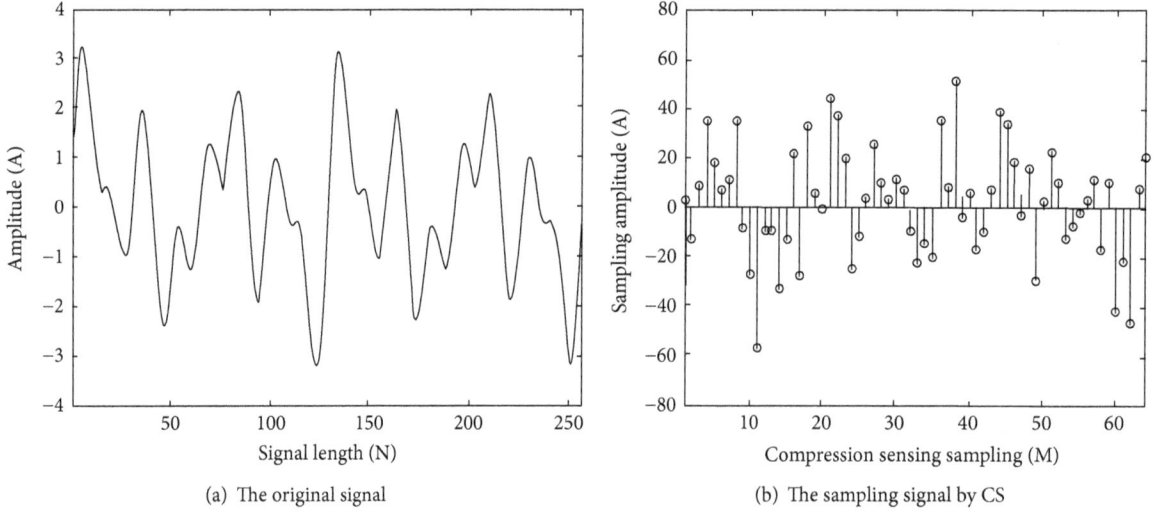

FIGURE 5: Sampling process of CS.

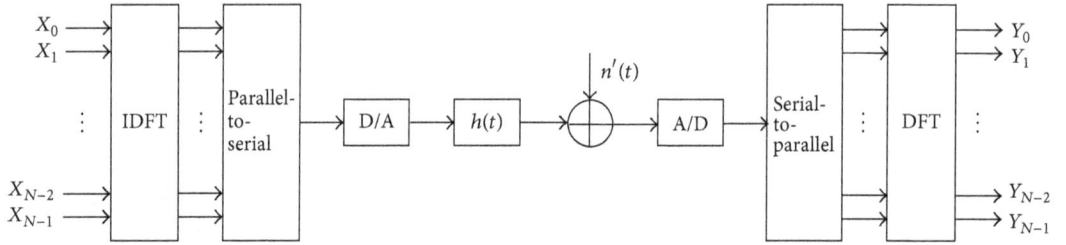

FIGURE 6: Pilot point channel estimation model.

Therefore, we just focus on the model and disturb signal by noise in the receiving end through couplers as described in

$$
\begin{bmatrix} y_0 \\ y_1 \\ \vdots \\ y_{M-2} \\ y_{M-1} \end{bmatrix}_{M \times 1} = \begin{bmatrix} c_0 & & c_1 \\ c_1 & \ddots & c_2 \\ \vdots & & \vdots \\ c_{M-1} & \ddots & c_M \\ c_M & & c_0 \end{bmatrix}_{M \times N} \begin{bmatrix} h_0 \\ h_1 \\ \vdots \\ h_{N-2} \\ h_{N-1} \end{bmatrix}_{N \times 1}
$$
$$
+ \begin{bmatrix} n_0 \\ n_1 \\ \vdots \\ n_{M-2} \\ n_{M-1} \end{bmatrix}_{M \times 1}. \tag{9}
$$

Assume the available power line communication channel characteristics y are at receiving end; we employ the sensing reconstruction algorithm (OMP, Orthogonal Matching Pursuit Algorithm) to estimate the power line channel impulse response $h(t)$.

OMP algorithm is the improved algorithm of MP algorithm. OMP algorithm selects and observes the most closely

matched atom signal from the atom library as OM algorithm. The orthogonality will make the selected atom not be repeated in the OMP algorithm iterative process, which ensures the optimality of iteration and thereby reduces the number of iterations and good reconstruction.

The restored reconstruction of OMP is as follows:

(1) Initialization is as follows: residual value $r_0 = y$, index set $\Lambda_0 = \Phi$, and iterations $i = 1$, $\Gamma_0 = \Phi$.

(2) Determine index value: $\lambda_{i+1} = \arg\max|\langle r_i, \tau_j \rangle|$, where τ_j is column j on matrix Φ; determine the position of the corresponding atom, that is, the position of nonzero element: $\{\Lambda_{i+1} = \Lambda_i \cup \lambda_{i+1}\}$, $\{\Gamma_{i+1} = \Gamma_i \cup \tau\lambda_{i+1}\}$.

(3) LS algorithm is used to obtain new estimates:

$$
\hat{x}_{i+1} = \arg\min \| y - \Gamma_{i+1}\hat{x} \| = \Gamma_{i+1}{}^+ x. \tag{10}
$$

in which + is devoted pseudoinverse.

(4) Calculate the new residual value: $r_{i+1} = y - \Gamma_{i+1}\hat{x}_{i+1}$.

(5) Optimize the iterative process: construct loop $i = i + 1$, and then repeat the indexing process until the completion of the required number m of iterations to terminate the iteration.

(6) Complete signal reconstruction: the calculated estimated value satisfies equality as $\hat{x}_{\Lambda_m} = \hat{x}_m$ and $\hat{x}_{\{1,\ldots,2N\}-\Lambda_m} = 0$.

Signals recovered percentage correctly (Gaussian)

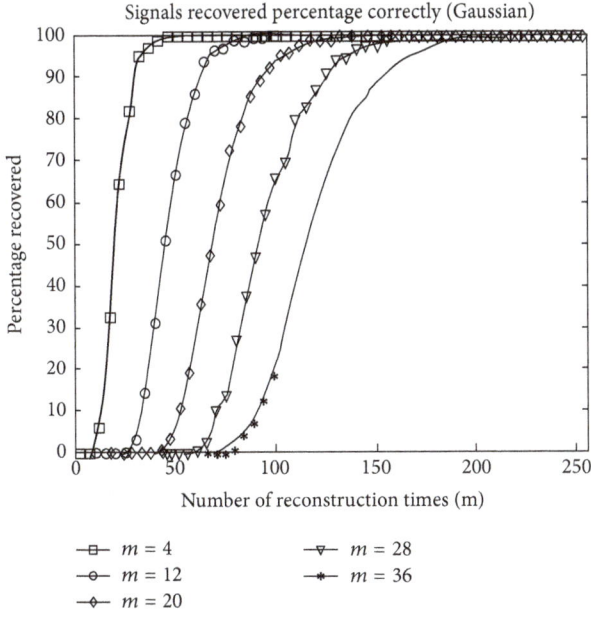

Power line channel amplitude frequency characteristic

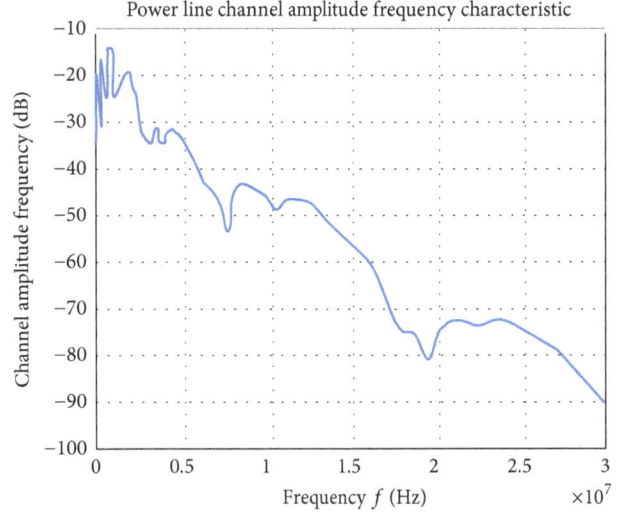

FIGURE 8: Amplitude frequency response of reference channel.

FIGURE 7: OMP multireconstruction.

TABLE 2: Parameters list.

Parameters	Value
Bandwidth simulation	$B_W = 30\,\mathrm{MHz}$
Sampling frequency	$f_s = 60\,\mathrm{MHz}$
Sampling time	$t = 10\,\mu s$

In each interaction, OMP algorithm obtains a nonzero element corresponding position in x and calculates the value of this element. After m iterations, it can get the estimation value $\hat{x}$, and then it estimates the power line channel impulse response. The reconstruction is as shown in Figure 7.

4. Simulation Result

Based on Matlab simulation platform, we built the simulation environment and simulated the power line channel impulse response. PLC channel frequency domain response is in 15 paths, and the longest path is 1000 m in the channel model. And based on the least squares channel estimation of the pilot sequences and CS-based power line channel estimation, we analyze the simulation results under the same conditions.

To build the simulation environment, we refer to the reference communication channel parameters and simulate the time-domain characteristics of power line reference communication channel. The parameters are shown in Table 2.

4.1. Different SNR (Signal to Noise Ratio). Power line noises are very complex; they can be roughly divided into five categories in the time domain: the colored background noise, narrowband noise, asynchronous to power frequency periodic impulse noise, synchronized to power frequency periodic impulse noise, and sudden impulse noise.

In implementation of the project, the average power of noises, the colored background noise, narrowband noise, and asynchronous to power frequency periodic impulse noise, are small. And these three noises time-varying characteristics are weak; they change slowly with time in the entire PLC carrier communication frequency band. Thus, they can be called background noise. Meanwhile, the average power synchronization frequency periodic pulse noise and sudden impulse noise are relatively large. And there are two noises changes, frequently and randomly. They thus are called random impulse noise. Although the random impulse noise appears little, this type of noise greatly affects the quality of communication. It usually causes the narrowband communication interruption. Therefore, we just consider the superposition persistence of background noise in the channel in simulation.

Based on the impulse response of power line channel, we built the suitable pilot signal matrix. The dimensions of fixed reception signal are as follows: M 50 and N 200. And due to the interference of power line channel by background noise, the signal to noise ratio changes are from 5 dB to 30 dB.

As shown in Figures 8 and 9, compared to the traditional least squares channel estimation, CS channel estimation has a better estimation by the compressed sensing technology in the low SNR and hostile environment. The estimation error is less than tenfold the traditional least squares channel estimation algorithm. As the SNR increases and the channel environment tends to be better, the least squares channel estimation and compressive sensing-based channel estimation can achieve good estimation. However, compared to compressive sensing-based channel estimation, the least squares estimation needs longer pilot signal, more data computation, and longer calculation time.

4.2. Different Compression Dimension. Based on the impulse response of power line channel, we built the suitable pilot signal matrix again. For the dimensions of fixed reception signal, we increase M from 50 to 150 gradually and N is at 200.

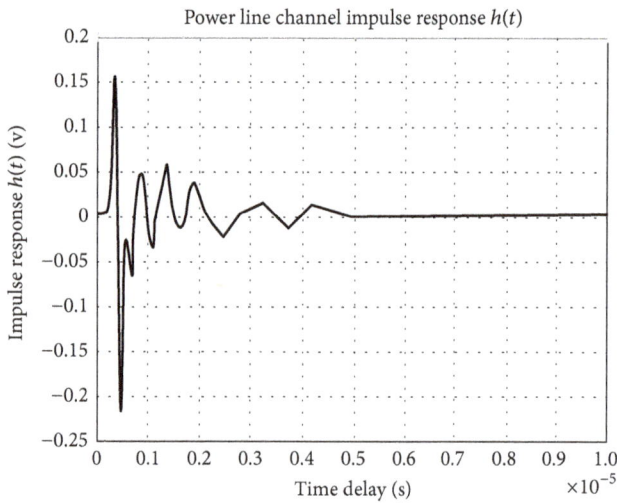

FIGURE 9: Impulse response of reference channel.

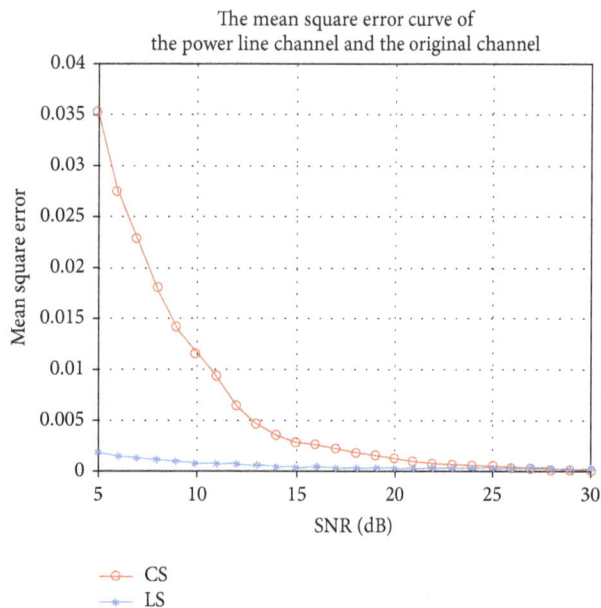

FIGURE 10: Performance comparison of CS and LS (SNR).

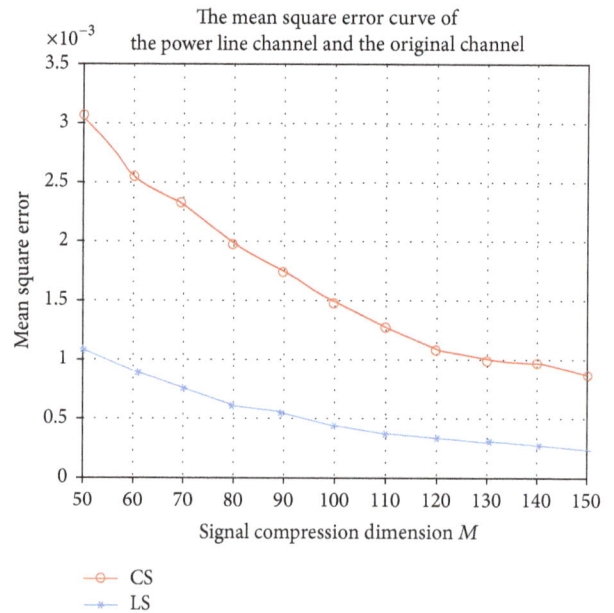

FIGURE 11: Performance comparison of CS and LS (M).

estimation algorithm has advantages and achieves compression objective data.

5. Conclusion

In this paper, we analyzed the sparsity of power line channel firstly. On this basis, we proposed a method of power line channel estimation based on compressive sensing. The simulation results show that this method used less pilot signal and fewer storage resources. However, the performance of the algorithm is better than the least square channel estimation algorithm. Therefore, the algorithm proposed in this paper has better application prospect.

In the future, we will still research various algorithms to improve the efficiency in compression sensing.

Conflicts of Interest

The authors declare that there are no conflicts of interest regarding the publication of this paper.

References

[1] M. Yigit, V. C. Gungor, G. Tuna, M. Rangoussi, and E. Fadel, "Power line communication technologies for smart grid applications: a review of advances and challenges," *Computer Networks the International Journal of Computer & Telecommunications Networking*, vol. 70, no. 10, pp. 366–383, 2014.

[2] S. K. Shen, "Power line communication system and control method thereof," 2016.

[3] A. J. Hicks Iii, B. Davis, G. Howell et al., "Power-line communications," US20160148499, 2016.

[4] B. M. Propp and D. L. Propp, "Power line communication apparatus," U.S. Patent 4,815,106. 1989.

Meanwhile, we added the noise power line channel and fixed the noise ratio at 15 dB.

As shown in Figures 10 and 11, in the unchanged situation of the power line signal to noise ratio, with the increasing of received signals' dimension of M, these two methods performance of power line channel estimation is improved. However, due to the sparsity of power line channel, CS-based power line channel estimation adopts the sparsity to improve the efficient and accurate estimation of the power line channel characteristics, which achieves better channel restoration reconstruction.

However, with the growing pilot sequences, the pilot signal-based least squares channel estimation method can achieve good channel estimation. However, compared to the former, the compressed sensing-based power line channel

[5] M. Götz, M. Rapp, and K. Dostert, "Power line channel characteristics and their effect on communication system design," *IEEE Communications Magazine*, vol. 42, no. 4, pp. 78–86, 2004.

[6] M. Zamani, Z. Zhang, and C. Li, "Channel estimation for optical orthogonal frequency division multiplexing systems," 2016.

[7] P. L. Zhang, H. X. Zhang, H. D. Liu, Y. J. Zhang, P. F. He, and X. L. Pang, "Particle filtering based channel estimation in OFDM power line communication," *Journal of China Universities of Posts & Telecommunications*, vol. 21, no. 5, pp. 24–30, 2014.

[8] B. Yunus and H. Li, "Analysis of power quality waveform for data transmission efficiency over IEC 61850 communication standard," in *Proceedings of the 1st International Power and Energy Conference (PECon '06)*, pp. 161–166, November 2006.

[9] C. Tse, "Power quality meter and method of waveform anaylsis and compression," 2015.

[10] A. Rahim Abdullah, H. T. N. Ahmad, A. N. Abidullah et al., "Performance evaluation of real power quality disturbances analysis using s-transform," *Applied Mechanics & Materials*, vol. 753, no. 2015, pp. 1343–1348, 2015.

[11] D. L. Donoho, "Compressed sensing," *IEEE. Transactions on Information Theory*, vol. 52, no. 4, pp. 1289–1306, 2006.

[12] R. G. Baraniuk, "Compressive sensing," *IEEE Signal Processing Magazine*, vol. 24, no. 4, pp. 118–124, 2007.

[13] J. A. Tropp and A. C. Gilbert, "Signal recovery from random measurements via orthogonal matching pursuit," *IEEE. Transactions on Information Theory*, vol. 53, no. 12, pp. 4655–4666, 2007.

[14] D. Tralic and S. Grgic, "Signal reconstruction via compressive sensing," in *Proceedings of the 53rd International Symposium (ELMAR '11)*, pp. 5–9, Zadar, Croatia, September 2011.

[15] S. C. Yan and J. Qi, "Research on service impact analysis for power communication network based on N-1 principle," *Advanced Materials Research*, vol. 846-847, pp. 396–399, 2014.

[16] J. Y. Shin and J. C. Jeong, "Power line channel model considering adjacent nodes with reduced calculation complexity due to multipath signal propagation and network size using infinite geometric series and matrices," *Transactions of the Korean Institute of Electrical Engineers*, vol. 58, no. 2, pp. 248–255, 2009.

[17] A. Tomasoni, R. Riva, and S. Bellini, "Spatial correlation analysis and model for in-home MIMO power line channels," in *Proceedings of the 16th IEEE International Symposium on Power Line Communications and Its Applications (ISPLC '12)*, pp. 286–291, March 2012.

[18] J. Matanza, S. Alexandres, and C. Rodríguez-Morcillo, "Advanced metering infrastructure performance using European low-voltage power line communication networks," *IET Communications*, vol. 8, no. 7, pp. 1041–1047, 2014.

[19] Y. Xiao, J. Zhang, F. Pan, and Y. Shen, "Power line communication simulation considering cyclostationary noise for metering systems," *Journal of Circuits, Systems and Computers*, vol. 25, no. 9, Article ID 1650105, 2016.

[20] Y. Wang, Q. D. Wang, X. Z. Hou, H. L. Sun, X. M. Chen, and X. J. Li, "Measurement and research on attenuation characteristics of low voltage power line communication channel," *Advanced Materials Research*, vol. 986-987, pp. 2068–2072, 2014.

10

Performance Analysis of Precoded MIMO PLC System based on Two-Sided Jacobi SVD

Hasna Kilani,[1] Mohamed Tlich,[2] and Rabah Attia[1]

[1]SERCOM Laboratory, Tunisia Polytechnic School, Carthage University, 2078 La Marsa, Tunisia
[2]CodinTek Company, El Ghazala Technological Park, 2088 Ariana, Tunisia

Correspondence should be addressed to Hasna Kilani; hasna.kilani@ept.rnu.tn

Academic Editor: Adam Panagos

This paper evaluates the performance of closed loop multiple input multiple output power line communication (CL MIMO PLC) system based on enhanced zero-forcing (ZF) equalizer. In this work, the two-sided Jacobi (TSJ) algorithm has been investigated for the computation of singular value decomposition of the channel matrix. Quantized parameters are feedback from the receiver to the transmitter for precoding process. Numerous simplifications are introduced for the reduction of the algorithm complexity. The performance of the CL MIMO PLC is evaluated in terms of bit error rate (BER), constellation error vector magnitude (EVM), and mean square error (MSE) between the constructed SVD matrices and Matlab computed ones.

1. Introduction

The PLC systems present the new trend for high level communication. The application of MIMO scenarios on PLC systems enhances the data throughput significantly. In MIMO wireline systems [1], the data streams can be demultiplexed into several substreams transmitted by different ports to improve the throughput performance of the overall communication system by utilizing the transmit diversity [2, 3].

Various architectures of receivers has been proposed in literature such as the zero-forcing (ZF) receiver, the minimum mean square error (MMSE) receiver, and the successive interference canceller (SIC) receiver. These techniques are investigated to decode the spatially multiplexed signals over MIMO systems [4–6]. Generally, the performance improvement from one type of the MIMO receiver to another comes at the price of higher implementation cost. For example, despite its reduced complexity of implementation, the ZF receiver is known to suffer from the effect of noise enhancement.

One can enhance the performance of the ZF receiver, by splitting the equalization algorithm among the transmitter and the receiver. The TSJ-SVD is used for the computing of precoding and decoding matrix. Several simplifications are introduced to reduce the hardware implementation of the precoding/decoding processes.

The remainder of the paper is organized as follows. In Section 2, the CL MIMO system is described. The precoding design scheme is then presented in Section 3 and introduced simplifications making computing complexity lower. Subsequently, the performance analysis of TSJ-SVD algorithm is given in Section 4. Finally, the paper is concluded in Section 5.

The superscript $(\cdot)^H$ denotes the conjugate transpose. In addition, $(\cdot)^+$ and $(\cdot)^T$ represent the pseudoinverse and transpose operations, respectively. $\mathbb{C}^{N_r \times N_t}$ denotes the set of $N_r \times N_t$ matrices over complex field.

2. System Description

For a MIMO PLC system composed of N_t transmission ports and N_r reception ports, the MIMO channel can be described by a complex matrix $H \in \mathbb{C}^{N_r \times N_t}$. The MIMO PLC model is then given by

$$r = Hx + n, \tag{1}$$

where $x = [x_1, x_2, \ldots, x_{Nt}]^T$ is the transmit signal, $r = [r_1, r_2, \ldots, r_{Nr}]^T$ is the received signal vector, and $n = [n_1, n_2, \ldots, n_{Nr}]^T$ is the noise at the receiver. In the remainder of the paper

$N_t = 2$ and $N_r = 4$. The channel matrix H has the following formula:

$$H = \begin{bmatrix} h_{11} & h_{12} \\ h_{21} & h_{22} \\ h_{31} & h_{32} \\ h_{41} & h_{42} \end{bmatrix}, \quad (2)$$

where $h_{ij} \in \mathbb{C}$ is the complex coefficient, $i \in [1, \ldots, 4]$ and $j \in [1, 2]$.

In conventional MIMO communication system based on zero-forcing (ZF) equalizer, the equalization matrix is given by

$$W_{\text{ZF}} = H^+ = \left(H^H H \right)^{-1} H^H. \quad (3)$$

The main idea of the closed loop MIMO system is based on decomposing the channel matrix H into precoding and decoding parts. The precoding matrix is then sent by the receiver to the transmitter in a feedback in order to precode the signal before being sent.

Here the decomposition is carried out by the SVD method largely used in MIMO. The channel matrix H can be decomposed into 2 parallel and independent SISO branches by SVD (see Figure 1):

$$H = UDV^H, \quad (4)$$

where $V \in \mathbb{C}^{2 \times 2}$ is the right-hand unitary matrix of the SVD, $U \in \mathbb{C}^{4 \times 2}$ is the left-hand unitary matrix, and $D \in \mathbb{R}^{2 \times 2}$ is a diagonal matrix containing the singular values of the channel matrix H.

In order to improve the MIMO equalization, rather than using simple ZF equalizer (2), the precoding matrix is incorporated in (1) by replacing the transmit symbol vector x by

$$s = Vx. \quad (5)$$

Equation (1) becomes

$$r = HVx + n = \left(UDV' \right) Vx + n = UDx + n. \quad (6)$$

If we denote W as the reception first-stage matrix, the received signal crossing W is

$$y = Wr = WUDx + Wn. \quad (7)$$

If $W = U^H$ then

$$y = Wr = Dx + U^H n. \quad (8)$$

The combination of precoding, channel, and reception first-stage matrices decomposes the equivalent channel into parallel streams reduced to the diagonal matrix D (see Figure 1). Then the ZF equalizer matrix becomes

$$W_{\text{ZF}} = W = U^H D^{-1}. \quad (9)$$

The precoding matrix V mixes the symbols of x, and thus the transmit symbol vector s contains all symbols of x; that is, each symbol is transmitted via each MIMO path. Thus, the full spatial diversity is achieved. The receiver side model is depicted in Figure 2. In this work, the 2×4 MIMO system is considered.

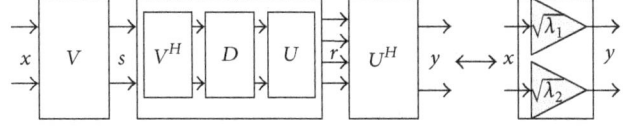

FIGURE 1: SVD based MIMO transmitter and receiver.

3. The Precoding Process

The proposed MIMO scheme is mainly based on the precoding process. After estimating the channel matrix H, its SVD is carried out using the TSJ method.

In this section, we introduce the TSJ-SVD algorithm. Then, we describe the main steps of the proposed simplified algorithm.

3.1. Basic Transformation on the Channel Matrix. In the literature, the TSJ algorithm is only used for real symmetric matrix [7]. In our case, the channel matrix H is complex. So to apply the TSJ algorithm we should transform the complex matrix into real symmetric matrix. We then firstly proceed by transforming the matrix $H \in \mathbb{C}^{2 \times 2}$ into Hermitian matrix A:

$$A = H^H H = \begin{bmatrix} a & be^{j\beta} \\ be^{-j\beta} & d \end{bmatrix}, \quad (10)$$

where a, b, and d are real values, β is an angle, and A is a Hermitian matrix such that

(i) $A^H = A$;

(ii) The diagonal elements of A are real; only the off-diagonals are complex conjugates:

$$C = \begin{bmatrix} 1 & 0 \\ 0 & e^{j\theta} \end{bmatrix}^H \begin{bmatrix} a & be^{j\beta} \\ be^{-j\beta} & d \end{bmatrix} \begin{bmatrix} 1 & 0 \\ 0 & e^{j\theta} \end{bmatrix} = \begin{bmatrix} a & b \\ b & d \end{bmatrix}, \quad (11)$$

where $\theta = -\beta$.

The obtained matrix C is real and symmetric and the Jacobi algorithm can be applied to decompose it into SVD form.

3.2. The Two-Sided Jacobi Transformation. As the MIMO system needs to feed back the precoding matrix to the transmitter, an optimization of the transferred parameters is fundamental to make the transferred parameters transfer overhead as little as possible. The transformation of H to C is carried out for this aim as will be detailed below.

The TSJ algorithm iteratively minimizes the off-diagonal elements of the matrix C expressed in

$$\text{Off}(C) = \sqrt{\sum_{m=1}^{2} \sum_{n=1, n \neq m}^{2} c_{mn}^2}. \quad (12)$$

The Jacobi transformations work by performing a sequence of orthogonal updates of C matrix described by

$$C_{i+1} \longleftarrow Q_i^T C_i Q_i, \quad (13)$$

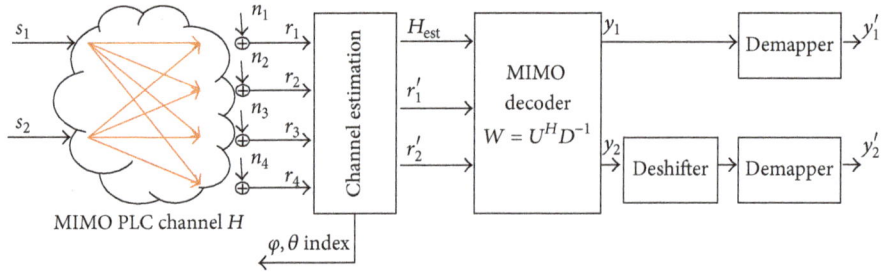

FIGURE 2: Receiver side of precoded MIMO PLC system.

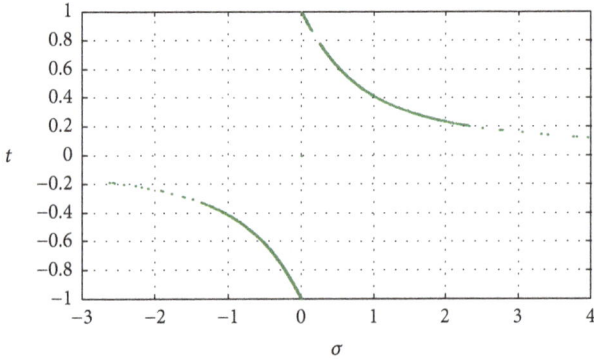

FIGURE 3: Variation of t as function of σ.

where C_i is the processed matrix at iteration i and Q_i is a Jacobi matrix at iteration i.

With C_{i+1} being "more diagonal" than its predecessor C_i, the off-diagonal values are compared to a threshold in order to decide whether the Jacobi algorithm converged or not. This threshold is proportional to the on-diagonal values of C and is defined by

$$\tau = \varepsilon \left(\frac{1}{2} \sqrt{\sum_{m=n=1,2} c_{mn}^2} \right), \tag{14}$$

where ε is a constant value.

The Jacobi matrix (named also given rotation) is defined by the following formula:

$$Q = \begin{bmatrix} \cos(\varphi) & \sin(\varphi) \\ -\sin(\varphi) & \cos(\varphi) \end{bmatrix}. \tag{15}$$

The rotation angle φ is chosen so that the off-diagonal elements of C are equal to zero [7]. φ should then satisfy the following:

$$\begin{bmatrix} \cos\varphi & \sin\varphi \\ -\sin\varphi & \cos\varphi \end{bmatrix}^T \begin{bmatrix} a & b \\ b & d \end{bmatrix} \begin{bmatrix} \cos\varphi & \sin\varphi \\ -\sin\varphi & \cos\varphi \end{bmatrix}$$

$$= \begin{bmatrix} a_1 & 0 \\ 0 & d_1 \end{bmatrix}. \tag{16}$$

Equation (16) can be simplified to

$$b \left(\cos\varphi^2 - \sin\varphi^2 \right) + (a - d) \cos\varphi \sin\varphi = 0. \tag{17}$$

Two cases arise:

(i) If $b = 0$ then C is already diagonal. Considering in this case $\cos\varphi = 1$ and $\sin\varphi = 0$ (rather than $\cos\varphi = 0$ and $\sin\varphi = 1$) makes no change to C. We have then $\varphi = 0$.

(ii) If $b \neq 0$ then (17) remains to be solved.

If $b \neq 0$, we define the parameters $\sigma = (a - d)/2b$ and $t = \cos\varphi/\sin\varphi$. Equation (17) becomes

$$t^2 + 2\sigma t - 1 = 0. \tag{18}$$

Equation (17) has the two roots:

$$t = \sigma \pm \sqrt{1 + \sigma^2}. \tag{19}$$

As a result, the cosine and sine values of the Jacobi matrix (see (15)) can be calculated as

$$\cos\varphi = \frac{1}{\sqrt{t^2 + 1}} \tag{20}$$

$$\sin\varphi = \cos\varphi \cdot t.$$

With C matrix being real symmetric and 2×2 sized, the theoretical maximum number of iterations i is thus equal to 1. In practice, the expression of t is quite complex to implement; it involves a division and calculation of square root, which is resource consuming when implemented on embedded platform. t should then be approximated, and the off-diagonal values of C are consequently no longer equal to 0 from the first iteration.

According to [8], the t parameter can be approximated to make the algorithm faster. The variation of t as function of σ is given in Figure 3.

It is easy to see that the error between t and its approximated value is limited when σ is close to 0. The value converges to zero when σ goes to infinity. The approximation of t will then develop for positive values of σ.

Using Taylor expansion the approximation of t is given by

$$\lim_{\sigma \to 0} \frac{1}{\sigma + \sqrt{1 + \sigma^2}} \approx 1 - \sigma + \frac{\sigma^2}{2},$$

$$\lim_{\sigma \to +\infty} \frac{1}{\sigma + \sqrt{1 + \sigma^2}} \approx \frac{1}{2\sigma}. \tag{21}$$

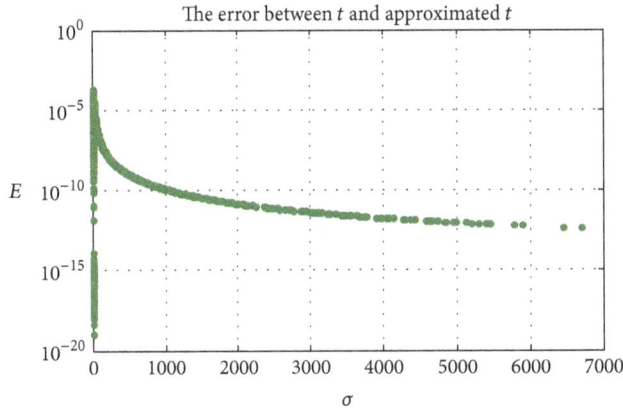

FIGURE 4: Approximation error of t.

Based on the piecewise approximation we get the following result:

$$
\tilde{t} = \begin{cases} 1 - \sigma + \dfrac{\sigma^2}{2}, & 0 \le \sigma < 1.25 \\[2mm] -\sigma + \sqrt{1 + \sigma^2}, & 1.25 \le \sigma < 5 \\[2mm] \dfrac{1}{2\sigma}, & \sigma \ge 5. \end{cases} \tag{22}
$$

The approximation error $E = |t - \tilde{t}|$ is reported in Figure 4. It is obvious that the approximation errors have been controlled within a precision almost equal to 10^{-4}, knowing that around $0t$ is very close to ± 1.

The Jacobi process is finished when the off-diagonal elements of the matrix become close to zero. Finally, the Jacobi transformation process determines the following:

(1) The right-hand singular vector matrix V is equal to the multiplication of successive matrix rotations Q by the transformation matrix applied to A (see (11)):

$$
V = \begin{bmatrix} 1 & 0 \\ 0 & e^{j\theta} \end{bmatrix} \begin{bmatrix} \cos(\varphi) & \sin(\varphi) \\ -\sin(\varphi) & \cos(\varphi) \end{bmatrix}, \tag{23}
$$

where

$$
Q = \begin{bmatrix} \cos(\varphi) & \sin(\varphi) \\ -\sin(\varphi) & \cos(\varphi) \end{bmatrix} = \prod_i \begin{bmatrix} \cos\varphi_i & \sin\varphi_i \\ -\sin\varphi_i & \cos\varphi_i \end{bmatrix}, \tag{24}
$$

$$
\varphi = \sum_i \varphi_i \tag{25}
$$

with φ_i being the rotation angle of the Jacobi matrix Q_i at iteration i.

(2) The singular values matrix $D = \sqrt{C_{i+1}}$.

(3) The left-hand U matrix is deduced by

$$
U = HVD^{-1}. \tag{26}
$$

In Figure 6 is reported the flowchart of the precoding process.

3.3. *Quantization of the Precoding Matrix Fed Back to the Transmitter.* As described above, performing the Jacobi based SVD on the channel matrix leads to 3 matrices: the diagonal matrix D, which contains the square root of the singular values of the channel matrix, the left-hand decoding matrix U, and the right singular vector matrix V also called precoding matrix.

In the transmitter, the precoding operation is performed on each carrier and consists of multiplying the output signals by the precoding matrix V. V is constructed using the two parameters θ and φ fed back from the receiver. To do this, the two parameters are uniformly quantized as follows:

$$
\varphi = \frac{k\pi}{2^{b_\varphi + 1}}, \quad k = 0, 1, \ldots, 2^{b_\varphi} - 1,
$$
$$
\theta = \frac{k\pi}{2^{b_\theta - 1}} - \pi, \quad k = 0, 1, \ldots, 2^{b_\theta} - 1, \tag{27}
$$

where b_φ is the number of b used to quantize φ and b_θ is the number of bits used to quantize θ.

4. Simulation Results

To evaluate the TSJ based SVD precoding/decoding process, we developed a MIMO communication system composed of the following:

(i) A transmitter chain that contains a random source delivering random streams containing $N = 13820$ bits, a splitter module to split the data into two MIMO signals, a mapping module using 16-QAM modulation, and a phase shifter which introduces a $\pi/2$ angle to the signal transmitted via the second antenna; the transmitter chain is shown in Figure 5.

(ii) A receiver chain that contains a channel estimation module, a MIMO decoder, a deshifter, and demapping, as reported in Figure 2.

The frequency band of the transmitted signal is [10 MHz, 86 MHz]. Simulations are carried out for 2×4 MIMO channel generated and based on the MIMO PLC model detailed in [9] and based on the Multiconductor Transmission Line theory (MTL) [10].

4.1. *Validation of the Proposed Precoding Algorithm.* The performance of the Jacobi transformation algorithm is evaluated according to the mean square error (MSE) between the computed eigenvalues denoted as λ_{svd}^l, $l = 1, 2$, obtained from the proposed Jacobi algorithm and the eigenvalues λ^l, $l = 1, 2$, obtained from the SVD function of Matlab (see Figure 7) where l is the antenna index. The MSE is defined by

$$
\text{MSE}_{\text{jacobi}}^l = \frac{1}{2} \left(\lambda_{svd}^l - \lambda^l \right)^2, \quad l = 1, 2. \tag{28}
$$

The transmitted streams constellations before and after precoding are presented in Figure 8.

The 16-QAM constellations undergo a specific rotation due to precoding matrix multiplication.

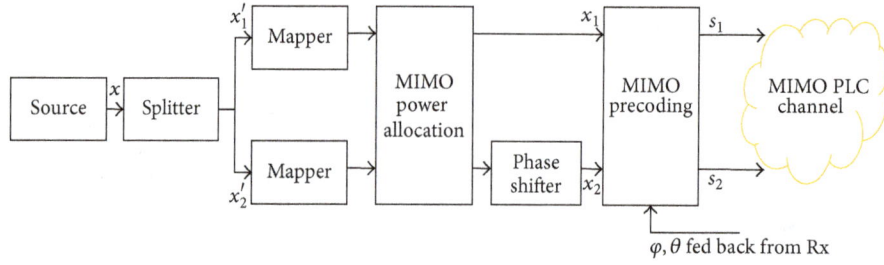

FIGURE 5: Transmitter chain of the precoded MIMO PLC system.

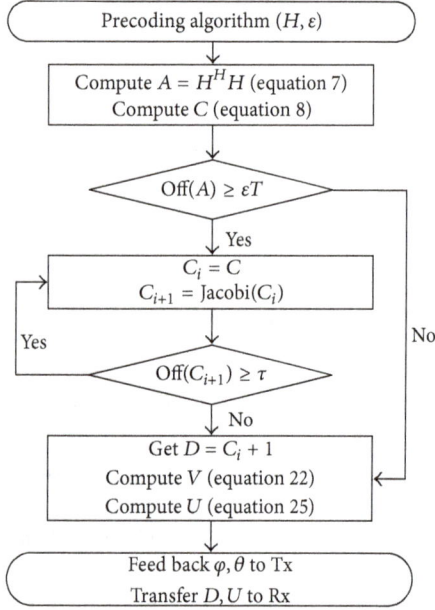

FIGURE 6: The precoding process flowchart.

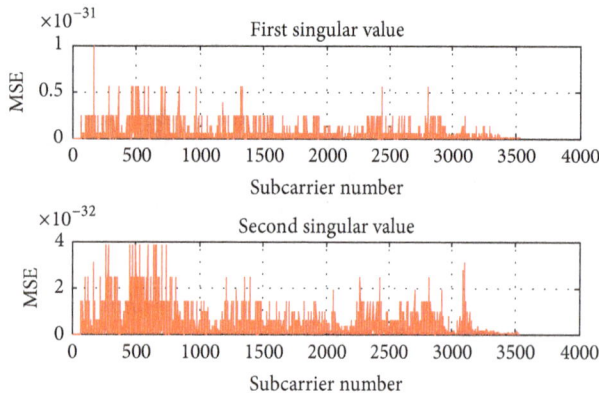

FIGURE 7: MSE of the two eigenvalues λ^1 and λ^2, respectively.

4.2. Evaluation of the MIMO System Performance.

In this section, the MIMO system performance evaluation is carried out in terms of energy conservation, Bit error rate (BER) variation, and the effect of the number of φ and θ quantization bits.

4.2.1. Energy Conservation of the System. The aim here is to verify the correctness of the TSJ-SVD algorithm in the decomposition of the estimated channel matrix H. The

MSE is calculated between the coefficients computed using the proposed SVD algorithm denoted by $h_{12,SVD}$ and the real values $h_{12,real}$ for 3 generated MIMO channels. The comparison results are shown in Figure 9.

According to Figure 9, the maximum value of MSE for the 3 different MIMO channels is below $6 \cdot 10^{-6}$. The same results are found for other matrix coefficients. This undeniably proves that the SVD algorithm based on Jacobi transform is energy conservative. This is explained by the fact that all matrix transformations applied to channel matrix are unitary.

4.2.2. Bit Error Rate as Function of SNR. In this section is evaluated the BER for different values of SNR, for bit streams coming from both MIMO antennas. Simulations are carried out without any error correcting encoding/decoding. BER is compared to theoretical BER for uncoded 16-QAM.

According to Figure 10, BER behavior of our MIMO system is close to theoretical BER. This proves that no additional noise is brought by our MIMO precoding/decoding processing.

4.2.3. The Error Vector Magnitude (EVM) as a Function of Quantization Bits. As described above, the receiver feeds back the two quantized parameters θ and φ to the transmitter in order to make the latter construct the precoding matrix V. In this section we study the effect of the number of quantization bits on the system performance. Simulations are carried out without adding any additive noise in order to only see the quantization noise effect. For evaluation purpose, we considered the constellation error vector magnitude (EVM) criteria. The EVM reflects the error between the received constellation after equalization and the original normalized 16-QAM constellation. In Figure 11 is reported the EVM (95th percentile point, which is the value where 95% of the individual symbol EVM values are below that point) as a function of the number of quantization bits, calculated on the two MIMO streams. Each EVM point is calculated for 3455 constellations.

We observe that the EVM converges to 0 when the number of quantization bits becomes high. In this case, the received constellation becomes closer to the original one. The EVM decreases below 0.1% from a quantization number equal to 16.

5. Conclusion

In this paper is described a method for linear precoded 2×4 MIMO PLC system based on TSJ-SVD algorithm.

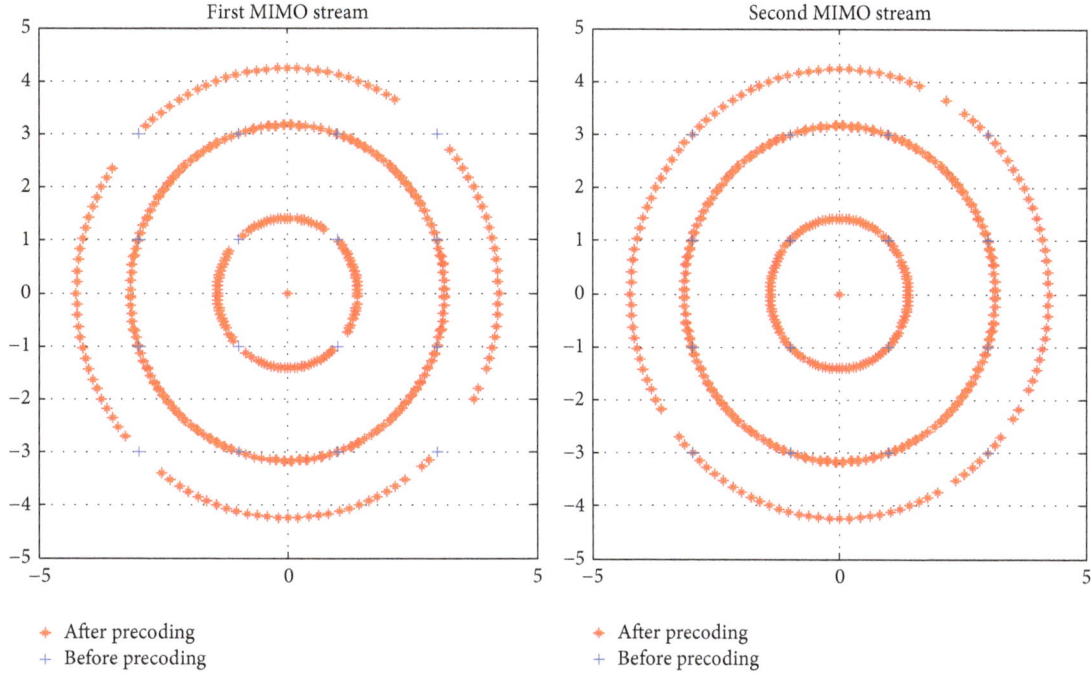

FIGURE 8: Transmitter chain of the precoded MIMO PLC system.

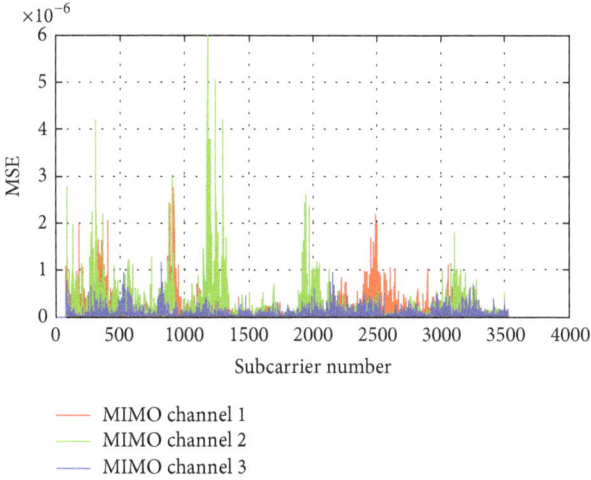

FIGURE 9: MSE between $h_{12,\text{SVD}}$ and $h_{12,\text{real}}$.

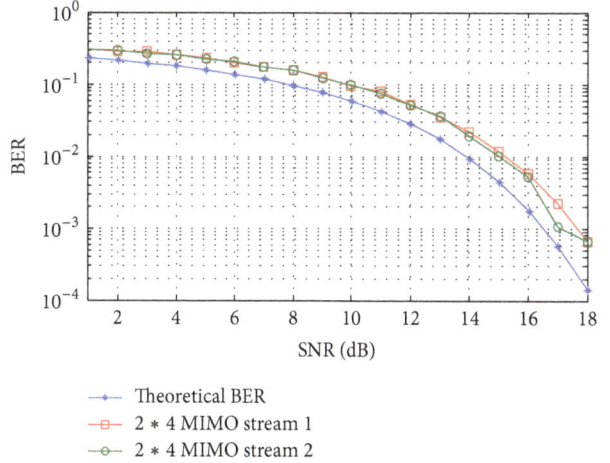

FIGURE 10: BER as function of SNR for bit streams from MIMO ports 1 and 2.

A simplified ZF equalizer is presented. The precoding matrix is compressed into 2 phase values which are quantified before being fed back from the receiver to the transmitter.

The proposed TSJ-SVD outputs were revalidated through comparison to Matlab SVD. Singular values λ^l, $l = 1, 2$, calculated from SVD and Matlab were juxtaposed.

The MIMO system performance was finally evaluated in terms of precoding matrix energy conservation, BER as function of SNR, and effect of quantization bits on EVM. It is proven that without error correcting code the proposed system BER is close to theoretical BER meaning that no additional noise is brought by our precoding/decoding processing. Also, based on the EVM 95th percentile point

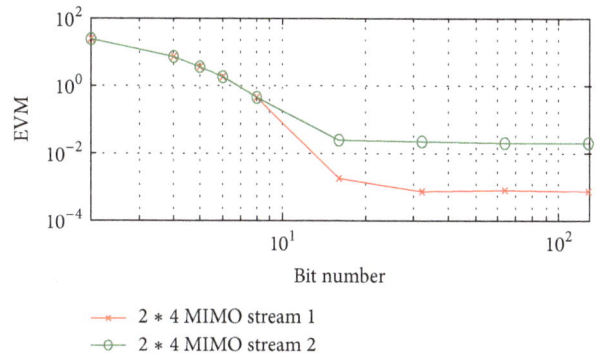

FIGURE 11: Effect of the number of quantization bits on EVM.

criteria, it was shown that the performance of the MIMO system is better for large number of quantization bits and that the EVM decreases below 0.1% from a quantization bits number equal to 16.

Conflict of Interests

The authors declare that there is no conflict of interests regarding the publication of this paper.

References

[1] R. Hashmat, P. Pagani, A. Zeddam, and T. Chonave, "MIMO communications for inhome PLC networks: measurements and results up to 100 MHz," in *Proceedings of the 14th Annual International Symposium on Power Line Communications and Its Applications (ISPLC '10)*, pp. 120–124, Rio de Janeiro, Brazil, March 2010.

[2] D. Schneider, *In home power line communications using multiple input multiple output principles [Dr. Ing. Dissertation]*, Verlag Dr. Hut, Munich, Germany, 2012.

[3] Seventh Framework Programme: Theme 3 ICT-213311OMEGA-Deliverable D3.2, "PLC Channel Characterization and Modeling," December 2008.

[4] G. J. Foschini, D. Chizhik, M. J. Gans, C. Papadias, and R. A. Valenzuela, "Analysis and performance of some basic space-time architectures," *IEEE Journal on Selected Areas in Communications*, vol. 21, no. 3, pp. 303–320, 2003.

[5] G. J. Foschini and M. J. Gans, "On limits of wireless communications in a fading environment when using multiple antennas," *Wireless Personal Communications*, vol. 6, no. 3, pp. 311–335, 1998.

[6] A. Scaglione, P. Stoica, S. Barbarossa, G. B. Giannakis, and H. Sampath, "Optimal designs for space-time linear precoders and decoders," *IEEE Transactions on Signal Processing*, vol. 50, no. 5, pp. 1051–1064, 2002.

[7] G. H. Golub and C. F. Van Loan, *Computation Matrix*, The Johns Hopkins University Press, 2013.

[8] T. Zhou, S. Fang, X. Yang, Z. Li, Q. Guo, and B. Jiang, "A Jacobi-based parallel algorithm for Matrix inverse computations," in *Proceedings of the International Conference on Wireless Communications & Signal Processing (WCSP '12)*, pp. 1–5, IEEE, Huangshan, China, October 2012.

[9] F. Versolatto and A. M. Tonello, "An MTL theory approach for the simulation of MIMO power-line communication channels," *IEEE Transactions on Power Delivery*, vol. 26, no. 3, pp. 1710–1717, 2011.

[10] C. R. Paul, *Analysis of Multiconductor Transmission Lines*, John Wiley & Sons, New York, NY, USA, 1994.

11

An Extraction Method of Weak Low-Frequency Magnetic Communication Signals based on Multisensor

Chao Huang,[1,2] **Xin Xu,**[1] **Dunge Liu,**[1,2] **Wanhua Zhu,**[1]
Xiaojuan Zhang,[1] **and Guangyou Fang**[1]

[1]*Key Laboratory of Electromagnetic Radiation and Detection Technology, Chinese Academy of Sciences, North 4th Ring Road West, Haidian District, Beijing 100190, China*
[2]*Graduate University of Chinese Academy of Sciences, No. 19A, Yuquan Road, Beijing 100049, China*

Correspondence should be addressed to Chao Huang; chaohuang0507@163.com

Academic Editor: John N. Sahalos

It is a technical challenge to effectively remove the influence of magnetic noise from the vicinity of the receiving sensors on low-frequency magnetic communication. The traditional denoising methods are difficult to extract high-quality original signals under the condition of low SNR (the signal-to-noise ratio). In this paper, we analyze the numerical characteristics of the low-frequency magnetic field and propose the algorithms of the fast optimization of blind source separation (FOBSS) and the frequency-domain correlation extraction (FDCE). FOBSS is based on blind source separation (BSS). Signal extraction of low SNR can be implemented through FOBSS and FDCE. This signal extraction method is verified in multiple field experiments which can remove the magnetic noise by about 25 dB or more.

1. Introduction

Because the electromagnetic wave of RF (radio frequency) attenuates in the water or on land, the application of low-frequency (LF, 30 Hz~300 kHz), ultralow-frequency (ULF, 300~3000 Hz), and extremely low-frequency (ELF, 3~30 Hz) electromagnetic spectrum receives widespread attention [1, 2]. The electromagnetic wave of low frequency is mainly used in submarine communication [2] and resource exploration [3].

Many scholars have made great efforts toward the theory of propagation of low-frequency electromagnetic field [4, 5]. They have created magnetic communication channel models [6] and analyzed the physical and statistical characteristics of channel noise [7]. Actually, many excellent signal processing methods of RF can be extended to that of low frequency. For the engineering application, numbers of signal denoising methods are proposed, such as the matched filtering [8], the optimization of receiving antenna [9], and denoising based on the noise distribution and empirical mode decomposition [10]. However, in the case of low signal-to-noise ratio or existing complex noise components, these methods are difficult to work effectively.

In the application of low-frequency magnetic field, not only the signal waveform should be obtained, but also the original signal should be acquired for locating magnetic source [11]. Magnetic communication receivers are often affected by the surrounding noise and signal distorts seriously. In this paper, the correlation of low-frequency magnetic field is analyzed, and the algorithms of the fast optimization of blind source separation and the frequency-domain correlation extraction are proposed. The extraction of signal can be implemented through these algorithms. FOBSS can obtain the signal similar to the original signal waveform. FDCE recovers the original signal combining with FOBSS. In essence, we acquire more information (or more relationships) through multisensor comparing with single sensor. We associate the advantages of the algorithms and the inner relationships to achieve the signal extraction.

In this paper, contents are arranged as follows. Section 1 introduces the background of the low-frequency field application and the present situation of low-frequency magnetic signal processing methods; Section 2 introduces the basic theory of blind source separation; Section 3 analyzes the numerical characteristics of low-frequency magnetic field sources and proposes FOBSS and FDCE; Section 4 presents the experimental design process and verifies algorithms through multiple experiments; finally, Section 5 summarizes the full text.

2. BSS

When signal is interfered by multiple noise sources under the condition of low SNR, the methods of signal decomposition (Fourier analysis, wavelet analysis, empirical mode decomposition, etc.) are not effective. BSS [12] can work well in some cases, but the requirements of BSS must be met.

In the signal processing of multisensor, assume that $\mathbf{x} = (x_1, x_2, \ldots, x_n)^T$ is the observation signal vector of n sensors, and each observation element of $\mathbf{x}$ is the linear mixture of m signal sources $\mathbf{s} = (s_1, s_2, \ldots, s_m)^T$. The linear model of x_i is written as follows:

$$x_i = \sum_{j=1}^{m} a_{ij} s_j, \quad i = 1, 2, \ldots, n, \quad (1)$$

where x_i is the measured signal of the ith sensor. a_{ij} is the unknown mixing coefficient between the jth source and the ith sensor. s_j is the jth signal source. Usually, m and n are equal. Equation (1) is expressed as matrix form:

$$\mathbf{x} = \mathbf{A} \cdot \mathbf{s}, \quad \mathbf{A} \in \left(a_{ij}\right)^{n \times m}. \quad (2)$$

Generally, before signal separation, $\mathbf{x}$ is preprocessed by spheroidizing decomposition. Blind source separation is to find the solution of inverse mixed matrix W and estimate the optimal source signals. The process can be shown as

$$\mathbf{y}(t) = \mathbf{W} \cdot \mathbf{x}(t) = \mathbf{W} \cdot \mathbf{A} \cdot \mathbf{s}(t) = \hat{\mathbf{s}}(t). \quad (3)$$

The basic requirements should be satisfied: (1) signal sources are statistically independent of each other; (2) the rank of the covariance matrix $\mathbf{x}$ is full. The results of separation are evaluated by the coherence. The time-domain coherence function between y_i and s_j is given by

$$\rho_{y_i s_j} = \frac{\mathrm{cov}\left(y_i(t), s_j(t)\right)}{\sqrt{\mathrm{cov}\left(y_i(t)\right) \mathrm{cov}\left(s_j(t)\right)}}, \quad (4)$$

where $\mathrm{cov}(\cdot, \cdot)$ is the covariance. The sequence and amplitude of the separated signal through BSS are uncertain. When the coherence is larger than 0.8, signal separation is successful. High coherence denotes that the waveform similarity of the separated result and the original signal is high.

3. Characteristics and Extraction Method of Low-Frequency Magnetic Field Signal

3.1. The Correlation Characteristics of Low-Frequency Magnetic Field. The propagations of low-frequency and high-frequency electromagnetic field are in different ways (e.g., the conduction current mainly contributes to the low-frequency field, and the displacement current plays a main role in high-frequency field). The propagation mode of low-frequency field is extremely complex. However, the near field of low-frequency magnetic field is approximately equivalent to the radiation field of magnetic source in electromagnetic field theory [13]. The vector potential and magnetic field excited by the changing current density source $\mathbf{J}(\mathbf{r}', t)$ are, respectively, written as

$$\mathbf{A}(\mathbf{r}, t) = \frac{\mu_0}{4\pi} \int_{V'} \frac{\mathbf{J}\left(\mathbf{r}', t - r/c\right)}{r} dV' = \mathbf{A}(\mathbf{r}) e^{-j\omega t}, \quad (5)$$

$$\mathbf{B}(\mathbf{r}) = \nabla \times \mathbf{A}(\mathbf{r}, t) = (\nabla \times \mathbf{A}(\mathbf{r})) e^{-j\omega t}, \quad (6)$$

where μ_0 is the permeability of air, c is the speed of light, and $\mathbf{A}(\mathbf{r})$ is expressed as $\mathbf{A}(\mathbf{r}) = (\mu_0/4\pi) \int_{V'} (\mathbf{J}(\mathbf{r}') e^{ikr}/r) dV'$. ω is the frequency. k is the wave number. $\mathbf{r}$ is the observation point. $\mathbf{r}'$ is the source point. $r = |\mathbf{r} - \mathbf{r}'|$. The frequency band of magnetic source is finite ($\omega_l \leq \omega \leq \omega_h$). Frequency is described as discrete form through the Fourier analysis. The field of observation point $\mathbf{r}_i$ can be given by

$$\mathbf{B}(\mathbf{r}_i) = \left(\nabla \times \mathbf{A}'(\mathbf{r}_i)\right) \sum_{\omega=\omega_l}^{\omega_h} e^{-j(\omega t - kr_i)}, \quad (7)$$

where $\mathbf{A}'(\mathbf{r}_i)$ is a factor related to the distance. Considering the wavelength of low-frequency field that is much longer than the distance between the observation point and the source point (the ratio of wavelength to distance tends to be infinite), the phase of (7) can be expressed as

$$\omega t - kr_i = \omega t - 2\pi \frac{r_i}{\lambda_\omega} \longrightarrow \omega t, \quad (8)$$

where λ_ω is the wavelength of ω. So the ratio of magnetic field at different observation points is

$$\frac{\mathbf{B}(\mathbf{r}_i)}{\mathbf{B}(\mathbf{r}_j)} = \frac{\nabla \times \mathbf{A}'(\mathbf{r}_i)}{\nabla \times \mathbf{A}'(\mathbf{r}_j)}. \quad (9)$$

And the coherence between the magnetic field of $\mathbf{B}(\mathbf{r}_i)$ and $\mathbf{B}(\mathbf{r}_j)$ is

$$\rho_{\mathbf{B}(\mathbf{r}_i)\mathbf{B}(\mathbf{r}_j)} = 1. \quad (10)$$

In theory, the magnetic fields of different observation points are highly relevant. The covariance matrix of $\mathbf{B}(\mathbf{r}_i)$ ($i = 1, 2, \ldots, m$) is $E(\mathbf{B}\mathbf{B}^T)$ ($\mathbf{B} = (\mathbf{B}(\mathbf{r}_1), \mathbf{B}(\mathbf{r}_2), \ldots, \mathbf{B}(\mathbf{r}_m))^T$). The singular value decomposition of $E(\mathbf{B}\mathbf{B}^T)$ is represented as $E(\mathbf{B}\mathbf{B}^T) = \mathbf{Q}\Sigma\mathbf{Q}^T$. If the rank of matrix Σ is full, the locations of the observation points meet the requirements of signal separation. If the rank of matrix Σ is not full, the locations of the observation points need to be adjusted (usually, placing the sensors randomly is appropriate; they need not to be relocated).

3.2. The Fast Optimization of Blind Source Separation.
Because matrix $\mathbf{W}$ (see Section 2) contains m^2 unknown elements, BSS is the process of $\mathbf{W}$ rotating $\mathbf{x}$ (see also Section 2). $\mathbf{W}$ is a unit orthogonal matrix. Now there is a method that reduces the numbers of the unknown elements of $\mathbf{W}$, so the iteration time of finding $\mathbf{W}$ can be shortened in calculation. According to the law of Givens rotation transformation [14], the mathematical process is given by

$$\mathbf{T}\boldsymbol{\xi}_k = \mathbf{T}_{1m}\mathbf{T}_{1(m-1)}\cdots\mathbf{T}_{12}\boldsymbol{\xi}_k = |\boldsymbol{\xi}_k|\,\mathbf{e}_j, \tag{11}$$

where $\mathbf{T}_{1i}$ is the Givens matrix, $\boldsymbol{\xi}_k$ is an arbitrary nonzero vector, and $\mathbf{e}_j$ is an arbitrary standard unit vector. If $\boldsymbol{\xi}_k$ is an arbitrary unit vector, there is a result: $\boldsymbol{\xi}_k = \mathbf{T}^{-1}\mathbf{e}_j$. Assume that $\mathbf{T}^{-1} = \mathbf{T}'$ ($\mathbf{T}^{-1}$ denotes reverse rotation). The rotation process of $\mathbf{e}$ ($\mathbf{e}$ is a standard orthogonal basis) can be expressed as $\mathbf{T}'\mathbf{e} = \mathbf{T}'_{12}\mathbf{T}'_{13}\cdots\mathbf{T}'_{1m} = \boldsymbol{\xi}$ ($\boldsymbol{\xi}$ is an arbitrary orthogonal basis). So the arbitrary orthogonal matrix of m-dimensional space can be acquired through the rotation of a standard unit matrix, and the number of rotations is at most $m-1$ times. Therefore $\mathbf{W}$ can be expressed as $\mathbf{W}(\theta_1, \theta_2, \ldots, \theta_{m-1})$ by $m-1$ unknown elements. The iteration time of $\mathbf{W}(\theta_1, \theta_2, \ldots, \theta_{m-1})$ can be defined as $t(\theta_1, \theta_2, \ldots, \theta_{m-1})$ (θ_i is the ith rotation angle). The iteration time of basic BSS is defined as $t(w_{11}, w_{12}, \ldots, w_{ij} \ldots, w_{mm})$ (w_{ij} is the element of $\mathbf{W}$). Under the condition of the same computational resource, the iteration time of $m-1$ unknown elements is statistically shorter than that of m^2 unknown elements. Therefore the ratio of mean iteration time between $t(\theta_1, \theta_2, \ldots, \theta_{m-1})$ and $t(w_{11}, w_{12}, \ldots, w_{ij} \ldots, w_{mm})$ is

$$\frac{E\left[t\left(\theta_1, \theta_2, \ldots, \theta_{m-1}\right)\right]}{E\left[t\left(w_{11}, w_{12}, \ldots, w_{mm}\right)\right]} \approx \frac{m-1}{m^2}. \tag{12}$$

3.3. The Frequency-Domain Correlation Extraction. FOBSS can only acquire the waveforms similar to the original signals. The recovery of original signals requires excellent algorithms of signal amplitude extraction. In signal extraction, the parameters of the adaptive filtering [15] are difficult to be adjusted and the adaptive filter cannot work effectively for nonperiodic signal. The signal extraction can be implemented in the frequency domain based on the correlation function. The advantage of frequency-domain extraction contains the process of power spectrum estimation. The correlation reflects in each frequency point. $u(t)$ and $v(t)$ are, respectively, defined as the time series of signals. $u(t)$ consists of the convolutions of various signals. $v(t)$ is the known waveform similar to a component of $u(t)$. The basic model is simply described as

$$u(t) = \sum_{i=1}^{n} h_i(t) * u_i(t), \tag{13}$$

$$v(t) = u_k(t), \quad k = 1, 2, \ldots, n,$$

where different components of $u(t)$ are statistically independent and $h_i(t)$ ($i = 1, 2, \ldots, n$) is an unknown kernel function of linear convolution. Now in order to extract $h_k(t) * u_k(t)$ of $u(t)$, we get the cross-correlation function $R_{uv}(\tau)$ between

$u(t)$ and $v(t)$. $R_{uv}(\tau)$ can be transformed to the corresponding frequency domain as $S_{uv}(\omega)$ by the Fourier transform. The mathematical process is described as

$$R_{uv}(\tau) = h_k(\tau) * R_{u_k u_k}(\tau),$$

$$S_{uv}(\omega) = h_k(\omega) S_{u_k u_k}(\omega), \tag{14}$$

$$h_k(\omega) = \frac{S_{uv}(\omega)}{S_{u_k u_k}(\omega)} = \frac{S_{uv}(\omega)}{S_{vv}(\omega)},$$

where $S_{uv}(\omega)$ is the cross power spectrum density function between $u(t)$ and $v(t)$, $h_k(\omega)$ is the Fourier transform of $h_k(t)$, and $S_{u_k u_k}(\omega)$ is the power spectrum density function of $u_k(t)$. Therefore the result of extraction is

$$h_k(t) * u_k(t) = F^{-1}\left(h_k(\omega) V(\omega)\right), \tag{15}$$

where $V(\omega)$ is the Fourier transform of $v(t)$. We give a deviation ratio λ_e of the extracted signal and the original signal such that

$$\lambda_e = \frac{|E\left(\hat{z}(t) - z(t)\right)|}{|E\left(z(t)\right)|} \times 100\%, \tag{16}$$

where $z(t)$ is the original signal, $\hat{z}(t)$ is the extracted signal, and $E(\cdot)$ is the mean value. If the mean values of $z(t)$ and $\hat{z}(t)$ are zero, $z(t)$ and $\hat{z}(t)$ are, respectively, replaced as the peak-to-peak values $z_{p-p}(t)$ and $\hat{z}_{p-p}(t)$.

4. Processing of the Experimental Data

4.1. Experiment Design. We have carried out multiple sets of magnetic communication experiment. Experimental equipment consists of the transmitter, the receivers, and the magnetic noise sources. The transmitter is an audio power amplifier AE7224 and the output impedance of the amplifier is less than $1 \times 10^6\ \Omega$. The transmitter antenna is a coil loaded by a magnetic core that it looks like a cylinder which is about 1 meter long. Besides, the diameter of antenna is 0.07 meter and the coil has 300 windings. The transmitting current is 1.5 A. The receiving antenna is also a magnetic induced coil and its conversion factor is 100 mV/nT. The AD of the receiver is a 24-bit converter and its LSB (Least Significant Bit) is 20 bits. The quantization noise is about 0.048 pT. The experiment sites are Kangxi Prairie in Beijing and An Guli Prairie in Hebei province, China. The scene is shown in Figure 1: the magnetic signal (the signal bandwidth is about 15 Hz) is transmitted with the carrier frequency of 480 Hz, and the transmitter is located at 100 meters in the Cartesian coordinate system; the receivers (multisensor) distribute in the range of 2 meters~4 meters away from the original point, and the sampling rate f_s is 2400 Hz. The actual sampling rate should be higher than Nyquist sampling rate at a suitable value; the magnetic noise is from the car engines, and the engines are in the range of 2 meters~4 meters away from the receivers. Magnetic noise is stochastic, and its frequency bandwidth covers the signal's bandwidth. So the signal contains a certain amount of in-band noise. The received signals are intensely interfered by the magnetic noise in time domain and frequency domain.

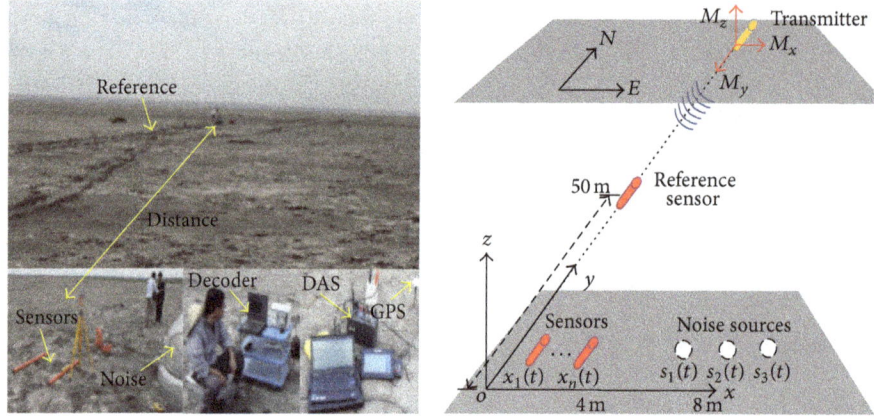

FIGURE 1: Field experiment.

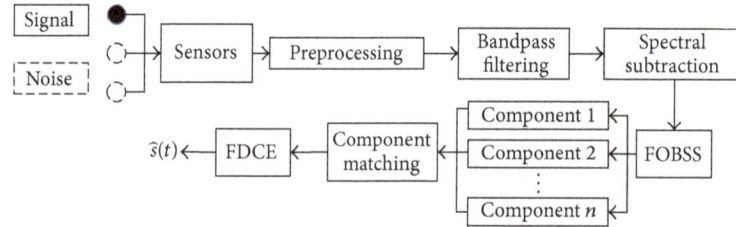

FIGURE 2: The procedure of signal extraction.

Only the signals of the multisensor close to the original point are applied to signal separation and signal extraction.

We have verified that the signal of the sensor at 50 meters is not affected by magnetic noise. When the noise sources and the transmitter are open in Figure 1, the coherence between the signal of sensor at 75 meters and that of sensor at 50 meters is 0.98. The coherence between the signal of sensor at 75 meters and that of sensor at original point nearby is 0. The coherence of the signal of sensor at 50 meters and that of sensor at original point nearby is 0. But when only the transmitter is open, the coherence of different sensors is 0.98. This phenomenon verifies that the reference sensor is not interfered by magnetic noise sources. The signal of the sensor (the reference sensor) at 50 meters is to prove the effectiveness of algorithms.

4.2. Data Processing. In Section 2, the signal model is theoretical. Under the condition of reality, the sensor itself contains micro noise, and the signal model is modified as another form:

$$x_i = \sum_{j=1}^{n} a_{ij} s_j + N_i, \quad i = 1, 2, \ldots, n, \qquad (17)$$

where a_{ij} is the mixing coefficient, s_j is the jth signal source, and N_i is the micro noise of the ith sensor (N_i is statistically independent). The procedure of signal extraction is shown in Figure 2, mainly including the preprocessing, the spectral subtraction [16], FOBSS, and FDCE. Firstly, whether the rank of the covariance matrix of the received signals is full or not should be confirmed; secondly, remove

the micro noise of sensors through spectral subtraction; thirdly, signal separation is executed by FOBSS; finally, the corresponding signal waveforms are obtained by component matching, and the original signal is obtained by FDCE from the row data (the signals after band-pass filtering). The main parts are FOBSS and FDCE. Component matching is to select the required component after FOBSS (not matched filter). Besides, computing correlation function through the waveform of code is to decide which component is the required component. The required component has a similar waveform to the original signal.

Experiments are carried out by the combinations which consist in whether transmitter and magnetic noise sources are open or not, and the reference sensor at 50 meters is used to validate the correctness of algorithms. In order to illustrate data, we transform (4) to frequency domain (the frequency-domain coherence function):

$$\rho_{y_i s_j}(\omega) = \frac{S_{y_i s_j}(\omega)}{\sqrt{S_{y_i y_i}(\omega) S_{s_i s_i}(\omega)}}, \qquad (18)$$

where $S_{y_i s_j}(\omega)$ is the cross power spectrum density function between y_i and s_j, $S_{y_i y_i}(\omega)$ is the power spectrum density function of y_i, and $S_{s_i s_i}(\omega)$ is the power spectrum density function of s_j.

As shown in Figure 1, magnetic noise sources are within the range of 2 meters~5 meters away from receivers. When only one magnetic noise is open, the coherence between the signals of different receivers is above 0.96 and there is no time delay between the signals of different receivers

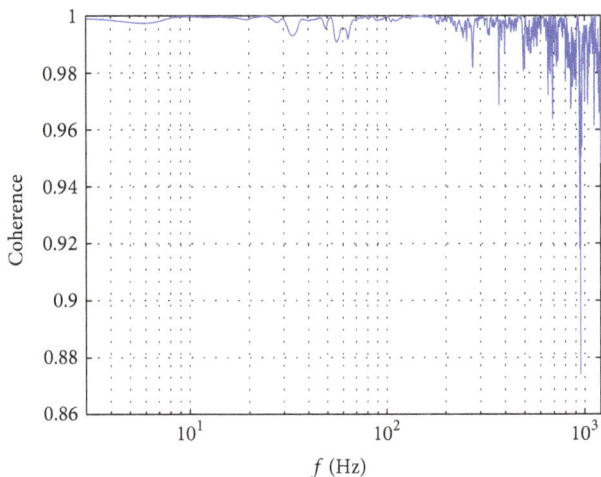

FIGURE 3: The frequency-domain coherence between magnetic noise signals of different receivers.

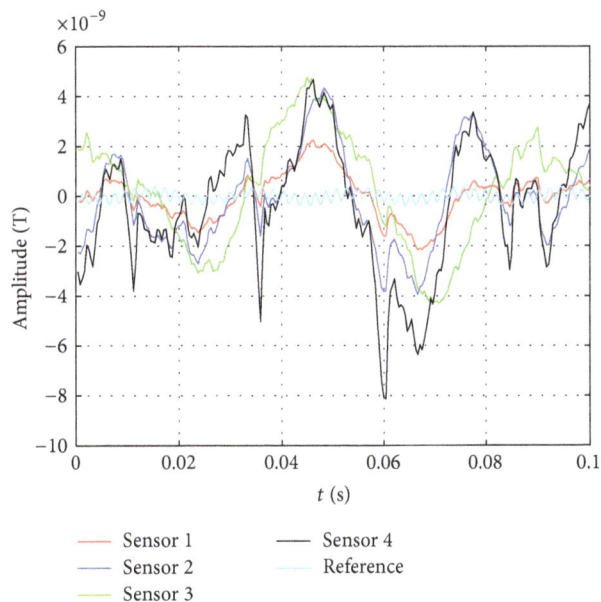

FIGURE 5: Frequency-domain waveforms of the received signals (three noise sources).

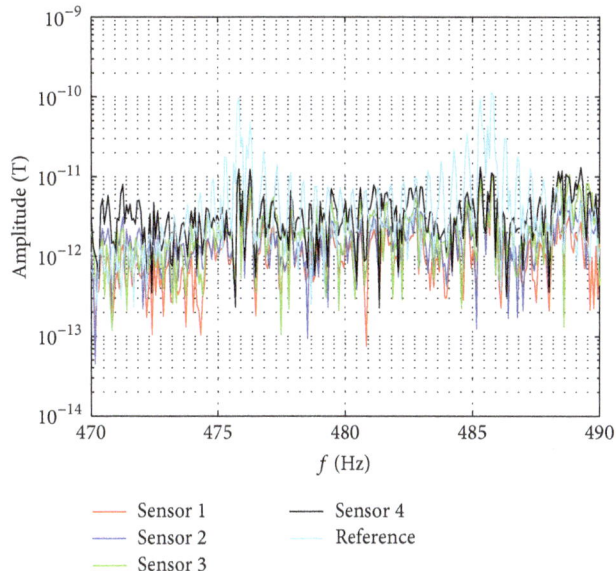

FIGURE 4: Time-domain waveforms of the received signals (three noise sources).

(these characteristics are consistent with the analysis of Section 3.1). Figure 3 shows the frequency-domain coherence, this feature meets the correlation characteristics of low-frequency magnetic field. The reference sensor at 50 meters cannot be affected by magnetic noise, and we only use it to verify the correctness of algorithms (the reference sensor is not used for signal separation and signal extraction). The coherence between different noise sources is zero. There is no correlation between different physical sources, so components separation is possible.

When one noise source and the transmitter are open, the SNR is about −9 dB. When three noise sources and the transmitter are open, the SNR is about −20 dB. Figure 4 shows the time-domain waveforms of different receivers in the case

that three magnetic noise sources and the transmitter are open, and the receiver at 50 meters is not influenced by noise sources. The sequence of sensors is arranged in Figure 1. The reference sensor is at 50 meters. Figure 5 shows the frequency spectra of Figure 4. The actual locations of sensors 1, 2, 3, and 4 have a little deviation compared with the locations shown in Figure 4. According to basic electromagnetic theory, the low-frequency magnetic field value decreases inversely proportional to distance when the source can be deemed as one point source. Because three noise sources are distributed near the sensors, three sources are not equivalent to one point source when the distance and the range of source are approximately at the same level. So field values in Figure 4 are not strictly consistent with that of one point source. There maybe exists offset when peaks and troughs appear at the same time. Since the signals are intensely interfered in time domain and frequency domain, the signal processing methods should be based on components separation but not on the synthetic filtering (it maybe contains the matched filtering, the optimal notch filtering, etc.) based on signal decomposition. The signal separation performance is generally influenced by the noise of sensor itself, industrial interference, and the experiment errors.

After preprocessing and independent noise removing, the received signals will be processed by FOBSS. We make a comparison between the separated results and the reference signal. Figure 6 shows a comparison between the separated results and reference signal (the separation results and reference signal are adjusted to the same scale), and the coherence calculated by (4) is 0.89 (in the case that three magnetic noise sources and the transmitter are open). The coherence of the separated signal under multiple sets of experiments is higher than 0.8. The communication channel may be influenced slightly by industrial frequency interference. Since the data of Figures 6(a) and 6(b) are obtained at different time and places,

(a) The comparison of separated results in time domain (one noise source)

(b) The comparison of separated results in time domain (three noise sources)

(c) The comparison of separated results in frequency domain (one noise source)

(d) The comparison of separated results in frequency domain (three noise sources)

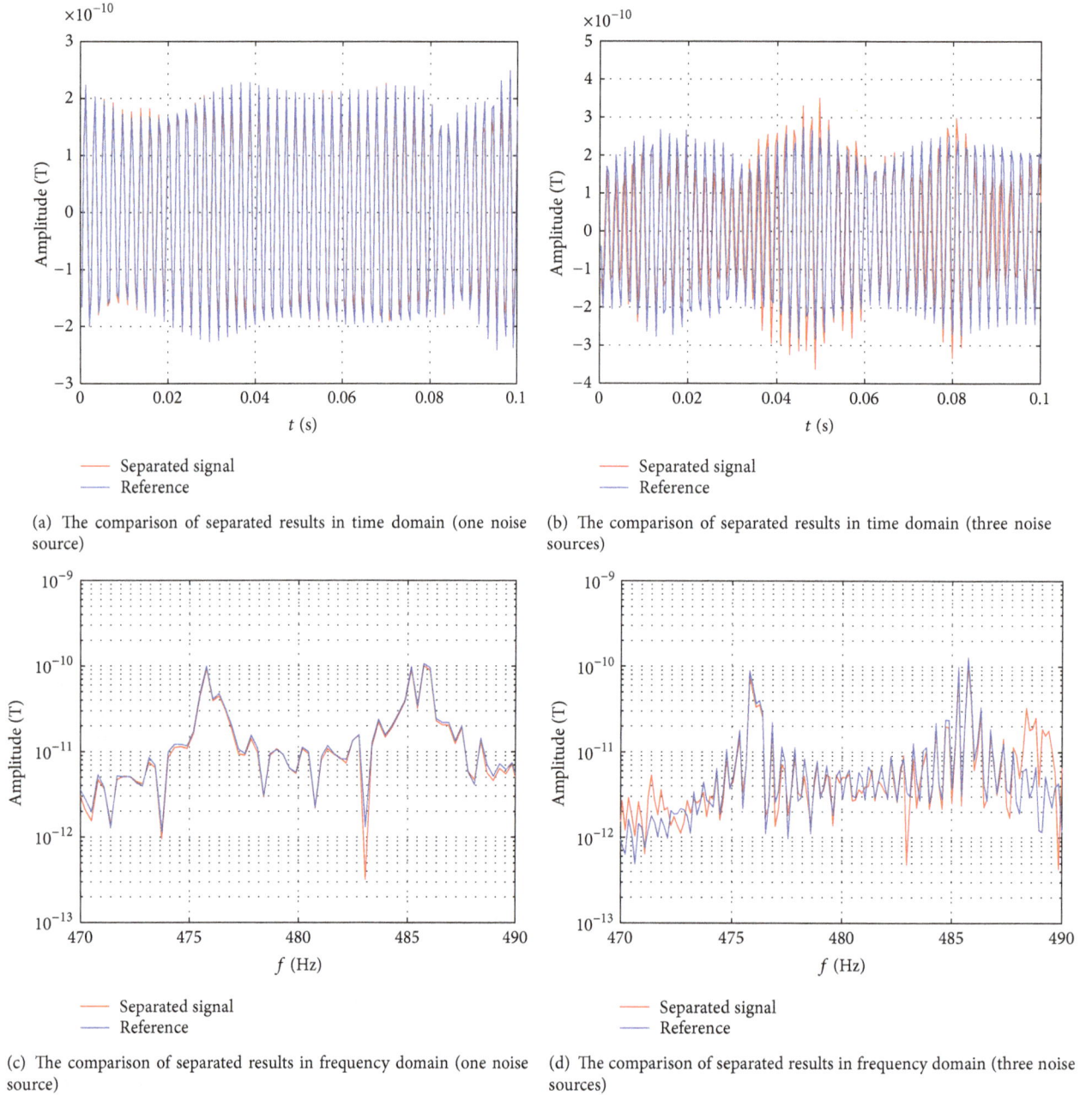

FIGURE 6: The separated results.

it is reasonable that their envelope causes slight difference in a larger image display. Both the separated signal and the signal of the reference sensor can be decoded correctly in BPSK modulation.

The signal separation algorithm (FOBSS) can only acquire the waveforms similar to the original signals. The signal extraction is realized by (15) (FDCE). The processed results of multiple experiments are shown in Table 1; the time-domain coherence is given by (4), and the deviation ratio is given by (16).

The key purpose of this paper is to extract weak true signal masked by strong noise. The "weak" (SNR is too low) is

TABLE 1: The results of signal processing.

Experiment site	Number of noise sources	Time-domain coherence ρ	Deviation ratio λ_e
An Guli Prairie	1	0.9924	3.42%
An Guli Prairie	1	0.9533	4.18%
An Guli Prairie	1	0.9501	4.22%
Kangxi Prairie	3	0.8945	7.51%
Kangxi Prairie	3	0.8184	9.03%

the main problem. The extracted signal through FOBSS and FDCE can be decoded errorless and it has the same waveform

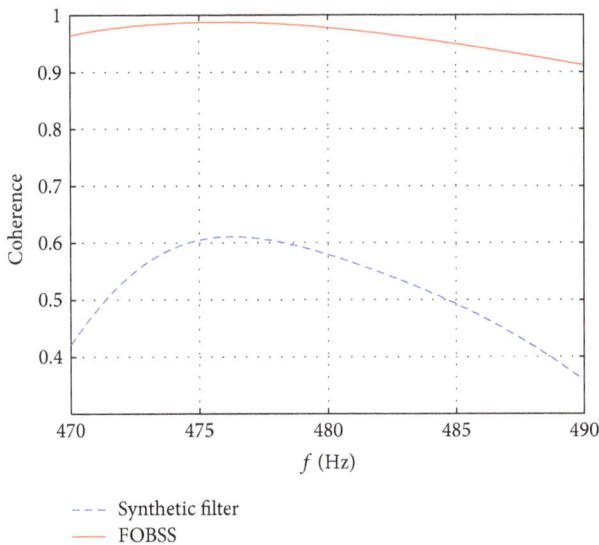

- - - Synthetic filter
— FOBSS

FIGURE 7: The comparison of frequency-domain coherence.

as the reference signal. They also obtained the true amplitude. In the actual environment, the signal in the transmitter has minute difference from the signal in the air. In order to verify algorithms, the reference signal only can be the signal in the air but not that in the transmitter. Even though there is a slight influence factor, both the receiving end and the reference at 50 meters are affected (other algorithms can easily solve usually). The effectiveness cannot be influenced.

4.3. Data Analysis. The coherence larger than 0.8 indicates high correlation in statistical theory. The synthetic filtering based on the time-frequency domain is difficult to filter out in-band noise; the ability of empirical mode decomposition to overcome in-band noise is limited and it is difficult to identify the mode; when the situation is more complex, it needs signal separation through building conditions. We regard the similarity (the frequency-domain coherence) to the reference signal as the standard to evaluate the ability of different signal processing methods. There is a comparison between FOBSS and the synthetic filter in Figure 7. The coherence of FOBSS tends to 1, and the coherence of the synthetic filter is low. Simultaneously, because design of standard narrowband filter is not easy, the synthetic filter is difficult to solve the problem.

The signal preprocessing and independent noise removal are important work. Compared with basic BSS, FOBSS can effectively improve the computation time in multiple experiments. Even though (12) is a statistical result, FOBSS is more suitable for engineering application. In the original signal extraction, the deviation ratio of FDCE is about 6 percent. In short, according to the performance of FOBSS and FDCE, the proposed signal extraction method is effective.

5. Conclusion

In this paper, we have analyzed the characteristics of low-frequency magnetic field and have proposed FOBSS and

FDCE. FOBSS improves processing efficiency. Even though FDCE is essentially from Fourier transform, it can be simply implemented and works well. A systemic method of extraction has been formed through the combination of FOBSS and FDCE.

This method has been effectively verified in multiple field experiments. Its advantage is that it can extract the waveform and the true amplitude under the condition of low SNR. It can remove the magnetic noise by about 25 dB or more. In future work, we will study the statistic characteristics of magnetic noise and give the mathematical relationship between the performance of the algorithms and independent noise.

Weak signal extraction is a difficult problem in signal processing. For the engineering application of low-frequency magnetic field, this paper has made preliminary progress in narrowband signal processing. In order to expand the application range of low-frequency signal, such as oil exploration and mining exploration, we need to increase the signal bandwidth. Because the processing of wideband signal has more difficulties, we will make further study.

Conflict of Interests

The authors declare that there is no conflict of interests regarding the publication of this paper.

Acknowledgments

This research is sponsored by the National High Technology Research and Development Program of China (Grant no. 2014AA093407) and the National Natural Science Foundation of China (Grant no. 41374186).

References

[1] A. Abubakar, T. M. Habashy, V. L. Druskin, L. Knizhnerman, and D. Alumbaugh, "2.5D forward and inverse modeling for interpreting low-frequency electromagnetic measurements," *Geophysics*, vol. 73, no. 4, pp. F165–F177, 2008.

[2] S. A. Wolf, J. R. Davis, and M. Nisenoff, "Superconducting extremely low frequency (ELF) magnetic field sensors for submarine communications," *IEEE Transactions on Communications*, vol. 22, no. 4, pp. 549–554, 1974.

[3] R. E. Grimm, "Low-frequency electromagnetic exploration for groundwater on Mars," *Journal of Geophysical Research: Planets*, vol. 107, no. E2, p. 1, 2002.

[4] W. Pan and K. Li, *Propagation of SLF/ELF Electromagnetic Waves*, Springer, 2014.

[5] L. O. Løseth, H. M. Pedersen, B. Ursin, L. Amundsen, and S. Ellingsrud, "Low-frequency electromagnetic fields in applied geophysics: waves or diffusion?" *Geophysics*, vol. 71, no. 4, pp. W29–W40, 2006.

[6] H. E. Rowe, "Extremely low frequency (ELF) communication to submarines," *IEEE Transactions on Communications*, vol. 22, no. 4, pp. 371–385, 1974.

[7] E. C. Field Jr. and M. Lewinstein, "Amplitude-probability distribution model for VLF/ELF atmospheric noise," *IEEE Transactions on Communications*, vol. 26, no. 1, pp. 83–87, 1978.

[8] J. E. Evans and A. S. Griffiths, "Design of a sanguine noise processor based upon world-wide extremely low frequency (ELF)

recordings," *IEEE Transactions on Communications*, vol. 22, no. 4, pp. 528–539, 1974.

[9] P. R. Bannister, "Orbiting transmitter and antenna for space-borne communications at ELF/VLF to submerged submarines," in *AGARD, ELF/VLF/LF Radio Propagation and Systems Aspects 14 p (SEE N93-30727 11-32)*, vol. 1, 1993.

[10] Y. Jeng, M.-J. Lin, C.-S. Chen, and Y.-H. Wang, "Noise reduction and data recovery for a VLF-EM survey using a nonlinear decomposition method," *Geophysics*, vol. 72, no. 5, pp. F223–F235, 2007.

[11] A. P. Nickolaenko and I. G. Kudintseva, "A modified technique to locate the sources of ELF transient events," *Journal of Atmospheric and Terrestrial Physics*, vol. 56, no. 11, pp. 1493–1498, 1994.

[12] M. T. Akhtar, T.-P. Jung, S. Makeig, and G. Cauwenberghs, "Recursive independent component analysis for online blind source separation," in *Proceedings of the IEEE International Symposium on Circuits and Systems (ISCAS '12)*, pp. 2813–2816, IEEE, May 2012.

[13] C. H. Papas, *Theory of Electromagnetic Wave Propagation*, Courier Dover Publications, 2013.

[14] F. Ling, "Givens rotation based least squares lattice and related algorithms," *IEEE Transactions on Signal Processing*, vol. 39, no. 7, pp. 1541–1551, 1991.

[15] S. T. Smith, *Geometric optimization methods for adaptive filtering [PhD thesis]*, Harvard University, 1993.

[16] T. Inoue, H. Saruwatari, Y. Takahashi et al., "Theoretical analysis of musical noise in generalized spectral subtraction based on higher order statistics," *IEEE Transactions on Audio, Speech, and Language Processing*, vol. 19, no. 6, pp. 1770–1779, 2011.

12

A Variable Weight Privacy-Preserving Algorithm for the Mobile Crowd Sensing Network

Jiezhuo Zhong,[1] **Wei Wu,**[1,2] **Chunjie Cao,**[1,3] **and Wenlong Feng**[1]

[1]*College of Information Science and Technology, Hainan University, Haikou, Hainan, China*
[2]*Institute of Deep-Sea Science and Engineering, Chinese Academy of Sciences, Sanya, Hainan, China*
[3]*State Key Laboratory of Marine Resource Utilization in the South China Sea, Hainan University, Haikou, Hainan, China*

Correspondence should be addressed to Chunjie Cao; chunjie_cao@126.com

Academic Editor: Zhuo Lu

Mobile crowd sensing (MCS) network collects scenario, environmental, and individual data within a specific range via the intelligent sensing equipment carried by the mobile users, thus providing social decision-making services. MCS is emerging as a most important sensing paradigm. However, the person-centered sensing itself carries the risk of divulging users' privacy. To address this problem, we proposed a variable weight privacy-preserving algorithm of secure multiparty computation. This algorithm is based on privacy-preserving utility and its effectiveness and feasibility are demonstrated through experiment.

1. Basic Theories

1.1. Architecture of Mobile Crowd Sensing Network. Mobile crowd sensing (MCS) network [1] takes the ordinary mobile terminals as the basis sensing units. The sensing task distribution and sensing data collection are achieved through collaboration via the mobile Internet. This represents a large-scale complex social sensing task. "Crowd" refers to the aspect of mobilizing the power and intelligence of the general public, and "sensing" is the process of acquiring the users' behavioral data under different scenarios using the sensors.

Figure 1 shows a typical MCS framework, which consists of the mobile users and the sensing platform. Mobile users are millions of mobile intelligent terminals, into which sensors are embedded (e.g., GPS, gravity sensor, temperature sensor, camera, microphone, and acceleration sensor). These sensors collect various sensing data, which are updated to the sensing platform via the mobile network or short-range wireless communication network. Upon receiving the data, the sensing platform will commence data analysis and processing. The processed data will be directly applied to a diversity of universal social sensing services. After the data analysis and processing are finished, each parcel of data will be evaluated. The mobile users participating in the sensing tasks

will be awarded based on the specific incentive mechanism, so as to attract more users into the large-scale sensing task. Liu proposed schemes based on both the Monopoly and Oligopoly models enhancing the location privacy of MCS applications by reducing the bidding and assignment steps in the MCS cycle [2]. Jin proposed a differentially private incentive mechanism that preserves the privacy of each worker's bid against the other honest-but-curious workers [3]. Furthermore, many researchers focused on the detailed information extraction processing in MCS including Hybrid Deep Learning Architecture [4] and Fog Computing and Data Aggregation Scheme [5, 6].

1.2. Application of the MCS Network. The MCS network comprising the mobile intelligent terminals and the mobile sensors is capable of large-scale, complex, fine-grained, and thorough data sensing and collection. For example, the use of MCS network for the collection, analysis, and fusion of the urban traffic flow information can provide highly efficient and convenient path planning and driver assistance system for the mobile users. The MCS network can also provide the decision-making support for urban transport planning and for the formulation of a safe and highly efficient urban transportation network. The MCS network-based sensing

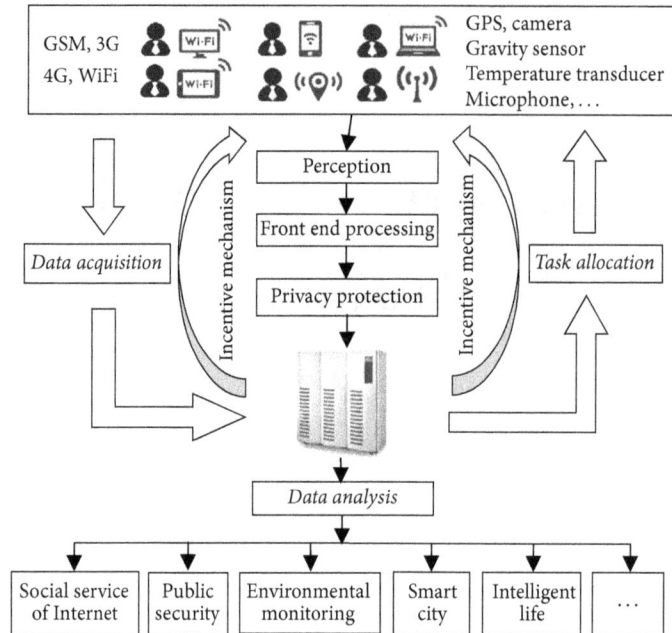

FIGURE 1: System structure diagram of mobile crowd sensing network.

and monitoring of urban domestic infrastructures offer convenient life services for the local residents. The wide prevalence of the mobile intelligent terminals is a solid guarantee for the high-efficiency and low-cost and large-scale monitoring of natural environment in the cities.

2. Privacy Protection Mechanism for MCS Network Users

2.1. Privacy Preserving in MCS Network. The sensing data collected by the MCS network are largely user privacy. Location data usually contain the sensitive information such as users' address, scope of activity, and transportation route. The mining of users' state of motion can obtain the sensitive information of users' living habits and health conditions. The biological data collected contains the information of users' voice, fingerprints, and basic physiological characteristics. The routine usage data of the mobile intelligent terminals are associated with the user privacy of a deeper level, including the users' hobbies and behavioral traits. Once the user privacy is divulged, there may be violation of privacy, harassment, fraud, or even direct economic loss. Therefore, designing the data security architecture for dynamic privacy protection under the MCS network is an urgent issue.

2.2. Related Technology of Privacy Preserving. The major privacy-preserving techniques used for MCS network are divided into the following types.

(1) Generalized Privacy-Preserving Algorithm. Anonymization is performed while sharing the sensing data, so that the sensitive information about the user's identity is removed without harming the meaningful deduction based on the anonymized sensing data. However, the currently used anonymization methods are usually greedy algorithms which have low execution efficiency.

(2) Perturbation-Based Privacy-Preserving Algorithm. The raw sensing data are perturbed by adding a random number, noises, and exchanges, so that the other party cannot mine the raw sensing data and privacy policies. The main difficulty with data perturbation is how to strike a balance between data correctness, privacy, and security.

(3) Secure Multiparty Computation (SMPC). This technique integrates data encryption and multiple parties are involved in the computation and mining. Because none of the parties have access to complete data, the users' privacy can be ensured. SMPC is now used for collaborative computing among a group of untrusted parties. Many researches have been carried out over the SMPC problem. In 2000, Lindell proposed the method of secure multiparty decision tree (ID3) to protect the data privacy of users [7]. Asharov proposed the threshold homomorphic encryption scheme to improve efficiency of the privacy protection algorithm [8]. In 2014, the threshold-based encryption of K-means outsourcing computing proposed by Liu is a more efficient privacy protection algorithm [9].

Proper application of information technology and algorithm design are the two major concerns in privacy protection. However, the users' attitudes towards privacy are generally neglected. A survey [10] indicates that 17% of the Internet users are still unwilling to provide their authentic information even under privacy protection; 56% of the Internet users are more willing to provide their authentic information in the presence of proper privacy protection; the remaining 27% of the Internet users do not particularly care about their privacy and will provide the authentic information with or without privacy protection. It is obvious that the users' attitudes towards privacy affect their willingness to share the personal information. Users may react differently to the prospect of disclosing different personal information. But under some incentive mechanisms, the psychological response of the users to the disclosure of different sensitive information may vary.

This study constructed an MCS network-based privacy-preserving algorithm by reference to SMPC. The weight function of privacy preference was built by combining the analysis of the users' sensitivity to the disclosure of privacy and classification of the privacy level of the sensing data. This proposed algorithm can effectively prevent the divulging of privacy information while achieving a maximal acquisition and analysis of the sensing data.

3. Variable Weight Privacy-Preserving Algorithm

3.1. Measure of User Privacy Sensing

3.1.1. User Multiattribute Assumption.
Suppose there exists Euclidean space, in which n dimensions represent n solutions to one problem; f_j denotes the attribute j, and G is a set of attributes, $G = \{f_1, f_2, \ldots, f_n\}$. x_i denotes one solution, and X is a set of one solution, $X = \{x_1, x_2, \ldots, x_m\}$. $x_{ij} = f_j(x_i)$ denotes the attribute value of the solution under attribute f_j. $D = (x_{ij})_{m \times n}$ denotes a decision-making matrix of solution X under the attribute G:

$$D = \begin{pmatrix} x_{11} & x_{12} & \cdots & x_{1n} \\ x_{21} & x_{22} & \cdots & x_{2n} \\ \vdots & \vdots & \ddots & \vdots \\ x_{m1} & x_{m2} & \cdots & x_{mn} \end{pmatrix}. \tag{1}$$

Considering the varying sensitivity to privacy, the users show different willingness to share their privacy in the MCS network. The influence factors of this willingness are divided into profit factors and risk factors, each of which is measured differently. Let $M = \{1, 2, \ldots, m\}$ be the set of the profit attributes, and $N = \{1, 2, \ldots, n\}$ be the set of risk attributes. The two sets are normalized by multiple attribute decision-making using the following formula:

$$y_{ij} = \frac{(x_{ij} - \min_i x_{ij})}{(\max_i x_{ij} - \min_i x_{ij})} \quad i \in M, \ j \in N. \tag{2}$$

After the transform, the synthetic matrix is $Y = (y_{ij})_{m \times n}$.

$$Y = \begin{pmatrix} y_{11} & y_{12} & \cdots & y_{1n} \\ y_{21} & y_{22} & \cdots & y_{2n} \\ \vdots & \vdots & \ddots & \vdots \\ y_{m1} & y_{m2} & \cdots & y_{mn} \end{pmatrix}. \tag{3}$$

3.1.2. Weight Determination of Privacy Perception Attributes.
As the users differ in privacy perception, each attribute will carry the information of different user preferences. Therefore, the given user preference can be expressed as the weight of the individual, and the weight of each attribute is expressed as

$$W = (w_1, w_2, \ldots, w_n)^T,$$
$$\sum_{j=1}^{n} w_j = 1. \tag{4}$$

The utility of each user is expressed as the sum of the weighted attributes. Hence, the user utility U_i is

$$U_i = \sum_{j=1}^{n} y_{ij} \cdot w_j, \quad (j = 1, 2, \ldots, n). \tag{5}$$

The utility analysis of users' privacy perception will provide not only the weight parameters for the SMPC, but also some suggestions for the collection modes of the sensing data under the MCS network. For example, the privacy information sensitive to most users will be prevented and a reasonable incentive mechanism can be designed on this basis. This is very important for increasing the confidence and participation level of users with a lower utility of privacy perception.

3.2. Variable Weight SMPC-Based Privacy-Preserving Algorithm

3.2.1. SMPC-Based Algorithm.
SMPC can be conceptualized by the following mathematical model: n participants $P_1, P_2, \ldots, P_n$ of the protocol jointly implement the function $f(x_1, x_2, \ldots, x_m)$. $S_{\text{input}} = \{x_1, x_2, \ldots, x_m\}$ is the set of input variables. The set of input variables S_{P_i} provided by the participant P_i ($i \in \{1, 2, \ldots, n\}$) is a subset of S_{input}, which satisfies $\bigcup_{P_i} S_{P_i} = S_{\text{input}}$, $S_{P_i} \cap S_{P_j} = \phi$ ($i \neq j$). It is required in the computing of the function that the input S_{P_i} from any participant P_i ($i \in \{1, 2, \ldots, n\}$) is not known to other participants P_j ($j \neq i$).

The essence of SMPC is a data encryption algorithm using the encryption scheme so as to ensure data privacy. Rivest et al. [11] proposed the concept of fully homomorphic encryption in 1978, aiming to construct an encryption mechanism that supports ciphertext retrieval. Goldwasser [12] studied the strategies used by mobile attackers in the secure channel model. They generalized the threshold mechanism to the ordinary SMPC. The plaintext will be revealed only when at least t participants are involved in the collaborative decryption. This effectively restricts the access to the final SMPC output and the participants will not disclose the data.

3.2.2. Weighted Threshold Secret Sharing Scheme Based on Mignotte Sequence.
The weighted threshold secret sharing scheme refers to that each participant assumes a different role, based on which different weights are assigned. The conventional weighted threshold secret sharing schemes achieve only works on the premise of assigning more secret shares

to those who are given special permission. However, this will increase the insecurity of key management and transmission. In this study, we adopted the weighted threshold secret sharing scheme based on *Mignotte* sequence. Regardless of the weight, each participant is only allowed one private key and there is no transmission of secret information between the participants and the dealer. Therefore, the cost of key transmission and storage is spared.

Mignotte sequence is defined as follows [13]:

Let $k, n \in Z$, $n \geq 2$, $2 \leq t \leq n$. If the integer sequence $m_1, m_2, \ldots, m_n$ satisfies

(1) $m_1 < m_2 < \cdots < m_n$;

(2) $(m_i, m_j) = 1$, where $1 \leq i < j \leq n$;

(3) $\prod_{i=0}^{t-2} m_{n-i} < \prod_{i=1}^{t} m_i$,

then sequence $m_1, m_2, \ldots, m_n$ is called a (t, n)-*Mignotte* sequence.

The weighted threshold secret sharing scheme based on the *Mignotte* sequence is designed.

(1) Parameter Configuration. In this scheme, the dealer assigns the weights to each participant using a digit with a length of large prime. The secret to be shared is determined and the relevant system parameters are configured. There are n participants and they constitute the set $U = \{u_1, u_2, \ldots, u_n\}$. The weight vectors of the participants are correspondingly $W = (w_1, w_2, \ldots, w_n)$. The threshold is t, and the secret to be shared is s.

(2) Construction and Expansion of Mignotte Sequence. The dealer needs the system parameters to construct an expanded Mignotte sequence fit for the weighted threshold secret sharing scheme. Meanwhile, the converted scheme should be equivalent to the original scheme. A (t, n)-*Mignotte* sequence is constructed as $m_1', m_2', \ldots, m_n'$, which is expanded into

$$\underbrace{m_1', \ldots, m_1'}_{w_1}, \underbrace{m_2', \ldots, m_2'}_{w_2}, \ldots, \underbrace{m_n', \ldots, m_n'}_{w_n}. \tag{6}$$

The above sequence is a sequence of n' primers, where $n' = \sum_{i=1}^{n} w_i$ which makes the sequence satisfy the following conditions:

(a) The product β of the last $t - 1$ numbers is smaller than the product α of the first t numbers.

(b) $\beta < s < \alpha$.

Let $m_1 = (m_1')^{w_1}, m_2 = (m_2')^{w_2}, \ldots, m_n = (m_n')^{w_n}$.

From above, it can be known that sequence $m_1, m_2, \ldots, m_n$ has the following property: $(m_i, m_j) = 1$ $(1 \leq i \leq j \leq n)$.

When $w_n + w_{n-1} + \cdots + w_j < t < w_1 + w_2 + \cdots + w_i$, $m_n m_{n-1} \cdots m_j < s < m_1 m_2 m_i$, $(1 \leq i \leq j \leq n)$.

Thus sequence $m_1, m_2, \ldots, m_n$ is the expanded Mignotte sequence, denoted as (W, t, n)-*Mignotte* sequence. This sequence is revealed.

(3) Generation of Secret Shares. The dealer computes the secret shares of each participant according to the *Mignotte* sequence $m_1, m_2, \ldots, m_n$ and the shared secret:

$$s_1 = s \bmod m_1,$$

$$s_2 = s \bmod m_2,$$

$$\vdots \tag{7}$$

$$s_n = s \bmod m_n.$$

This S_i is sent to the participant u_i via the secret channel.

(4) Secret Restoration. Suppose there are k participants who constitute the set A, $A = (u_1, u_2, \ldots, u_k)$, and restore the secret. The vector weights for each participant in A constitute the set $W = (w_1, w_2, \ldots, w_n)$.

When the sum of the weight vectors of each participant in A is above or equal to the threshold, that is, $\sum_{i=1}^{t} w_i \geq t$, the following congruence equations are constructed:

$$x = s_1 \bmod m_1,$$

$$x = s_2 \bmod m_2,$$

$$\vdots \tag{8}$$

$$x = s_k \bmod m_k.$$

$x = \sum_{i=1}^{k} s_i M_i^{-1} M_i \pmod{m}$, where $M_i = m \mid m_i$, $M_i^{-1} M_i \equiv 1 \pmod{m_i}$, $(1 \leq i \leq k)$ and the solution x is the shared secret s.

4. Implementation and Deployment

MapReduce System was used for the high-efficiency parallel processing in the large-scale matrix multiplication in the weighted threshold secret sharing scheme. On the simple data center comprising 5 host machines, the Hadoop distributed storage and computing environment was deployed as a mimic of the sensing platform in the MCS network. One host machine was the Master node, which was deployed with the roles of NameNode and JobTracker for the management of distributed data and task decomposition; 4 host machines were the Slaver and were deployed with the roles of DataNode and TaskTracker for the distributed data storage and task execution. The implementation and deployment (Figure 2) are illustrated below.

(1) The initialization program at the data center would preset the system parameters. The threshold t was determined. The weight vectors of each participant were initialized. The key management system as the trusted third party generated n pairs of homomorphic public and private keys. The public keys hom_PK were the same, and the n private keys were distributed to different participant nodes.

(2) The block function MR_Splitter( ) in the MapReduce System was responsible for dividing the sensing data files submitted by the clients in the MCS network into blocks. Each block was 64 M. The data blocks were encrypted using the

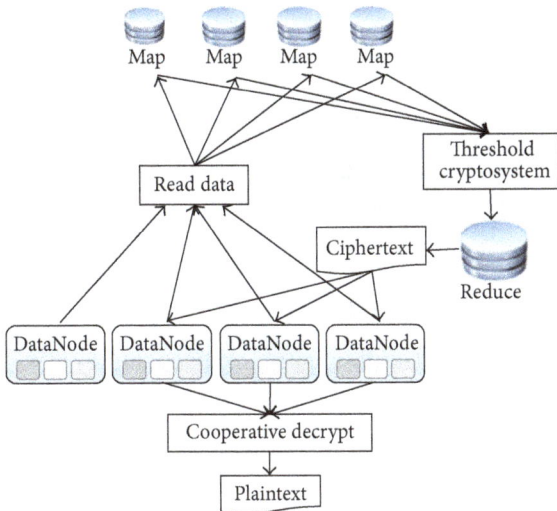

FIGURE 2: Flowchart of deployment.

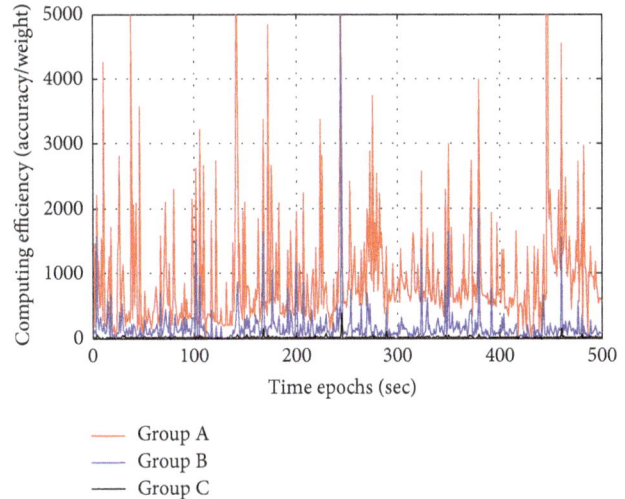

FIGURE 3: Privacy computing efficiency iteration results of all groups.

public key hom_PK. The encrypted data block file is stored in the distributed file system of the DataNode.

(3) An intermediate ⟨key, value⟩ pair was computed during the matrix multiplication in the privacy-preserving algorithm. The map nodes were allocated to each operation. Before the mapper output the ⟨key, value⟩ pair, the ciphertext for each participant was generated using formulae (7).

(4) Reducer replicated the intermediate output of the corresponding division from the mapper output terminal to the local file system.

(5) At least t participants were involved in the decryption of the ciphertext using the decryption algorithm in formulae (8). These participants would share the decrypted information with other participants. The information decrypted by the t participants was then combined with the information decrypted by the remaining participants to obtain the final result.

For users in network society, we divide them into three groups according to the weights aforementioned in Sections 2 and 3, that is, privacy careless person (group A), practical privacy person (group B), and the group who protect their privacy strictly (group C). In group A, they are not so sensitive with privacy and willing to share their true information. In group B, they may share personal files while policies and regulations are carefully learned. In group C, they are not interested in any sharing information activities at any circumstances.

In a certain survey, the percentage results of groups A, B, and C are obtained as 33.1%, 57.4%, and 9.5% from 352 users on the Internet, and we can initiate the weights of sharing by 0.9, 0.5, and 0.1. These parameters are easily adjusted during privacy protection mechanism proposed here.

As is shown in Figure 3, the three groups in privacy iteration results are given. Group A indicates that since they are not concerned about their information, those provided data are true and the efficiency is acceptable. Group B is matching data on the condition that they believe the privacy is protected, so that their efficiency is not stable and high.

Group C are not willing to share their information, and their provided information is not all correct, which influences the computing fundamentally.

5. Algorithm Performance Analysis

5.1. Security Analysis. The private information of each participant is randomly divided into m fragments in a certain way. Each participant selects one fragment randomly and preserves it. The remaining fragments are randomly allocated to other participants. After the fragments are reallocated according to the protocol, each participant will own an equal amount of fragments. Each participant owns one fragment of his or her information plus one fragment transmitted from another participant. Therefore, even if participants P_{i-1} and P_{i+1} conspire, they can only infer the reallocated information of participant P_i and do not know other private information N_i. Any two conspiring participants can only infer the reallocated information of the third party. Then, combining with the information fragments owned by themselves, they can infer the private information of the third party. But when there are more than 3 participants, it will be very difficult to infer all information of the other participants by conspiracy. When there are more than 4 participants and when most participants are honest, the possibility of information leak will approach 0.

5.2. Complexity Analysis. Computational complexity: each round of computation consists of m operations (different from the m aforementioned), and m rounds involve m^2 operations. Thus the computational complexity $S(m)$ is expressed as $S(m) = m^2$, as shown in Figure 4.

Communication Complexity. Each participant needs to transmit $m - 1$ fragments to other participants. Therefore, in the fragment transmission stage, $m(m - 1)$ communications will occur. In the computing stage, each participant needs

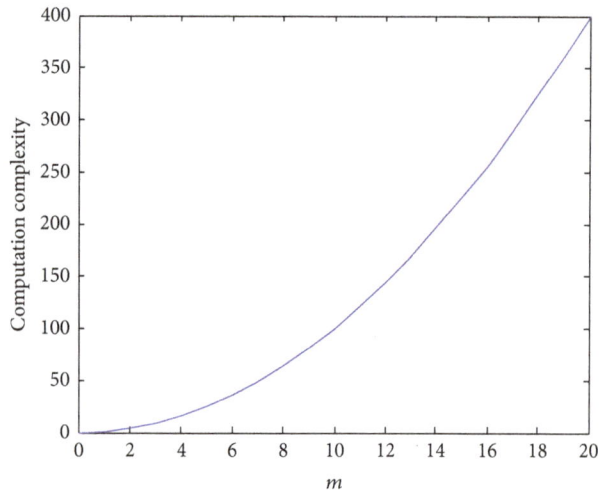

FIGURE 4: Computation complexity.

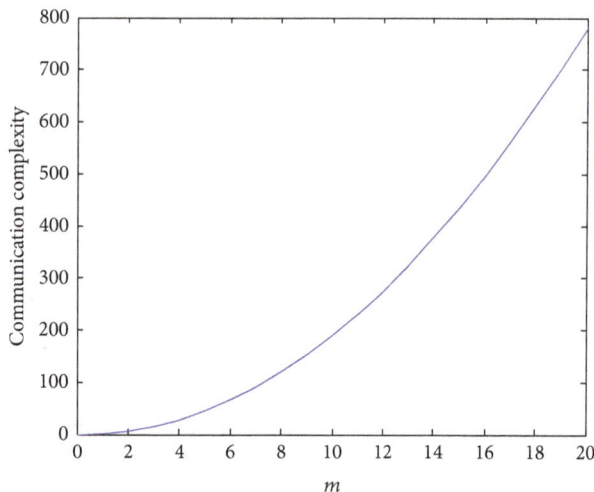

FIGURE 5: Communication complexity.

to transmit the summation of some fragments to other participants over the ring structure. Therefore, each round of computation consists of m communications, and m rounds involve m^2 communications. The overall communication complexity $C(m)$ of the algorithm is expressed as $C(m) = m(m - 1) + m^2 = 2m^2 - m$, as shown in Figure 5.

6. Summary and Forecast

To protect against privacy violation in the MCS network, we proposed a variable weight SMPC-based privacy-preserving algorithm. The weighted threshold secret sharing scheme based on Mignotte sequence was applied for the encryption of the sensing data and private key management. Considering the different attitudes of users towards the disclosure of the private information, the privacy of the information was graded. Thus the weight parameters of the privacy-preserving algorithm were determined based on the utility analysis of the users' privacy perception. The proposed model was

deployed in the Hadoop distributed environment to verify its effectiveness and validity. The implementation of the SMCP protocol requires several participants, among which communications are necessary. This will incur significant communication and computational costs. How to enhance the reliability of channel communication and to increase the efficiency of sensing data encryption are issues awaiting resolution.

Conflicts of Interest

The authors declare that there are no conflicts of interest regarding the publication of this paper and confirm that the mentioned received funding in the Acknowledgments did not lead to any conflicts of interest regarding the publication of this manuscript.

Acknowledgments

This work was supported by the Natural Science Foundation of Hainan Province (no. 20166216 and no. 617033) and Education and Reaching Research Project of Hainan University (no. hdjy1325) investigated by Jiezhuo Zhong; National Natural Science Foundation of China (no. 61661019), the Major Science and Technology Project of Hainan Province (no. ZDKJ2016015), the Natural Science Foundation of Hainan Province (no. 20156217), and the Higher Education Reform Key Project of Hainan Province (no. Hnjg2017ZD-1) by Chunjie Cao; National Science and Technology Support Program (no. 2015 BAH55F01-5) and Natural Science Foundation of Hainan Province (no. 614232) investigated by Wenlong Feng.

References

[1] R. K. Ganti, F. Ye, and H. Lei, "Mobile crowdsensing: current state and future challenges," *IEEE Communications Magazine*, vol. 49, no. 11, pp. 32–39, 2011.

[2] B. Liu, W. Zhou, T. Zhu et al., "Invisible hand: a privacy preserving mobile crowd sensing framework based on economic models," *IEEE Transactions on Vehicular Technology*, vol. 66, no. 5, pp. 1-1, 2017.

[3] H. Jin, L. Su, B. Ding et al., "Enabling privacy-preserving incentives for mobile crowd sensing systems," in *Proceedings of the IEEE 36th International Conference on Distributed Computing Systems (ICDCS '16)*, pp. 344–353, Nara, Japan, June 2016.

[4] S. A. Ossia, A. S. Shamsabadi, A. Taheri et al., "A Hybrid Deep Learning Architecture for Privacy-Preserving Mobile Analytics," https://arxiv.org/abs/1703.02952.

[5] S. Basudan, X. Lin, and K. Sankaranarayanan, "A privacy-preserving vehicular crowdsensing based road surface condition monitoring system using fog computing," *IEEE Internet of Things Journal*, no. 99, pp. 1-1, 2017.

[6] C. Xu, R. Lu, H. Wang, L. Zhu, and C. Huang, "PAVS: a new privacy-preserving data aggregation scheme for vehicle sensing systems," *Sensors*, vol. 17, no. 3, p. 500, 2017.

[7] Y. Lindell and B. Pinkas, "Privacy preserving data mining," in *Advances in Cryptology—CRYPTO 2000, 20th Annual International Cryptology Conference, Santa Barbara, California, USA,*

August 20–24, 2000, vol. 1880 of *Lecture Notes in Computer Science*, pp. 36–54, 2000.

[8] G. Asharov, A. Jain, A. López-Alt, E. Tromer, V. Vaikuntanathan, and D. Wichs, "Multiparty computation with low communication, computation and interaction via threshold FHE," *Lecture Notes in Computer Science (including subseries Lecture Notes in Artificial Intelligence and Lecture Notes in Bioinformatics)*, vol. 7237, pp. 483–501, 2012.

[9] L. Liu and M. Tamer Özsu, *Encyclopedia of Database Systems*, Springer, New York, NY, USA, 2017.

[10] D.-H. Shin, "The effects of trust, security and privacy in social networking: A security-based approach to understand the pattern of adoption," *Interacting with Computers*, vol. 22, no. 5, pp. 428–438, 2010.

[11] R. L. Rivest, L. Adleman, and M. L. Dertouzos, *On Data Banks And Privacy Homomorphism Proc of Foundations of Secure Computation*, Academic Press, New York, NY, USA, 1978.

[12] S. Goldwasser, "Multi party computations: past and present," in *Proceedings of the sixteenth annual symposium on Principles of distributed computing (ACM '97)*, pp. 1–6, August 1997.

[13] M. Mignotte, "How to share a secret," *Lecture Notes in Computer Science*, vol. 149, no. 2, pp. 371–375, 1983.

Dynamic Antenna Alignment Control in Microwave Air-Bridging for Sky-Net Mobile Communication using Unmanned Flying Platform

Chin E. Lin and Ying-Chi Huang

Department of Aeronautics and Astronautics, National Cheng Kung University, Tainan, Taiwan

Correspondence should be addressed to Chin E. Lin; chinelin@mail.ncku.edu.tw

Academic Editor: Sing Kiong Nguang

This paper presents a preliminary study on establishing a mobile point-to-point (P2P) microwave air-bridging (MAB) between Unmanned Low Altitude Flying Platform (ULAFP) and backhaul telecommunication network. The proposed Sky-Net system relays telecom signal for general mobile cellphone users via ULAFP when natural disaster sweeps off Base Transceiver Stations (BTSs). Unlike the conventional fix point microwave bridging application, the ULAFP is cruising on a predefined mission flight path to cover a wider range of service. The difficulty and challenge fall on how to maintain antenna alignment accurately in order to provide the signal strength for MAB. A dual-axis rotation mechanism with embedded controller is designed and implemented on airborne and ground units for stabilizing airborne antenna and tracking the moving ULAFP. The MAB link is established in flight tests using the proposed antenna stabilizing/tracking mechanism with correlated control method. The result supports backbone technique of the Sky-Net mobile communication and verifies the feasibility of airborne e-Cell BTS.

1. Introduction

Ever since 1990s, mobile communication had been gradually becoming the most dependent livelihood communication means on handy sets. Base Transceiver Stations (BTSs) for mobile communications are usually built on the ground for services. On the natural disasters, such as earthquake, typhoon, or flood, the BTSs as well as power systems are easily affected or destroyed and thus black out communications from outside forces to inside victims. The most important part for disaster rescue would be the recovery of communications. Any common methods to establish mobile communication for any users within the disaster area would be most appreciated even just for a short period.

The Sky-Net project addresses the establishment and recovery of a temporary mobile phone service for general users by using Unmanned Low Altitude Flying Platform (ULAFP) carrying with an airborne e-Cell BTS to relay mobile signal to distant ground BTS. The proposed Sky-Net connects mobile communication into backbone network of Chunghwa Telecom Company in Taiwan.

The concept of relaying wireless signal via flying platform had been studied in the past decade. Literatures categorized different flying platforms from high or low flight altitudes. Widiawan and Tafazolli [1] introduced the necessary part of the wireless communication relay using High-Altitude Platform (HAP), including platform type, energy consumption for long time mission, and communication loading for the system. Because of the Line-of-Sight (LoS) characteristics of the flying platform, this kind of airborne relay equipment can provide more coverage or increase the link capacity for heavy loading area. The coverage emulation studied by Feng et al. [2] also adopted multi-HAP on signal relay missions. The increment of link capacity was proved by Grace et al. [3] and Elshaikh et al. [4]. The airborne relay equipment uses various wireless link types. Because of the wide usage and application of personal electronic devices, like mobile phone or laptop, relaying wireless signal with Internet service

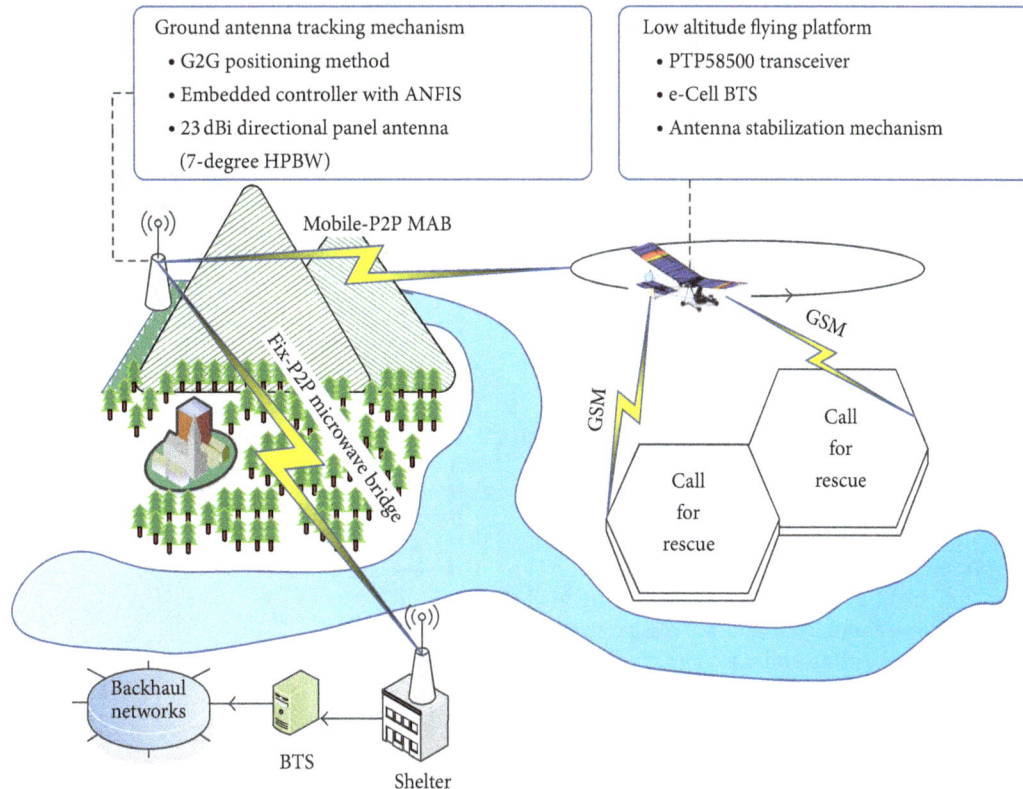

FIGURE 1: The proposed Sky-Net mobile signal relaying concept.

becomes the major issue on such research. Hariyanto et al. [5] introduced a study on Emergency Broadband Access Network (EBAN) application using balloon to hover at altitude range from 100 to 500 meters. Hariyanto sorted out three major alternate solutions for wireless Internet service, such as WiMAX, 3G, and WiFi. Qiantori et al. [6] demonstrated the result of airborne WiFi service using two balloons to carry transceiver module. They work on launching a balloon hovering at 440 m (AGL) to provide service area up to 3.88 km. The maximum throughput of this system is limited under 0.8 Mbps, which is just enough for transmitting voice data for user [6]. The other hybrid type of relaying is introduced by Guo et al. [7] to propose a WiFi link between users and the drone. The latter part of airborne decode-and-forward (DaF) device exchanges data pack between the drone and the nearest BTS. However, in the Sky-Net project, system architecture is proposed as directly relayed signal into backbone networks of Telecom Company, which is operating under E1 protocol in this paper. Besides, instead of using balloon as LAFP, the Sky-Net proposed a fix-wing unmanned ultralight aircraft as flying platform like HeliNet project in Europe [8] in order to fly against the unpredictable weather conditions. Under Sky-Net concept, mobile signal is carried by microwave with frequency at 5.8 GHz in order to achieve high bandwidth for steady link. The relay mechanism can be sorted into two parts as Fix-P2P and Mobile-P2P, separately. Like general microwave bridging, Fix-P2P relays signal from functional BTS to ground antenna tracking station for Mobile-P2P, as shown in Figure 1.

In order to establish a microwave air-bridging (MAB), a directional antenna is adopted with advantage on highly narrowed power beam. It can also support farther wireless link establishment and higher link quality. Yet, the narrow antenna pattern comes with directional restriction; once the flying platform is out of the coverage area of the ground directional antenna, signal strength would drop and therefore affect the integrity of MAB. On the other hand, if the flying platform would meet inevitable attitude change during cruise flight, the polarization or beam direction of airborne antenna is changing in whole flight time, which may affect wireless link quality too.

For the ULAFP and ground unit, a dual-axis rotation platform is proposed in this paper in order to provide well directional antenna alignment from the flying platform and maintain steady attitude for airborne gimbal on mounting antenna. Both ground and airborne rotation mechanisms have to apply proper control method to achieve the required MAB performance. Flight tests for verifying the performance and integrity of the proposed MAB for mobile signal relaying operation are demonstrated. It supports the evidence for the reliable backbone technique for further developments and realizations of the Sky-Net project.

2. Antenna Stabilization/Tracking Mechanism

In consideration to maintain the most antenna pattern overlap ratio for wireless link, airborne antenna needs to be covered under the coverage area of the ground directional antenna, and vice versa. Smart antenna with beam forming performance is an ideal solution for such condition; Falletti et al. [9] did the analysis of Bit Error Rate (BER) for bot high speed train and HAP antenna. Falletti verified that the smart antenna showed suitable wireless relaying to mobile target. On the other hand, several antenna types for relaying different carrier frequency were studied by Miura and Suzuki [10]. Both electrical and mechanical scanning techniques are compared. However, by reducing system complexity, the Sky-Net project proposes a dual-axis rotation mechanism on both airborne and ground units as directly connected with backbone networks in mobile telecommunication. While the purpose ensures the coherence of polarization of airborne antenna during the flight mission, the airborne dual-axis mechanism needs to compensate the dynamic changes during flights. The ULAFP attitude in roll and pitch motion varies due to flight control or aerodynamic disturbance. Consequently, it is most required to control the ground antenna tracking unit to fit its performance on aiming at the ULAFP in Sky-Net operation. Both rotation mechanisms have correlated control algorithm and mechanism design, separately, for microwave alignment.

2.1. Airborne Antenna Stabilization.

In the Sky-Net project, a 10 dBi omnidirectional linearly polarized antenna is selected to match with the other 2.5 dBi circularly polarized antenna to fit the multiple-input-multiple-output (MIMO) performance for Motorola PTP58500 transceiver, which is telecommunication signal relaying equipment for long range Ethernet bridging. The major task of airborne mechanism should maintain the antenna's polarization for ULAFP during cruise flight. In other words, the antenna in vertical polarization should be installed to point to the ground regardless of ULAFP's attitude and flight performance. This kind of gimbal has been widely studied for applications to aerial photography from airborne systems, such as fix-wing or multirotor Unmanned Aerial Vehicle (UAV) [11]. Figure 2 illustrates the dimension and coordinates arrangement of the proposed antenna stabilization mechanism.

Using the microelectromechanical system (MEMS) technique, the Inertial Measurement Unit (IMU) now could be small enough to merge into microcontroller inside a small piece of Printed Circuit Board (PCB). It is therefore widely used on embedded control applications in robotics. UAV and the proposed stabilization mechanism are typical uses. Studies have been made for attitude estimation to control the gimbal mechanism [12]. Madgwick et al. [13] introduced the real-time attitude estimation in microcontroller with algorithm as less calculation cost as possible in dedicate realization. In this paper, Madgwick et al.'s method is adopted as attitude information for controlling the gimbal with modifications. The proposed stabilization block diagram is shown in Figure 3.

In Figure 3, the control core in each control loop solves the command to manipulate the mechanism attitude. Either pitch angle or roll angle in the gimbal should be maintained under designed thresholds. Otherwise the control law will

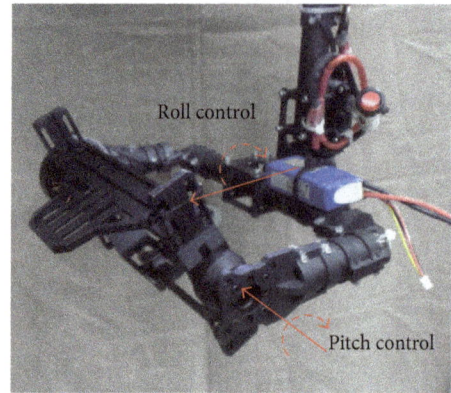

FIGURE 2: Airborne dual-axis antenna stabilization gimbal.

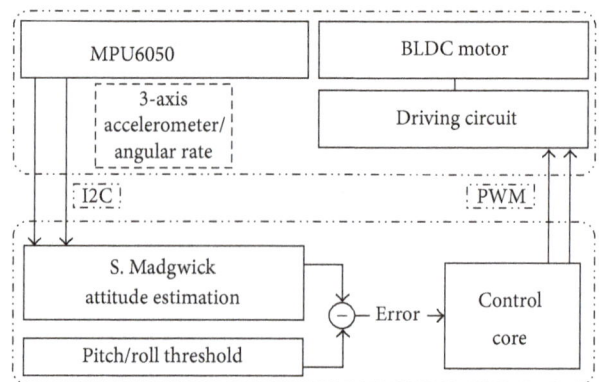

FIGURE 3: Block diagram of antenna stabilization control.

output the driving signal for Brushless DC Motor (BLDC) in sequent of roll and pitch axes to compensate the angle offset due to ULAFP motion or external disturbance. In the flight tests to verify the stabilization performance of designed gimbal, the data are logged as roll and pitch angles of the gimbal referring to ULAFP attitude, correspondingly, as shown in Figure 4.

Blue line in Figure 4 represents the attitude changing status of ULAFP during the flight stage. The glitch on blue line data reveals the inevitable vibration when the antenna mechanism is directly mounted onto ULAFP. It may cause the antenna polarization offset and lead MAB throughput to drop down temporarily or even lose connection. The proposed gimbal stabilization mechanism should be effectively maintained for both pitch and roll angles in good alignment to around zero degrees while ULAFP is flying. This is shown by the red line in Figure 4.

2.2. Ground Antenna Tracking Mechanism.

In addition to airborne stabilization mechanism, the ground tracking unit mounting with highly directional panel antenna needs even more accurate positioning control on elevation and azimuth angle for tracking the ULAFP. The ground antenna tracking mechanism should cover the airborne antenna under Half-Power Beam Width (HPBW) area. The ground antenna tracking mechanism is driven by stepper motor with gear

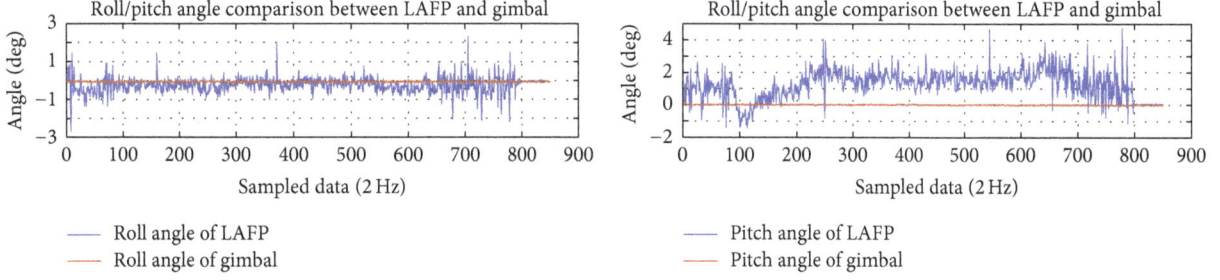

FIGURE 4: Gimbal roll and pitch angles referring to ULAFP attitude.

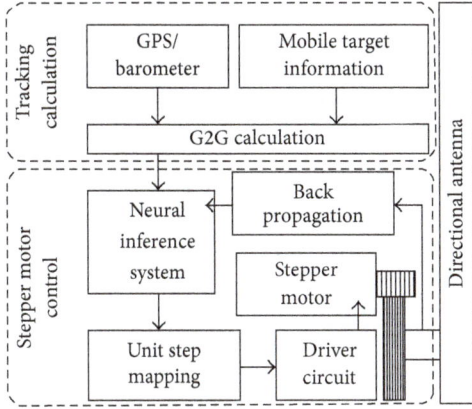

FIGURE 5: Block diagram of ground antenna tracking unit.

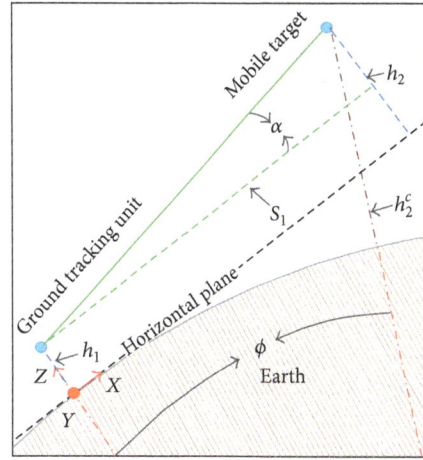

FIGURE 6: Elevation angle error due to earth curvature.

mechanism. Therefore, the tracking mechanism can be categorized into tracking calculation and stepper motor control, as shown in Figure 5.

For applications like target tracking out of visual range, the related angle displacement between the ground tracking unit and the ULAFP is small. The tracking mechanism needs very high accuracy feature in perturbation control much more than its rotation speed. And also, as the Line-of-Sight (LoS) distance between the ULAFP and the ground tracking unit is long enough, GPS error becomes negligible in calculation. A simple but efficient tracking method based on GPS-to-GPS (G2G) method is adopted in the Sky-Net project. By using GPS and barometric data, the related azimuth and elevation angles can be solved easily. However, as distance increases, the calculation error appears significantly because of the earth curvature, especially on the elevation angle. G2G method assumes that the ground tracking unit and the ULAFP are on the same surface of coordinates. Both sides are located in the same Cartesian Coordinates system. Figure 6 indicated the error accumulation as distance S_1 increases.

If the ground tracking unit and the ULAFP are close enough, angle calculation can be simplified into 3-axis Cartesian Coordinates directly. By transforming the GPS position into another coordinates system, such as TM2 in TWD97, distance between the ground antenna tracking unit and the ULAFP can be obtained, marked as S_1 in Figure 6. In addition to coordinate transformation, the altitude difference can be

calculated from barometric data and GPS altitude data. The elevation angle α can be expressed as

$$\alpha = \arctan\left(\frac{h_2 - h_1}{S_1}\right). \tag{1}$$

However, as S_1 is far from both terminals, the curvature of the earth will evidently affect the accuracy of elevation angel α. Figure 6 also shows how the curvature makes the error bigger as S_1 increases. In TWD97 data, the earth model is an ellipsoid and its radius scale is relatively huge compared to ULAFP flight path. Within a small region, the earth is reasonably assumed as a sphere not an ellipsoid for simplicity. Therefore between the ground and the airborne terminals in Figure 6, the earth radius is R and the inclination angle between them toward earth center is ϕ; the elevation angle α of the ground terminal can be recalculated as

$$d_1 = R^2 + \left(h_2^c + R\right)^2 - 2R\left(h_2^c + R\right)\phi,$$
$$d_2 = \left(h_1 + R\right)^2 + \left(h_2^c + R\right)^2 - 2\left(h_1 + R\right)\left(h_2^c + R\right)\cos\phi, \tag{2}$$
$$\alpha = \cos^{-1}\left(\frac{h_1^2 + d_1^2 - d_2^2}{2 - h_1 d_1}\right).$$

The relative azimuth and elevation angles obtained from G2G calculation output control parameters for stepper motor

TABLE 1: Step division of stepper motor (full-step@1.8°).

Mode	Resolution	Mode	Resolution
1	1/1	9	1/40
2	1/2	10	1/50
3	1/2.5	11	1/100
4	1/5	12	1/125
5	1/8	13	1/200
6	1/10	14	1/250
7	1/20	15	1/500
8	1/25	16	1/1000

mechanism. Instead of direct driving stepper motor to the desired angle displacement, a neural inference system with back propagation updating mechanism is proposed in this paper. Its merit mainly shows the advantage unit step switching using stepper motor in the mechanism drive. Unit step division of the adopted stepper motor is shown in Table 1. Combining with the division factor and its gear mechanism ration, the resolution of the ground antenna tracking unit hence becomes 0.18 degrees to 0.00018 degrees. It makes the antenna mechanism in microstep control.

As ULAFP flying along the designated mission path, cruising speed or actual path may be different due to the characteristic differences of flying vehicle, pilot, or even flight control computer. Hence, even though the flight path is redesigned and designated before taking off, the instant angle changing rate of elevation and azimuth axis is still different in the whole flight mission. The unit step switching feature of stepper motor is suitable in this implementation on angle positioning application with wide range angle displacement. Controller needs to be designed with capability on making appropriated unit step selection in order to consider both tracking accuracy and response speed. The proposed Adaptive Neural Fuzzy Inference System (ANFIS) plays the role of selecting the output unit step of stepper motor of each axis.

Figure 7 shows the typical 1st order Sugeno Fuzzy Model [14] adopted in this paper. Two input variables and one output system structure are proposed. Therefore, a MISO fuzzy logic can be expressed as the following human describing sentences:

$$\text{Rule}_{1,1}\text{: If } x_1 \text{ is } A_1 \text{ AND } x_2 \text{ is } B_1 \text{ Then } y_{1,1} \text{ is } C_{1,1}$$
$$\text{Rule}_{1,2}\text{: If } x_1 \text{ is } A_1 \text{ AND } x_2 \text{ is } B_2 \text{ Then } y_{1,2} \text{ is } C_{1,2}. \quad (3)$$

The describing sentences use fuzzy linguistic label of membership function to express logic inference in human language, where $A = \{A_1, A_2, A_3, \ldots, A_m\}$ and $B = \{B_1, B_2, B_3, \ldots, B_n\}$ are linguistic label set. Subscripts m and n of each fuzzy set vector represent the vector dimension, separately. As shown in Figure 7, the proposed ANFIS architecture is divided into four layers. Nodes in Figure 7 are defined as fix node for circle shape, which are predesigned and will not be changed in the whole inference process. On the other hand, the rectangle nodes are adaptive ones, which are the updating target for ANFIS in back propagation (BP) training.

The proposed neural fuzzy inference process can be classified into four layers: input, fire strength calculation, rule formation, and output, as below.

Layer 1 (Inputs). In layer one, two input variables are directly mapped into membership functions to get the corresponding membership grade:

$$O^1_{1,i} = \mu_{A_i}(x_1), \quad i = 1, 2, 3, \ldots, m,$$
$$O^1_{2,j} = \mu_{B_j}(x_2), \quad j = 1, 2, 3, \ldots, n. \quad (4)$$

For arbitrary crisp input data, the ith membership functions for input x will be generated as $\mu_{A_i}(x)$ and $\mu_{B_i}(x)$. Superscript of O^e represents the output of layer e.

Layer 2 (Fire Strength). Fire strength factor is defined as multiplication product of two input membership grades in order to represent the logical AND operation. This indicates the implementation strength of the corresponding rule. Therefore, in Layer 2, the output terms and the fire strength matrix $[\alpha]_{m,n}$ can be presented as

$$O^2_{i,j} = \mu_{A_i}\mu_{B_j}$$
$$[\alpha]_{m,n} = \mu_A{}^T \cdot \mu_B = \begin{bmatrix} \alpha_{1,1} & \cdots & \alpha_{1,n} \\ \vdots & \ddots & \vdots \\ \alpha_{m,1} & \cdots & \alpha_{m,n} \end{bmatrix}, \quad (5)$$

Layer 3 (Rule Formulation). In this layer, the core of Fuzzy Logic Controller (FLC) is shown as Rule Base Matrix (RBM). RBM is the decision maker to generate the appropriate output results. However, unless the designer is familiar with the control target, RBM could generate the region logical confliction or lack of system response. The Error Correction Mechanism (ECM) is set in this layer with the purpose of training and updating the value of each node until the output error scale converges into the desired margin. In this paper, a rule function is specially designed and set for generating the corresponding value of nodes in Layer 3, as shown below:

$$O^3_{i,j} = f^{m,n}_{i,j} = (-m + 2i - 1)p_{i,j} + (-n + 2j - 1)q_{i,j}, \quad (6)$$

where $p_{i,j}$ and $q_{i,j}$ are the *learning factor* for nodes belonging to inputs X_1 and X_2, respectively. The rule function is designed for general stepper motor driven platform. The formulated rule values carry with +/− signs to represent motor driving direction. According to the network in Figure 7,

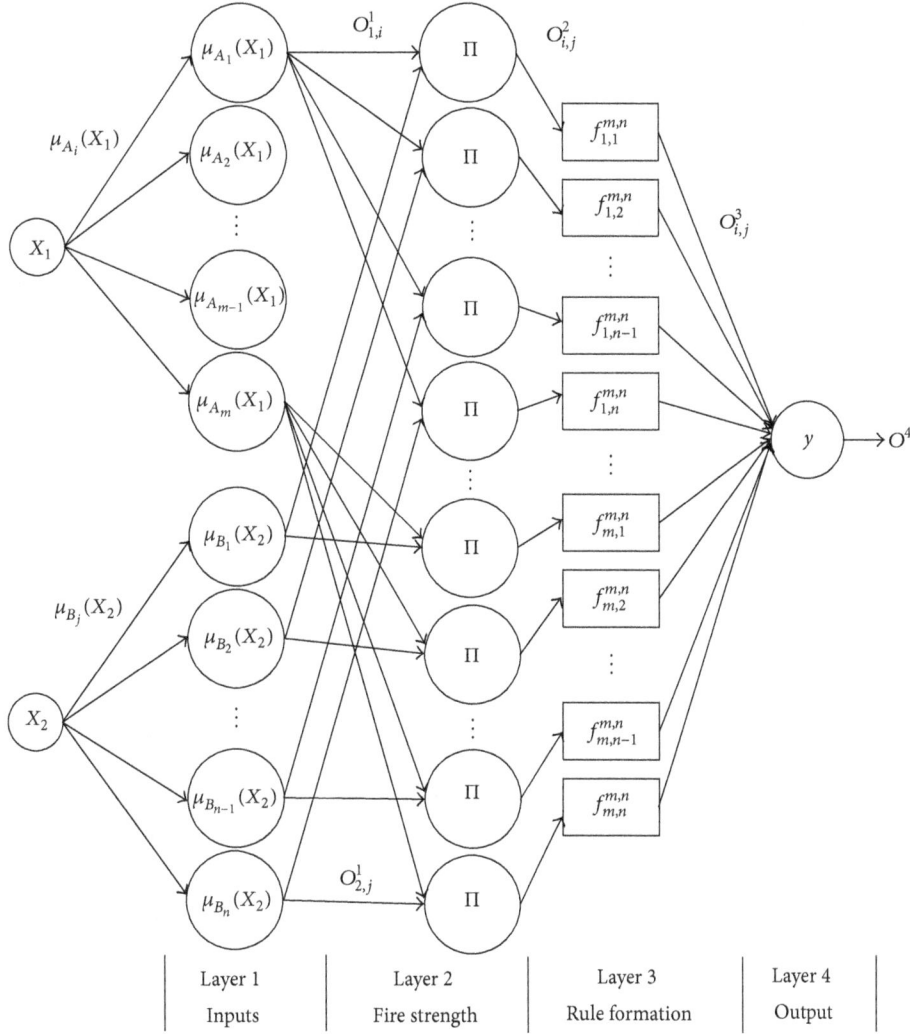

FIGURE 7: Architecture of adopted ANFIS.

output result from Layer 3 can be represented in matrix form as

$$[RBM]_{m,n} = [\alpha]_{m,n} \times \begin{bmatrix} f_{1,1}^{m,n} & \cdots & f_{1,n}^{m,n} \\ \vdots & \ddots & \vdots \\ f_{m,1}^{m,n} & \cdots & f_{m,n}^{m,n} \end{bmatrix}$$

$$= \begin{bmatrix} O_{1,1}^3 & \cdots & O_{1,n}^3 \\ \vdots & \ddots & \vdots \\ O_{m,1}^3 & \cdots & O_{m,n}^3 \end{bmatrix}. \qquad (7)$$

Layer 4. Output result in Layer 4 is acquired using the gravity center method. The overall crisp output of fuzzy inference process is calculated in this single node. This is also called the defuzzification stage. Consider

$$O^4 = \frac{\sum_{i=1}^m \sum_j^n \alpha_{i,j} f_{i,j}^{m,n}}{\sum_{i=1}^m \sum_j^n \alpha_{i,j}}. \qquad (8)$$

In order to improve the result form of the neural fuzzy inference system, back propagation is adopted for adjusting rule formation parameters in Layer 3 of Figure 7. For a single output system, the goal of training reduces the value of the following cost function:

$$E(k) = \frac{1}{2}(d(k) - y(k))^2 = \frac{1}{2}(e(k))^2. \qquad (9)$$

At time k, $d(k)$ and $y(k)$ are the desired and actual system outputs, separately. The update law applied on arbitrary parameters $C_{i,j}$ in the neural network at time $k + 1$ is the combination of its value at time k. The learning result and the momentum part are shown as

$$C_{i,j}(k+1) = C_{i,j}(k) - \eta^c \frac{\partial E(k)}{\partial C_{i,j}} + \tau^c \Delta C_{i,j}(k),$$

$$\frac{\partial E(k)}{\partial C_{i,j}} = -e(k) \frac{\partial O_{i,j}^3}{\partial C_{i,j}}. \qquad (10)$$

Learning rate and momentum factor are represented as η^c and τ^c, separately. Therefore, the update law of $p_{i,j}$ and $q_{i,j}$ of each node in Layer 3 at time k can be expressed as

$$p_{i,j}(k+1) = p_{i,j}(k) + \eta^p e(k)\frac{\partial O_{i,j}^3}{\partial P_{i,j}} + \tau^p \Delta p_{i,j}(k),$$

$$q_{i,j}(k+1) = q_{i,j}(k) + \eta^q e(k)\frac{\partial O_{i,j}^3}{\partial q_{i,j}} + \tau^q \Delta q_{i,j}(k). \tag{11}$$

The partial differential parts can be expanded as follows:

$$\frac{\partial O_{i,j}^3}{\partial P_{i,j}}$$

$$= \frac{\partial}{\partial P_{i,j}}$$

$$\cdot \frac{\sum_{i=1}^m \sum_{j=1}^n \alpha_{i,j}\{(-m+2i-1)p_{i,j} + (-n+2j-1)q_{i,j}\}}{\sum_{i=1}^m \sum_j^n \alpha_{i,j}(k)}$$

$$= \frac{(\partial/\partial P_{i,j})\sum_{i=1}^m \sum_{j=1}^n \alpha_{i,j} f_{i,j}^{m,n}}{\overline{\alpha}(k)},$$

$$\overline{\alpha}(k) = \sum_{i=1}^m \sum_j^n \alpha_{i,j}(k) = \frac{\alpha_{i,j}(-m+2i-1)}{\overline{\alpha}(k)}. \tag{12}$$

Similarly, for $q_{i,j}$, the partial differential part in update law is

$$\frac{\partial O_{i,j}^3}{\partial q_{i,j}} = \frac{\alpha_{i,j}(-n+2j-1)}{\overline{\alpha}(k)}. \tag{13}$$

According to the above equations, the change scale of BP process at time k is determined by three factors: (a) the summation of fire strength of each of the membership functions in Layer 1, (b) labeled number of the corresponding nodes, and (c) the output error magnitude. The *learning factor* affects $O_{i,j}^3$ result directly. However, in the BP process, output error scale is possibly huge and leads to the overtraining result. It could cause the oscillation or divergence in training. A threshold is added into the BP before updating new value of target parameter in order to avoid the overtraining condition. The maximum value of each *learning factor* is determined according to the corresponding labeled value:

$$p_{i,j}^{\max} = \frac{O_{i,j\max}^3}{(-m+2i-1)},$$

$$q_{i,j}^{\max} = \frac{O_{i,j\max}^3}{(-n+2j-1)}, \tag{14}$$

$$O_{i,j\max}^3 = \mathrm{abs}(-m+1) + \mathrm{abs}(-n+1).$$

Therefore, the update law can be rewritten into the following equations:

$$\overline{p}_{i,j}(k+1) = \min\left(p_{i,j}^{\max}, p_{i,j}(k+1)\right),$$

$$\overline{q}_{i,j}(k+1) = \min\left(q_{i,j}^{\max}, q_{i,j}(k+1)\right). \tag{15}$$

In the beginning of the training process, the initial value of learning factor is set to 1. Instead of using a random forming fuzzy surface, a rough established fuzzy surface is generated, and therefore the proposed method has ability to provide system with a degraded output, instead of leading system output with a conflict decision.

Verification of the proposed ANFIS control method for antenna tracking mechanism control is presented with flight test data in this study. Control core adopted by this paper is a microcontroller with Cortex-M4 architecture running at 180 MHz as CPU speed. Figures 8 and 9 indicate the rule updating with the proposed rule formation method. Figures 10 and 11 show the difference on angle displacement error and online tracking result versus time between predesign and random set fuzzy surface under the proposed ANFIS method.

Both of predesigned and random set fuzzy surface training results are shown in Figures 8 and 9. Even though BP process can update the fuzzy surface at each calculation and makes the frequently used rule be well-trained, there are some decision confliction parts hidden by untrained blocks. On the other hand, the predesigned fuzzy surface has some regions that have not been trained in flight mission, too. Because of the logic coherence of fuzzy surface, the system can still generate suitable output decision and shorten training time compared with random set fuzzy surface method.

According to Figures 10 and 11, the tracking error reveals that the random set fuzzy surface has slower response on adapting a tracking mechanism. In tracking error chart, glitches from Figure 10 are the part that control core "step" into random set fuzzy surface part. Therefore, the potential logic confliction generates the inappropriate output for stepper motor. The tracking error is relatively larger than the proposed predesign fuzzy surface method, as shown in Figure 11. There is noticeable glitch difference on error chart in Figure 11 compared with Figure 10. Besides, because the tendency of fuzzy surface of the proposed method is coherent, it needs less number of training times for adjusting $p_{i,j}$ and $q_{i,j}$ on each node in neural networks. Under the above introduced antenna tracking accuracy, MAB establishment can be realized between ground antenna tracking mechanism and ULAPF.

3. Verification of MAB Establishment

In the flight test, an Eipper ultralight aircraft with license number JJ-2071 is converted in unmanned ultralight (UUL) aircraft using ULAFP for Sky-Net project [15, 16]. In order to achieve 360-degree communication in the whole cruising path, two omnidirectional antennas with circular and linear polarization are installed on airborne antenna stabilization mechanism, as shown in Figure 12.

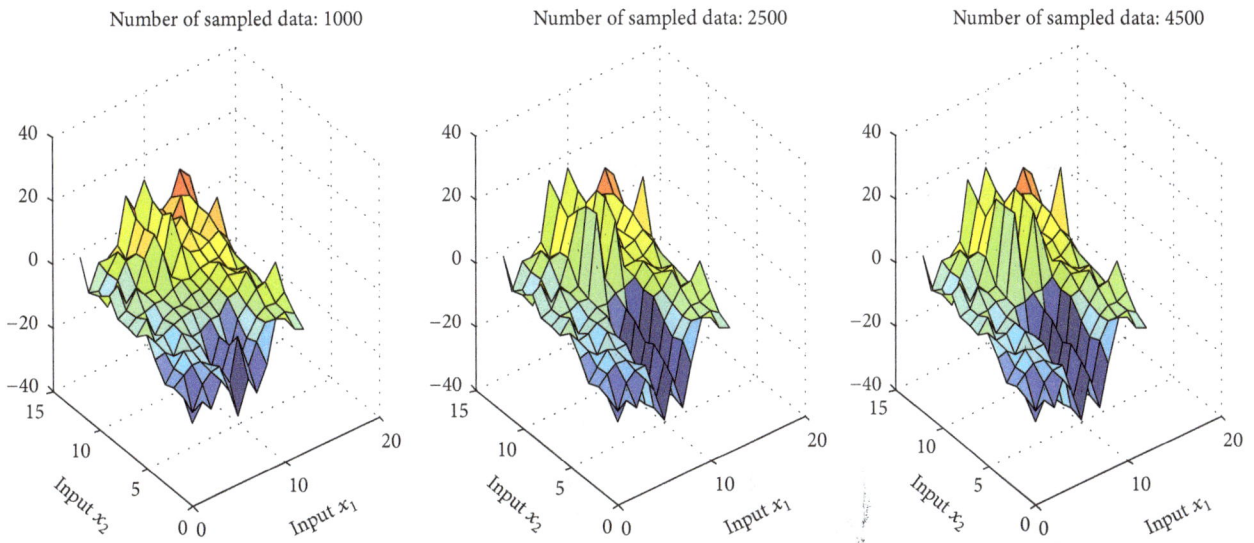

FIGURE 8: Training result of random set fuzzy surface.

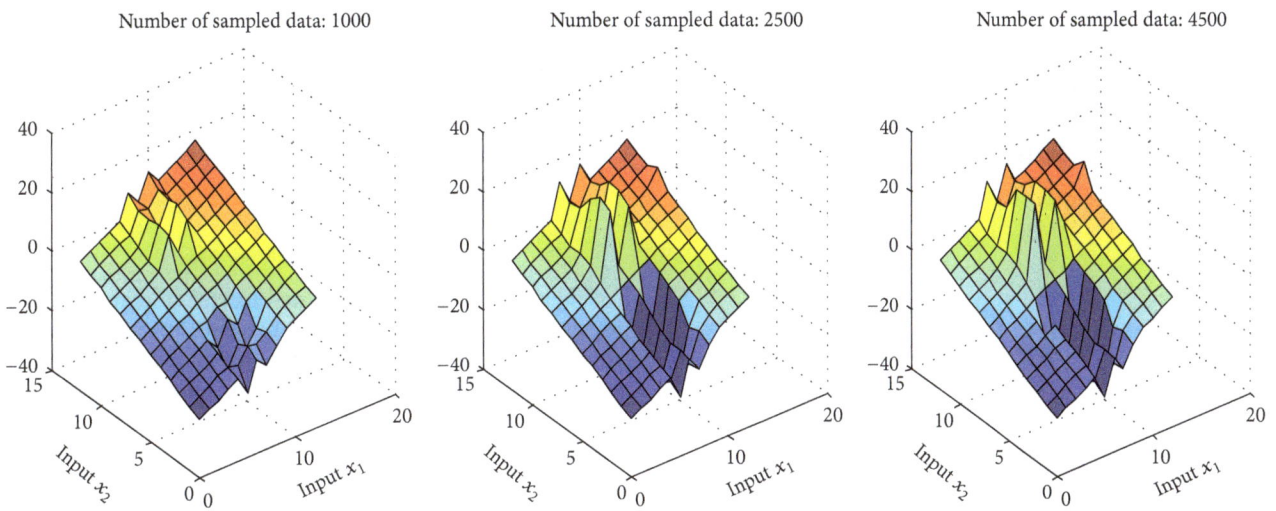

FIGURE 9: Training result of predesigned fuzzy surface.

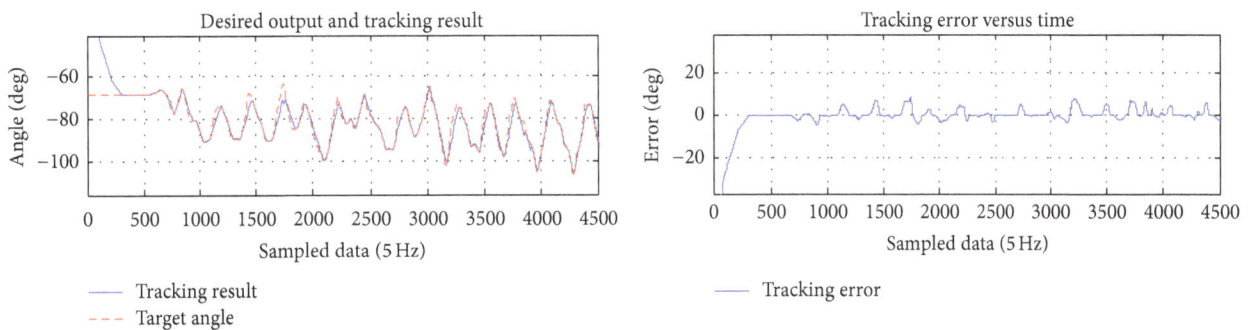

FIGURE 10: Tracking error of random set fuzzy surface with proposed ANFIS.

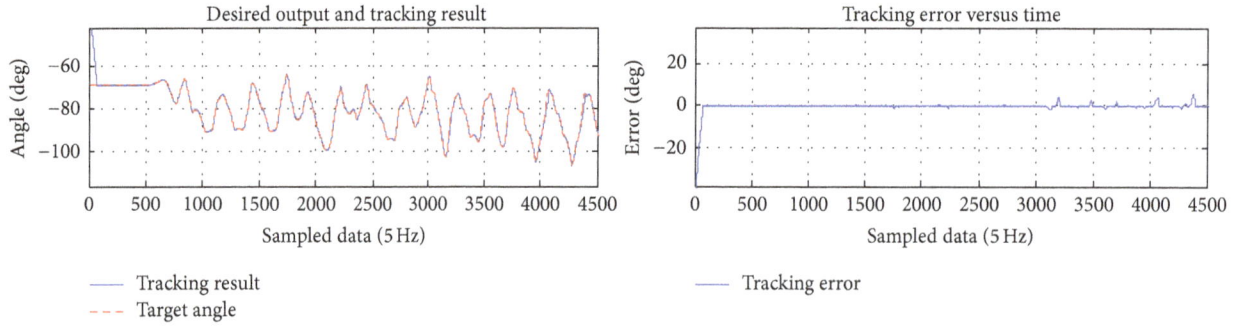

FIGURE 11: Tracking error of predesigned fuzzy surface with proposed ANFIS.

FIGURE 12: Airborne antenna on JJ2071 for ULAFP test.

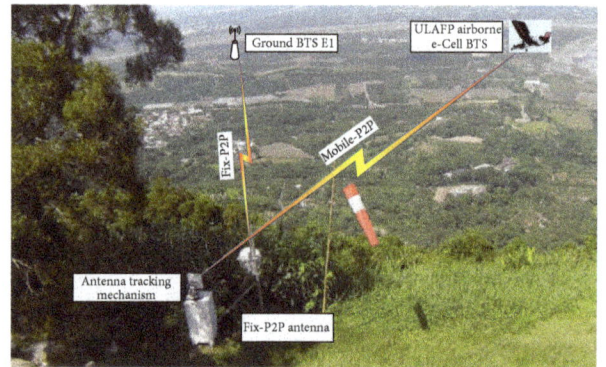

FIGURE 14: Test site for Sky-Net microwave air-bridging at 440 m (MSL).

FIGURE 13: Ground antenna tracking mechanism with microwave module.

Ground antenna tracking mechanism is set up at mountainside in southern Taiwan, as shown in Figure 13, at above ground level (AGL) 334 m (or about 440 m MSL). A PTP58500 transceiver with integrated panel antenna is mounted at 2-axis rotation mechanism with well alignment on both azimuth and elevation axes. The test site is shown in Figure 14.

Flight test path was designed to ensure that both sides of microwave module antenna pattern can be overlapped. In order to verify the tracking system being capable of maintaining constant microwave link, the fight paths and distances are predetermined to check the microwave link. The flying platform test altitude is fixed at 440 m~450 m (MSL) as shown with flight path in Figure 15.

Telecom data package in backbone network of Chunghwa Telecom is transmitted in E1 format, and therefore MAB system has a bandwidth threshold (3.5 Mbps) in order to satisfy the basic telecom data exchanging to establish a stable mobile communication service. Table 2 shows that both system signal strength and vector errors are factors which affect system maximum throughput. As the wireless link status is changed due to antenna misalignment or weather condition, microwave module would change the modulation in order to fit current wireless link condition. Therefore, the sticking point of maintaining the QoS of MAB falls on keeping the system signal strength higher than −87.96 dBm and vector error is lower than −11 dB, shown by red line in Figures 16 and 17. These are also the threshold to hold the QPSK 0.75 modulation to provide a steady 3.54 Mbps system bandwidth.

The tracking steps are classified into take-off, test phase in cruise, and landing. In order to ensure that the system signal strength scale is qualified for MAB, the ground tracking unit must maintain ULAFP under area coverage of conical

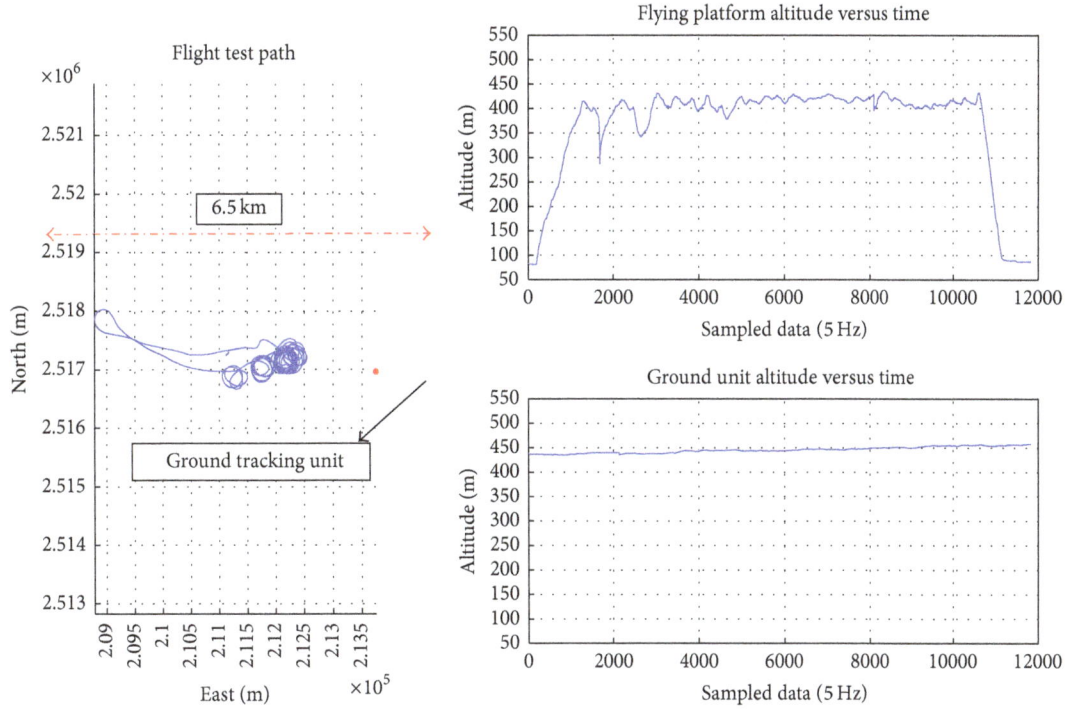

FIGURE 15: Flight test path and altitude.

TABLE 2: PTP58500 adaptive system modulation.

Modulation mode	Signal threshold (dBm)	Vector error (dB)	Bandwidth (Mbps)
16QAM 0.75	−81	−17.8	7.39
16QAM 0.50	−84.78	−14.5	5.45
QPSK 0.75	**−87.96**	**−11**	**3.54**
QPSK 0.50	−89.78	−9.5	2.04
BPSK 0.50	−99.02	−1.9	0.61

TABLE 3: Antenna/transceiver specification of ground tracking unit.

Antenna polarization	Linear
H-HPBW/V-HPBW	8 degrees
Gain	23 dBi
Transceiver Tx power	−18~27 dBm
Transceiver sensitivity	−69~−94 dBm

TABLE 4: Ground tracking mechanism specification and flying path.

Flying platform	
Cruising speed	60~72 km/hr
Flight path distance	2.5~6 km
Turning radius	500 m
Antenna tracking mechanism	
Azimuth/elevation rotary angle	360 degrees/−15~90 degrees
Azimuth/elevation tracking speed	10 dps/5 dps* (max.)

*dps: degree per second.

antenna pattern. According to the ground directional panel antenna pattern and microwave module specification in Table 3, the ground tracking mechanism is designed with predetermined flight path as Table 4 to match with the ideal MAB performance.

The preliminary flight test results with flight path in Table 3 are shown in Figures 16 and 17, for tracking mechanism using random set and predesigned fuzzy surface with the proposed ANFIS method.

Since Half-Power Beam Width (HPBW) of the directional panel antenna on the ground has 8 degrees for both vertical

and horizontal patterns, MAB is very sensitive on antenna alignment accuracy. Even a small tracking delay could lead the system bandwidth to drop drastically or be disconnected. The gap in Figure 15 is caused by control core "step" into the logic confliction region while tracking, in which rule is formed by random set fuzzy surface as neural networks just starting the training process. Even though the fuzzy surface would be trained appropriately for flight mission as time goes on, there are still some regions that never get updated which hide the potential output risk for the whole tracking system. On the other hand, the proposed rule formation methods with predesigned fuzzy surface have capability of being less time consuming on rule training and maintain coherent output result. This sustains MAB throughput for 4 Mbps more than the determined bandwidth threshold. This is shown in Figure 17. The steady MAB throughput, as shown in Figure 17, is therefore qualified for relaying telecom signal

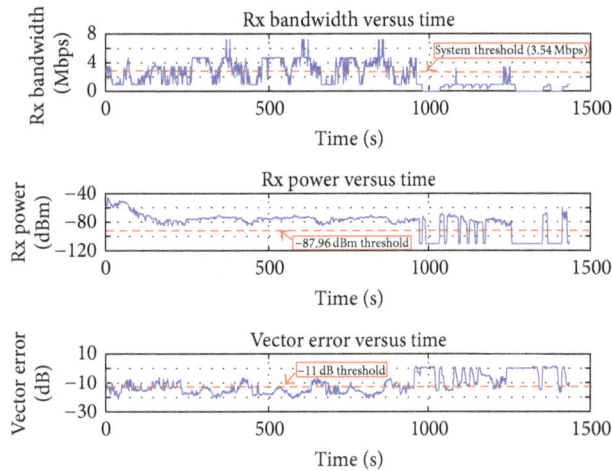

FIGURE 16: System bandwidth, Rx power, and vector error of MAB (random set fuzzy surface).

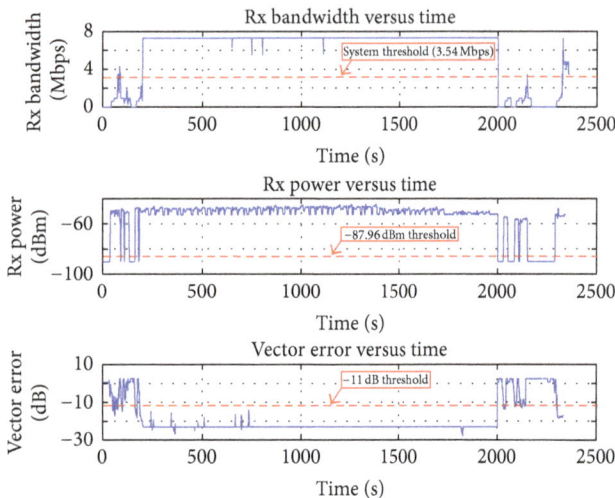

FIGURE 17: System bandwidth, Rx power, and vector error of MAB (predesigned fuzzy surface).

with E1 format. This is the key technique support for further applications with airborne *e-Cell* BTS in the Sky-Net project.

4. Conclusion

The Sky-Net project proposes a temporary airborne mobile communication service in disaster stroke area. A ULAFP is used to carry *e-Cell* BTS to relay mobile service via microwave transceiver for general users to the backbone network of the mobile communication providers. With the demand on high throughput and steady microwave wireless link, directional antenna is adopted for ground tracking device to establish reliable MAB between flying platform and ground E1. It is designed for aiming airborne antenna on ULAFP with circular and linear polarization, separately. In consideration of maintaining airborne antenna polarization during flight missions, an airborne self-stabilized gimbal is

introduced for ensuring the attitude of antennas regardless of ULAFP's dynamic motion. Meanwhile, performance of ground antenna tracking mechanism is planned to establish microwave link adapting different tracking scenario and ensure the control output being coherent. Therefore, an ANFIS with modified rule formation layer is implemented into the control core. Under the proposed control method, tracking result can maintain logical coherence and achieve accurate directional antenna alignment to mobile target. Both airborne stabilization and ground antenna tracking mechanism performance are verified via flight test in this paper. Test results have strongly been proven with the receiving bandwidth and power for MAB in high quality and good integrity by link status data log. This is the evidence to support the proposed Sky-Net architecture being feasible and effective for relaying telecommunication signal through ULAFP from the ground infrastructure of mobile system provider into designed areas.

Conflict of Interests

The authors declare that there is no conflict of interests regarding the publication of this paper.

Acknowledgment

This work is supported by National Science Council for Sky-Net Development under Contract NSC 102-2221-E-006-081-MY3.

References

[1] A. K. Widiawan and R. Tafazolli, "High Altitude Platform Station (HAPS): a review of new infrastructure development for future wireless communications," *Wireless Personal Communications*, vol. 42, no. 3, pp. 387–404, 2007.

[2] Q. Feng, J. McGeehan, and A. R. Nix, "Enhancing coverage and reducing power consumption in peer-to-peer networks through airborne relaying," in *Proceedings of the IEEE 65th Vehicular Technology Conference (VTC-Spring '07)*, pp. 954–958, April 2007.

[3] D. Grace, J. Thornton, G. Chen, G. P. White, and T. C. Tozer, "Improving the system capacity of broadband services using multiple high-altitude platforms," *IEEE Transactions on Wireless Communications*, vol. 4, no. 2, pp. 700–709, 2005.

[4] Z. E. O. Elshaikh, M. R. Islam, A. F. Ismail, and O. O. Khalifa, "High altitude platform for wireless communications and other services," in *Proceedings of the 4th International Conference on Electrical and Computer Engineering (ICECE '06)*, pp. 432–438, December 2006.

[5] H. Hariyanto, H. Santoso, and A. K. Widiawan, "Emergency broadband access network using low altitude platform," in *Proceedings of the International Conference on Instrumentation, Communication, Information Technology, and Biomedical Engineering (ICICI-BME '09)*, pp. 1–6, November 2009.

[6] A. Qiantori, A. B. Sutiono, H. Hariyanto, H. Suwa, and T. Ohta, "An emergency medical communications system by low altitude platform at the early stages of a natural disaster in Indonesia," *Journal of Medical Systems*, vol. 36, no. 1, pp. 41–52, 2012.

[7] W. Guo, C. Devine, and S. Wang, "Performance analysis of micro unmanned airborne communication relays for cellular networks," in *Proceedings of the 9th International Symposium on Communication Systems, Networks & Digital Signal Processing (CSNDSP '14)*, pp. 658–663, IEEE, Manchester, UK, July 2014.

[8] D. Grace, J. Thornton, C. Spillard, T. Konefal, and T. C. Tozer, "Broadband communications from a high-altitude platform: the european helinet programme," *Electronics and Communication Engineering Journal*, vol. 13, no. 3, pp. 138–144, 2001.

[9] E. Falletti, M. Laddomada, M. Mondin, and F. Sellone, "Integrated services from high-altitude platforms: a flexible communication system," *IEEE Communications Magazine*, vol. 44, no. 2, pp. 85–94, 2006.

[10] R. Miura and M. Suzuki, "Preliminary flight test program on telecom and broadcasting using high altitude platform stations," *Wireless Personal Communications*, vol. 24, no. 2, pp. 341–361, 2003.

[11] M. Quigley, M. A. Goodrich, S. Griffiths, A. Eldredge, and R. W. Beard, "Target acquisition, localization, and surveillance using a fixed-wing mini-UAV and gimbaled camera," in *Proceedings of the IEEE International Conference on Robotics and Automation (ICRA '05)*, pp. 2600–2606, April 2005.

[12] M. M. Abdo, A. R. Vali, A. R. Toloei, and M. R. Arvan, "Stabilization loop of a two axes gimbal system using self-tuning PID type fuzzy controller," *ISA Transactions*, vol. 53, no. 2, pp. 591–602, 2014.

[13] S. O. H. Madgwick, A. J. L. Harrison, and R. Vaidyanathan, "Estimation of IMU and MARG orientation using a gradient descent algorithm," in *Proceedings of the IEEE International Conference on Rehabilitation Robotics (ICORR '11)*, pp. 1–7, July 2011.

[14] T. Takagi and M. Sugeno, "Fuzzy identification of systems and its applications to modeling and control," *IEEE Transactions on Systems, Man and Cybernetics*, vol. 15, no. 1, pp. 116–132, 1985.

[15] C. E. Lin, T.-Y. Lu, H.-Y. Chen, and Y.-H. Lai, "System reengineering in mechanism design and implementation for unmanned ultra-light (part I)," *Journal of Aeronautics, Astronautics and Aviation Series A*, vol. 46, no. 2, pp. 124–131, 2014.

[16] C. E. Lin, H.-Y. Chen, T.-Y. Lu, and Y.-C. Huang, "System reengineering in flight control verification for unmanned ultra-light (part II)," *Journal of Aeronautics, Astronautics and Aviation Series A*, vol. 46, no. 2, pp. 132–140, 2014.

14

Cooperative Optimization QoS Cloud Routing Protocol based on Bacterial Opportunistic Foraging and Chemotaxis Perception for Mobile Internet

Shujuan Wang,[1] **Long He,**[2] **and Guiru Cheng**[1]

[1]*Changchun University of Technology, Changchun 130012, China*
[2]*Sinopharm A-THINK Pharmaceutical Co. Ltd., Changchun 130012, China*

Correspondence should be addressed to Guiru Cheng; chengguiru@ccut.edu.cn

Academic Editor: James Nightingale

In order to strengthen the mobile Internet mobility management and cloud platform resources utilization, optimizing the cloud routing efficiency is established, based on opportunistic bacterial foraging bionics, and puts forward a chemotaxis perception of collaborative optimization QoS (Quality of Services) cloud routing mechanism. The cloud routing mechanism is based on bacterial opportunity to feed and bacterial motility and to establish the data transmission and forwarding of the bacterial population behavior characteristics. This mechanism is based on the characteristics of drug resistance of bacteria and the structure of the field, and through many iterations of the individual behavior and population behavior the bacteria can be spread to the food gathering area with a certain probability. Finally, QoS cloud routing path would be selected and optimized based on bacterial bionic optimization and hedge mapping relationship between mobile Internet node and bacterial population evolution iterations. Experimental results show that, compared with the standard dynamic routing schemes, the proposed scheme has shorter transmission delay, lower packet error ratio, QoS cloud routing loading, and QoS cloud route request overhead.

1. Introduction

Research of QoS (Quality of Services) guarantees that routing protocol has been the focus issue in mobile Internet [1]. In particular, some research staff has combined the bionics research results of chemotaxis system [2, 3] and bacterial foraging optimization [4] with the above problem. However, how to improve the QoS routing protocol effect of mobile Internet and Cloud platform has not been in-depth study.

On the one hand, Szymanski [5] presented a Constrained Multicommodity Maximum-Flow-Minimum-Cost routing algorithm, which could compute the maximum-flow routings for all smooth unicast traffic demands within the Capacity Region of a network subject to routing cost constraints. Berger et al. [6] proposed the interdomain traffic routing for dealing with the generalization of the shortest path and path-trading problem. A model of Cascading Failures on the Interdomain Routing System was proposed by Liu et al. [7]. The Virtual Internet Routing Lab platform was shown

in article [8], which was used in the training, education, or research of Internet service. A delay-guaranteed energy profile-aware routing (DEAR) algorithm proposed for a green Internet was proposed to resolve the limitations of energy profiles and delay guarantees in article [9].

On the other hand, a variant of the bacterial foraging optimization algorithm was proposed by Tan et al. [10], which included the time-varying chemotaxis step length and comprehensive learning strategy. Naveen et al. [11] formulated the network reconfiguration problem as nonlinear objective optimization problem. A new long term scheduling for optimal allocation of capacitor bank in radial distribution system was presented by Devabalaji et al. [12], which can minimize the system power loss to resolve the equality constraints. The quantitative behavior of one-dimensional classical solutions for a hyperbolic-parabolic system describing repulsive chemotaxis was investigated in article [13]. Based on the knowledge of fuzzy mathematics, microeconomics, and swarm intelligence, the author proposed the flexible QoS unicast routing scheme with QoS satisfaction degree and utility introduced [14]. The novel bacterial chemotaxis optimization method (BCO) to QoS multicast routing scheme was proposed by Yong [15]. Chen et al. [16] proposed the bacterial colony optimization algorithm based on bacterial chemotaxis.

On the basis of the above research results, we studied the bacterial foraging model of mobile Internet and presented the chemotaxis perception cooperative routing protocol. Finally, the QoS cloud routing protocol for mobile Internet was proposed.

The rest of the paper is organized as follows. Section 2 describes the bacterial foraging model. In Section 3, we design the chemotaxis perception cooperative routing protocol. In Section 4, we proposed the QoS cloud routing protocol for mobile Internet. Experiment results are given in Section 5. Finally, we conclude the paper in Section 6.

2. Bacterial Foraging Model for Internet

In the mobile Internet, we set each mobile node as a bacterium. The behavior of mobile nodes searching for the next hop data receiving nodes is defined as bacterial foraging. Mobile nodes movement process is defined as a bacterial motility. When k bacterium is moving, the target was searching in the subnetwork of current bacteria. Set the distance to the d_{bk}. Then, the different directions of the current crawling angle as α were chosen. When k bacterium is in the direction of the movement direction of the best individual in the subnetwork, the attraction force of the bacteria b_{af} is determined by the following formula:

$$d_{bk} = \frac{t_k \sqrt{v_k \sin \alpha}}{\tan \beta},$$

$$b_{af} = \left[b_p - \sum_{i=1}^{k} \frac{\sin \beta}{d_{bk}} \right]^{-\cos \alpha}. \tag{1}$$

Here, t_k denotes the peristalsis delay of k bacteria. v_k denotes the peristalsis speed of k bacteria. b_p denotes the overall dynamics of bacterial populations. β is the angle of antenna and signal direction.

The mobile Internet is a multidimensional space field of the bacteria. Data transmission route selection problem is decomposed into the problem of the bacteria on the multidimensional field. The food delivery process may be accomplished by a number of bacteria in the same field, and it may be completed by a new structural field S_F, which is derived from the bacterial assemblage from different fields, as defined in

$$V_{S_F} = [v_1, v_2, \ldots, v_n],$$

$$S_F = \begin{bmatrix} b_1 \\ b_2 \\ \vdots \\ b_n \end{bmatrix} V_{S_F}. \tag{2}$$

Here, V_{S_F} is the bacterial movement velocity in the structure field. v is the mobile speed of node. b is the amount of food carried by the bacteria. The amount of information transmitted is denoted by S_F for the mobile communication of the structure field.

The dynamic changes of the multifield and the structure of the bacteria in the process of the movement and feeding of the bacteria are shown in Figure 1. There is bacteria foraging in multidimensional field, when the same field can not meet the needs of food delivery and other fields through mapping to establish the route, thus generating the structure, and there are various fields to select the best bacteria to join the structure of the field. By this mapping mechanism, it can prevent the local best bacterial ability to limit the population creep performance but also can shorten the time of route update and improve the convergence precision of the searching. The mapping results could be connected with the Internet by combining with the wireless cooperative networks.

Among them, from the dynamic evolution of the multidimensional field to the structural field, the principle of bacterial opportunity to feed is used. In Figure 3, three multidimensional fields and two structural fields are given. When the bacteria foraging from M_{F1} and M_{F2} to M_{F3} could not satisfy the transmission requirements, the opportunity structure field would be constructed based on the optimal direction and speed of the multidimensional field. Therefore, there is the different bacteria structure, the food principle, and the food amount. For example, S_{F1} is composed of good bacteria from M_{F1} and M_{F2}. The food principle is the XOR operation between M_{F1} and M_{F2}. The food amount is the sum of the food scales of M_{F1} and M_{F2}. S_{F2} would be obtained by the fusion of S_{F1} and M_{F1}, M_{F2}, and M_{F3}. The bacteria in the field are dynamic, which could be selected and optimized according to the requirements of food and information transmission requirements in the multidimensional field. The structural parameters of the above structures are shown in the following formula:

$$b_k^t = b_k(\alpha, \beta, b_k),$$

$$S_{F1} = \text{sub}(M_{F1}) \cup \text{sub}(M_{F2}) \cup \text{sub}(M_{F3}) \sum_{I=1}^{K} b_k^t, \tag{3}$$

$$S_{F2} = \delta(M_{F1} \cap M_{F2} \cap M_{F3}) \cup \text{sub}(M_F) \cup S_{F1}.$$

Here, let b_k^t denote transfer peristalsis from the multidimensional field to the structural field of k bacteria. δ is opportunity foraging factor of bacteria, which is determined by the bacteria in real time and the attraction force.

Hence, the bacteria opportunity foraging algorithm in the mobile Internet described is as follows.

Algorithm 1 (BOF ($d_{bk}, b_{af}, \alpha, \beta, \delta$)).

  BOF function Initialization of each Bacteria

  α and β to d_{bk}

  b_p to b_{af}

  Searching for optimal bacteria in M_F

Multidimensional field

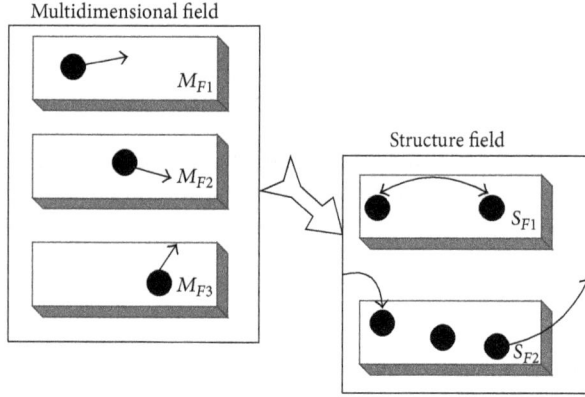

FIGURE 1: Multidimensional field and structure field of bacterial opportunistic foraging.

For $i = 1$ to k

{

Statistics V_{S_F};

Compute S_F;

$S_{F1} \leftarrow \text{sub}(M_{F1}) \cup \text{sub}(M_{F2}) \cup \text{sub}(M_{F3}) \sum_{I=1}^{K} b_k^t$;

$S_{F2} \leftarrow \delta(M_{F1} \cap M_{F2} \cap M_{F3}) \cup \text{sub}(M_F) \cup S_{F1}$}

return S_F

3. Chemotaxis Perception Cooperative Routing Protocol

In the process of bacterial foraging, the direction of bacterial motility and the structure of the field are characterized by chemotaxis. All the bacteria in the field can creep along the same direction and can also be spread to the food in the field of structural field after the food is carried and dispersed. The position of a trend of drug resistance is shown in the following formula:

$$b_k^{i+1} = b_k^i \left(\alpha, \beta, b_f \right) + \frac{\sum_{i=1}^{L} S_{F_i}}{\sqrt{\Delta S_F}}. \tag{4}$$

Here, ΔS_F is the update information of the structure field.

On the basis of chemotaxis perception, the behavior of the bacterial foraging population has the characteristics of dormancy, activation, and direction. When bacteria are crawling, the bacteria are perceived to be more resistant and opportunistically release the lure of the bacteria to the neighbor and move along the structural field for the next round. When the bacteria in the active state are all completed by the lure and the movement, the whole structure field becomes an active field, and the relationship between the

bacteria and the colony behavior of the two groups is shown in the following formula:

$$B_{S_F} \left(b_k, n, m \right) = \sum_{i=1}^{n} b_k^m B_{S_F}^i \left(b_k^i, i, m \right),$$

$$B_{S_F} = \sum_{j=1}^{m} \left[-\gamma \exp \left(\cos \left(|\alpha - \beta| \right)^2 \right) \right]^{m-j}. \tag{5}$$

Here, B_{S_F} denotes the bacterial population behavior of structure field. n is bacterial number of structure fields. m is the number of neighbors who have become more resistant to the bacteria. With the bacterial motility and structural field of the update, the attraction of the drug is gradually enhanced. Bacterial colony behavior becomes gradually mature. The position and direction of the bacteria are unified, as shown in the following formula:

$$B_{S_F}^{i+1} \left(b_k, i+1, m \right) = \cos \left(|\alpha - \beta| \right)^2 B_{S_F}^i \left(b_k^i, i, m \right),$$

$$b_k^{i+1} = b_k^i \left(\sin \alpha, \arctan \beta, \gamma \right). \tag{6}$$

After multiple iterations of the search, the driving of the chemotaxis and the total number of bacteria in the food are shown in the following formula:

$$T_{CH} \left(B_{S_F}^n \right) = \sum_{i=1}^{n} b_p^i \exp \left(\sin^2 \alpha - \arctan^2 \beta \right). \tag{7}$$

Here, T_{CH} denotes the total foraging amount of bacteria population. In order to make full use of the characteristics of the drug resistance to the bacteria, the mobile node route is established, and the angle parameters of the bacterial motility and the attraction force are calculated.

At the same time, in order to speed up the accumulation of bacteria and route establishment, the bacterial community would be sorted linearly and opportunistically according to the food principle and individual food amount. The bacteria with good performance would be produced through the drug resistance drive. In order to simplify the algorithm, in order to maintain the vitality of the population, the bacteria and the bacteria in the dormant state are consistent with the direction of the movement of the drug.

In the actual Internet communication environment, the transmission data of mobile nodes is affected by external forces or their resources are limited, and the process of routing communication based on bacterial chemotaxis may be destroyed. But the bacterial population is helping to find a more abundant area of food. Therefore, in view of the end-to-end communication, the mechanism of the drug delivery mechanism based on the bacterial cooperative control is beneficial to the bacterial foraging and population stability. In the cooperative routing algorithm, the individual behavior and the behavior of the bacteria are repeated after many iterations, and the bacteria can be a certain probability to the recombinant population or the structure field and will be the advantage of the spread of the food gathering area. The mobile node information fusion scheme can not only protect

TABLE 1: Internet settings.

Parameter	Value	Parameter	Value
Simulation time	2500 s	Perceived distance of mobile node	200 m
Internet area	100 m * 1200 m	Number of maximum active nodes	30
Concurrent service number	5	Packet size	1200 bytes
Cloud number	8	Packet sending speed	3 packets/s
Disk size of cloud	1 Tbytes	CPU number of cloud	4
Kernel operating frequency	2 GHz	Memory size of cloud	4 Gbytes

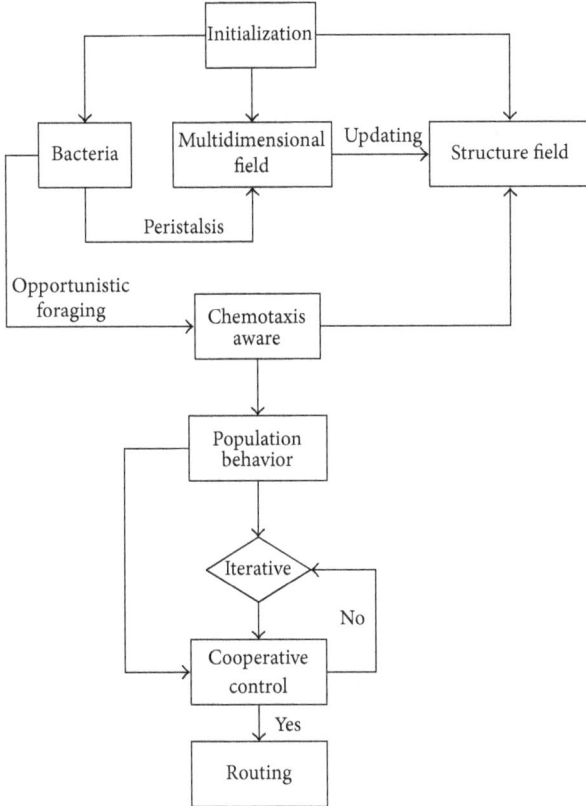

FIGURE 2: Route establishment based on bacterial opportunity to search for food and drug delivery.

the bacteria from the local optimal value but also enhance the global routing performance.

In summary, the establishment of a route based on bacterial opportunity foraging and the route of drug delivery is described in Figure 2.

4. QoS Cloud Routing Protocol for Mobile Internet

In the mobile Internet to join the cloud platform, the mobile node's bacterial opportunity foraging behavior can express the communication distance between the data sending node and the receiving node. Bacterial chemotaxis of mobile nodes can reflect the quality and stability of the communication link between the Internet and the cloud platform. In the

actual communication environment, by detecting the degree of attenuation of the bacterial foraging ability of the mobile nodes, the population behavior of the mobile node is analyzed, such as the direction and speed of the short time of the mobile node, and the probability of the interruption of the link is caused by the structural field update. In the cloud routing mechanism of QoS, the opportunistic optimization of the bacterial population is formed, which will lead to a large loss of multipath gain between the Internet and the cloud platform. Therefore, QoS cloud routing bacterial bionic optimization is based on multipath by establishing mechanism through hedge mapping between iterations in mobile Internet node physical layer of the antenna array and bacterial population evolution. QoS cloud routing path was selected and optimized.

The signal in the end-to-end communication with the bacterial opportunity foraging is shown in the following formula:

$$\text{Sig}_b = \sqrt{h_t^2 h_r^2} \frac{\sin^{\sqrt{G_t G_r}} \alpha}{b_{dk}^n}. \tag{8}$$

Here, G_t denotes antenna gain of sending node. G_r denotes antenna gain of receiving node. h_t and h_r denote the antenna height of the transmitter and the receiving nodes, respectively. If b_{dk}^n is larger than 1, the bacterial opportunity is not related to the attraction force of the population. In the QoS cloud routing request message in the bacterial population in the process, the activation of the neighbor bacteria opportunity type according to the movement speed and load information of the opportunity to calculate the cloud path of the opportunity weight W is shown in the following formula:

$$w_k = \prod_{k \in S_F} \left(\delta \frac{\rho - b_{dk}}{\rho} \right) \tan \beta. \tag{9}$$

Here, the evolution of the bacteria will occur, which belong to the same field. ρ denotes the peristalsis bacteria foraging radius of node.

As shown in Figure 3, bacteria nodes sent out a signal of the Internet data transmitting. After the opportunity for foraging peristalsis two multidimensional cloud paths are formed in the cloud platform. By formula (9), the opportunity weights of two multidimensional field cloud paths are $w1$ and $w2$. Here, $w1$ is less than $w2$. Although number of active bacteria nodes of dotted cloud routing path is less than one of the solid cloud path, the solid cloud path has more attractive

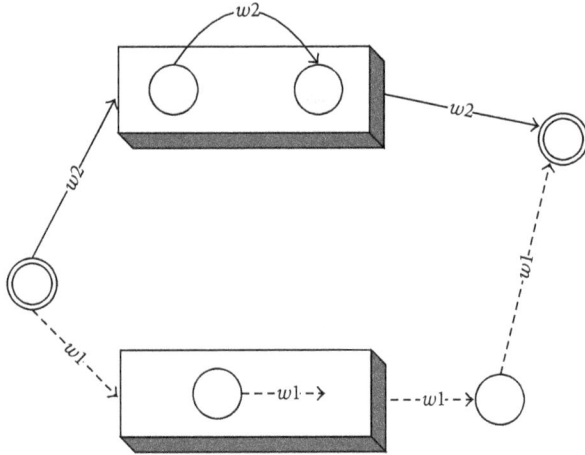

FIGURE 3: QoS cloud path selection.

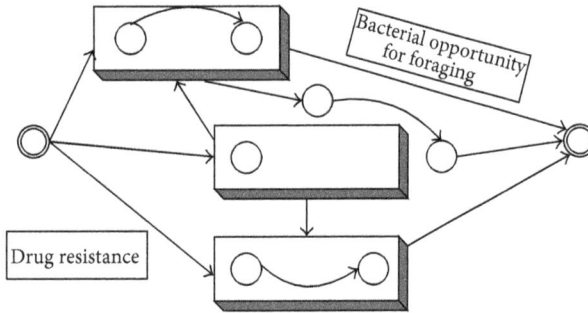

FIGURE 4: Process of QoS cloud route discovery.

force. In order to avoid breaking the path of the cloud, the current QoS cloud path should be the solid path.

QoS cloud path of cooperative maintenance process is shown in Figure 4. The RREP Format in DSR (Dynamic Source Routing) Options Header is shown in Figure 5. The bacterial neighbor node receives the request message from the data sending end node to the cloud path request message, and the bacteria spread the cloud path request to the bacteria population in the same multidimensional field through the opportunistic feed. There are multiple paths of mobile nodes in the bacterial population through collaborative construction, to maintain the cloud path through the computer.

In order to eliminate bacterial opportunity to feed into the dead cycle, the cooperative maintenance of the QoS cloud path requires the maintenance of the individual in the end to end cloud path of the chemotaxis and motility of the bacterial node sequence, to maintain its monotonic increasing characteristics, as shown in the following formula:

$$\left(-S_{F_{i+1}}, B_{S_F}^i\left(b_k^i, i, m\right)\right) > \left(S_{F_i}, B_{S_F}^{i-1}\left(b_k^{i-1}, i-1, m\right)\right)$$

$$R_{bd} = \sum_{i=1}^{L} S_{F_i} \frac{\cos \alpha}{\sqrt{\Delta S_F}}. \tag{10}$$

5. Performance Evaluation

In this section, we analyzed and evaluated the performance of the proposed COQCRP (Cooperative Optimization QoS Cloud Routing Protocol based on bacterial opportunistic foraging and chemotaxis perception) routing protocol. Two groups of simulation experiments were designed. The DRP (Dynamic Routing Protocol) with the Internet was compared with the proposed protocol, respectively, based on the mobile speed of the nodes and the size of the Internet in the case of the QoS performance of the Internet. Table 1 records the Internet parameter settings.

In order to evaluate the performance of COQCRP, four performance evaluation criteria of QoS cloud routing are statistically analyzed.

End-to-End QoS Cloud Communication Delay. It is the time interval of data packets transmitted from the Internet node to destination node through cloud.

QoS Cloud Route Load. It is the control packet number of QoS routing cloud cooperative optimizations divided by the number of correctly received data packets.

QoS Cloud Routing Overhead. It is the number of COQCRP cloud routing requests sent or forwarded in the unit time.

In Experiment 1, the analysis of the Internet based on the mobile speed of the mobile node of the mobile speed of the QoS cloud routing performance, the results are shown in Figure 6. The meaning of coordinates x and y in Figure 6(a) is mobile speed of node and average end-to-end delay. The meaning of coordinates x and y in Figure 6(b) is mobile speed of node and packet error ratio.

The COQCRP significantly improves the end-to-end cloud communication delay, as shown in Figure 6(a). The reason is that, based on QoS cloud, opportunistic bacterial foraging by selection mechanism improves the robustness of the path of the cloud and relationship of cloud routing size for the maintenance delay. COQCRP controlled the message broadcast of cloud path request cooperatively, which improved the Internet link bandwidth utilization and avoided the cloud computing error, as well as the data packet collision probability. With the increase of the node moving speed, the performance of COQCRP and DRP is similar. This is due to the probability of packet collision and the speed of the node of the Internet is proportional to the speed of the node. In Figure 6(b), the simulation results show the effect of the change of node moving speed on the packet loss rate. With the increase of the node moving speed, the path break frequency increases and the packet loss rate increases. COQCRP data transmission path quality is improved. COQCRP can provide an effective guarantee for data packet transmission and packet error ratio is lower than DRP.

In Experiment 2, the analysis of the impact of the scale of the Internet on the QoS cloud routing performance is shown in Figure 7. The meaning of coordinates x and y in Figure 7(a) is cloud scale and routing load. The meaning of coordinates x and y in Figure 7(b) is cloud scale and route request overhead.

32 bits			
Opportunity foraging weight	Chemotaxis perception	Length of multidimensional field	Length of structure field
Address of sending node			
Cooperative nodes			4–10 bytes
Address of receiving node			

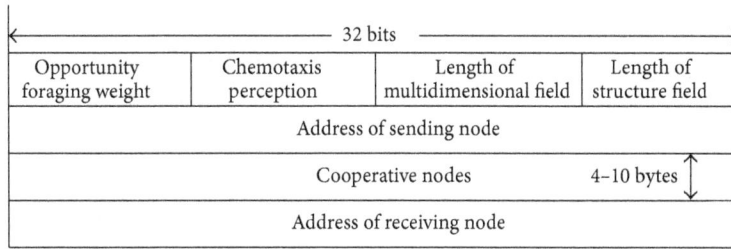

FIGURE 5: The RREP format in DSR options header.

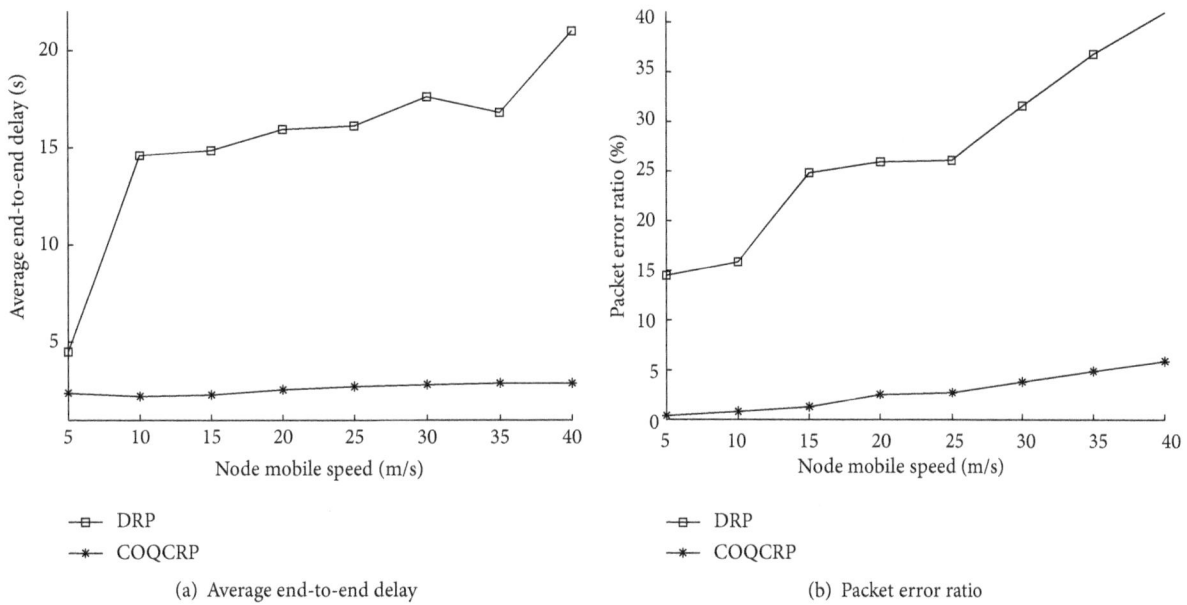

(a) Average end-to-end delay

(b) Packet error ratio

FIGURE 6: Performance analysis with speed.

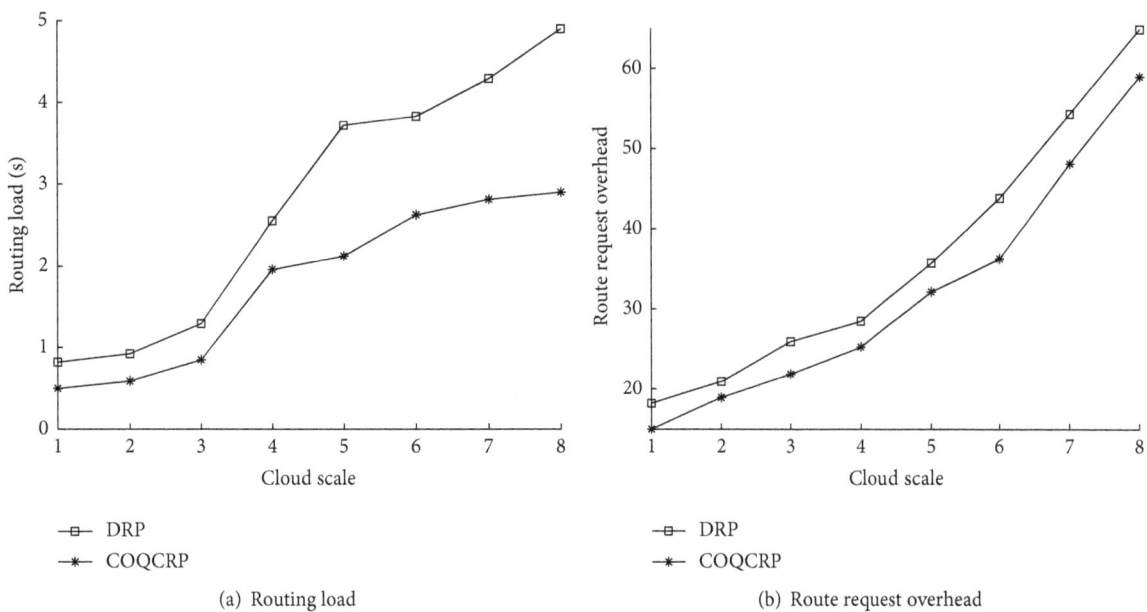

(a) Routing load

(b) Route request overhead

FIGURE 7: Performance analysis with cloud scale.

Due to the extension of the path lifetime and COQCRP QoS cloud routing load (Figure 7(a)) and cloud routing request overhead (Figure 7(b)) that significantly reduced, COQCRP effectively saved the Internet resources and improved the work efficiency of the cloud platform.

6. Conclusions

According to the mobility management requirements of mobile Internet and cloud platform resource constrained characteristics, a chemotaxis perception cooperative QoS cloud routing optimization mechanism was proposed based on the opportunistic bacterial foraging bionics. First, we assume that each mobile node of the mobile Internet is a bacterium. A data forwarding node group was established through the progress of the bacterial opportunistic feed and the characteristic of bacterial motility. Secondly, based on the analysis of the direction of the movement of the bacteria and the characteristics of the trend of the structural field, the establishment of the mechanism of the drug resistance is established by the distributed load control of all the bacteria in the field. Finally, the cloud platform joined the mobile Internet. We combined the mobile node opportunistic bacterium foraging behavior and bacterial chemotaxis. A cooperative optimization QoS cloud routing mechanism was proposed to guarantee the quality and stability of the communication link between the Internet and cloud platform. The experimental results show that the proposed mechanism has the advantages of real-time, reliability, routing load, and resource utilization of cloud platform in QoS communication.

Conflict of Interests

The authors declare that there is no conflict of interests regarding the publication of this paper.

References

[1] M. S. Siddiqui, D. Montero, R. Serral-Gracià, X. Masip-Bruin, and M. Yannuzzi, "A survey on the recent efforts of the internet standardization body for securing inter-domain routing," *Computer Networks*, vol. 80, pp. 1–26, 2015.

[2] Q. S. Zhang and Y. X. Li, "Global existence and asymptotic properties of the solution to a two-species chemotaxis system," *Journal of Mathematical Analysis and Applications*, vol. 418, no. 1, pp. 47–63, 2014.

[3] C.-H. Huang and P. A. Iglesias, "Cell memory and adaptation in chemotaxis," *Proceedings of the National Academy of Sciences of the United States of America*, vol. 111, no. 43, pp. 15287–15288, 2014.

[4] I. A. Mohamed and M. Kowsalya, "Optimal size and siting of multiple distributed generators in distribution system using bacterial foraging optimization," *Swarm and Evolutionary Computation*, vol. 15, pp. 58–65, 2014.

[5] T. H. Szymanski, "Max-flow min-cost routing in a future-internet with improved QoS guarantees," *IEEE Transactions on Communications*, vol. 61, no. 4, pp. 1485–1497, 2013.

[6] A. Berger, H. Roglin, and R. van der Zwaan, "Internet routing between autonomous systems: fast algorithms for path trading," *Discrete Applied Mathematics*, vol. 185, pp. 8–17, 2015.

[7] Y. J. Liu, W. Peng, J. S. Su, and Z. Wang, "Assessing the impact of cascading failures on the interdomain routing system of the internet," *New Generation Computing*, vol. 32, no. 3-4, pp. 237–255, 2014.

[8] J. Obstfeld, S. Knight, E. Kern, Q. S. Wang, T. Bryan, and D. Bourque, "VIRL: the virtual internet routing lab," *ACM SIGCOMM Computer Communication Review*, vol. 44, no. 4, pp. 577–578, 2014.

[9] E.-J. Lee, Y.-M. Kim, and H.-S. Park, "DEAR: delay-guaranteed energy profile-aware routing toward the green internet," *IEEE Communications Letters*, vol. 18, no. 11, pp. 1943–1946, 2014.

[10] L. J. Tan, F. Y. Lin, and H. Wang, "Adaptive comprehensive learning bacterial foraging optimization and its application on vehicle routing problem with time windows," *Neurocomputing*, vol. 151, no. 3, pp. 1208–1215, 2015.

[11] S. Naveen, K. S. Kumar, and K. Rajalakshmi, "Distribution system reconfiguration for loss minimization using modified bacterial foraging optimization algorithm," *International Journal of Electrical Power and Energy Systems*, vol. 69, pp. 90–97, 2015.

[12] K. R. Devabalaji, K. Ravi, and D. P. Kothari, "Optimal location and sizing of capacitor placement in radial distribution system using bacterial foraging optimization algorithm," *International Journal of Electrical Power & Energy Systems*, vol. 71, pp. 383–390, 2015.

[13] D. Li, R. Pan, and K. Zhao, "Quantitative decay of a one-dimensional hybrid chemotaxis model with large data," *Nonlinearity*, vol. 28, no. 7, pp. 2181–2210, 2015.

[14] X. Wang, D. Jiang, and M. Huang, *Flexible QoS Unicast Routing Scheme Based on Utility and BCC [EB/OL]*, Sciencepaper Online, Beijing, China, 2010.

[15] H. Yong, "QOS multicast routing algorithm based on bacterial chemotaxis optimisation," *Computer Applications and Software*, vol. 29, no. 11, pp. 269–271, 2012.

[16] L. Chen, L.-Y. Zhang, Y.-J. Guo, J.-Q. Zhao, and Q. Li, "Sequential blind signal separation algorithm based on bacterial colony chemotaxis," *Journal on Communications*, vol. 32, no. 4, pp. 77–85, 2011.

15

Communication Behaviour-Based Big Data Application to Classify and Detect HTTP Automated Software

Manh Cong Tran and Yasuhiro Nakamura

Department of Computer Science, National Defense Academy, 1-10-20 Hashirimizu, Yokosuka, Kanagawa 239-0811, Japan

Correspondence should be addressed to Manh Cong Tran; manhtc@gmail.com

Academic Editor: Jun Bi

HTTP is recognized as the most widely used protocol on the Internet when applications are being transferred more and more by developers onto the web. Due to increasingly complex computer systems, diversity HTTP automated software (autoware) thrives. Unfortunately, besides normal autoware, HTTP malware and greyware are also spreading rapidly in web environment. Consequently, network communication is not just rigorously controlled by users intention. This raises the demand for analyzing HTTP autoware communication behaviour to detect and classify malicious and normal activities via HTTP traffic. Hence, in this paper, based on many studies and analysis of the autoware communication behaviour through access graph, a new method to detect and classify HTTP autoware communication at network level is presented. The proposal system includes combination of MapReduce of Hadoop and MarkLogic NoSQL database along with xQuery to deal with huge HTTP traffic generated each day in a large network. The method is examined with real outbound HTTP traffic data collected through a proxy server of a private network. Experimental results obtained for proposed method showed that promised outcomes are achieved since 95.1% of suspicious autoware are classified and detected. This finding may assist network and system administrator in inspecting early the internal threats caused by HTTP autoware.

1. Introduction

Application layer attacks pose an ever serious threat to network security for years since it always comes after a technically legitimate connection has been established. Because of the flexibility and interoperability of HTTP since everything users need can be found through web services, its based communication is always allowed in most of network. Consequently, HTTP-based automated software (autoware) is blooming in utilizing in reaching Internet users. Unfortunately, besides normal autoware such as for operating system or software updating purpose, in recent years, cyber criminals turn to fully exploit web as a medium of communication environment to lurk a variety of forbidden or illicit activities through spreading HTTP malicious autoware such as fraudulent adware, spyware, or bot. HTTP traffic and autoware can be classified in some categories as in Figure 1:

(i) Human traffic is kind of traffic which is generated by users with their intention when they use normal software such as web browser to access their websites to get information they needed. In this kind of traffic, users clearly understand their accessed sites, who they contact to, and which information they obtain.

(ii) On the other side, the graph presents nonhuman traffic to which users unintentionally have access; they come from autoware. This traffic can be requested from normal software such as antivirus updater, mail client, browser's toolbar, greyware encompasses adware, spyware, joke programs, and malicious software acting as HTTP-based botnet and trojan horses.

Normal autoware can be controlled and beneficial for user; however, since greyware and malicious software penetrate into users' network, they turn out to be internal threats, from which attackers can conduct various types of application layer attacks through these agents, which are really difficult to prevent such as DoS/DDoS, malware distribution, or identity theft. The distinction between malicious and normal activities from HTTP traffic is becoming tougher because the malicious requests merges adequately with legitimate HTTP

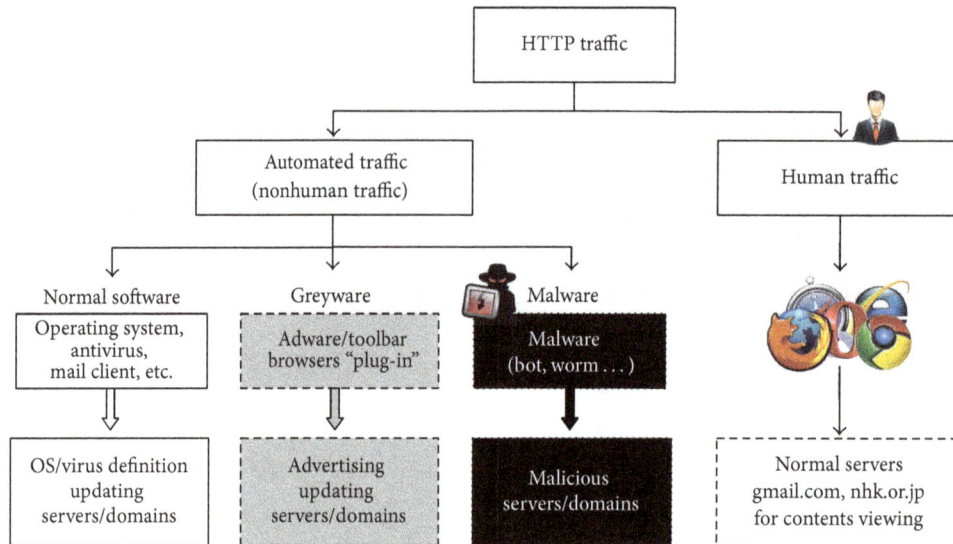

FIGURE 1: HTTP traffic and automated software categories.

traffic. Furthermore, in a large private network, detection and also classification between types of HTTP autoware traffic are really great challenge when huge requests are generated each day.

To maintain communication, perform updates, or receive commands, all kinds of HTTP-based autoware have common characteristics that they generate repetitively legal traffic and requests to their servers/domains. However, in detail, there are some sophisticated differences in the way of communication behaviour of autoware to their sites. In this paper, based on the analysis and study of autoware communication behaviour, a method in classification and detection of HTTP autoware at network level is proposed. To overcome the issue of handling huge of traffic each day, a big data based system proposal is implemented. In that, a combination between MapReduce of Hadoop [1] and MarkLogic NoSQL database [2] with xQuery supported [3] is suggested for experiment. The method is experimented with real traffic data generated from a university network, and a promised result is archived in classification and detection of malicious HTTP autoware communication.

The remainder of the paper is organized as follows. Related work is discussed in Section 2. In Section 3, features extraction and terminology which included autoware communication behaviour analysis and core terminologies are presented. Section 4 is about detailed description of proposed method which includes algorithms and all components responsible. Section 5 presents applied big data application, the evaluation for proposed method, and experiment results. Finally, conclusion and future work are summarized in Section 6.

2. Related Work

There were a considerable number of techniques which aim to protect users against malware; however, it continues to be a challenging problem. Traditional defense mechanisms such

as antivirus (AV) products are the most common content-based malware detection techniques. These types of AV software run on end-user systems and employ signature-based detection to identify variants of known malware. As a consequence, the signature generation and update cycle cause an inherent delay in protecting users against new variants of malware [4]. Additionally, with the aim of limiting AV engines effectiveness, malware authors have developed increasingly sophisticated evasion techniques such as packing and polymorphism, aimed at circumventing detection by AV engines [5, 6]. Oberheide et al. [7] figure many undetected malware binaries by using signature-based techniques, and major AV engines just detect only 30% to 70% of recent malware. As the same content, Rajab et al. [4] show that less than 40% of malicious binaries can be detected by four AV engines in their experiment.

Many botnet detection methods are presented in [8–11]. Ashley [8] has suggested a method for detecting potential HTTP C&C activity based on repeated HTTP connections to a website. According to this, an algorithm is proposed for detecting HTTP polling activity. Lu et al. in [9], using signature-based techniques, propose a hierarchical framework to automatically discover malicious bot on a large-scale Wi-Fi ISP network, in which the network traffic is classified into different application communities by using payload-signature. These signatures were used to separate known traffic from unknown traffic in order to decrease the false alarm rates. Eslahi et al. [10] proposed an approach to reduce the false alarm HTTP botnet detection; in this research, high access rate traffic, which might be other security threats, is filtered out. Basil AsSadhan and Moura [11] proposed a detection method in which it concentrates in C&C communication analysis and find that it exhibits a periodic behaviour. In [11], a method which applied discrete time series is analyzed to examine the aggregate traffic behaviour in order to detect botnet C&C communication channels traffic. These researches [8–11] focus on botnet communication to C&C server, but actually HTTP threats do not just come from malicious bots but also can be from other types of automated software such as HTTP spyware, adware, or unauthorized applications.

Shin et al. in [12] proposed a framework to detect bot malware at host and network level. At host level, they monitor human-process interactions by using hook technique to capture user mouse and keyboard activities. These hook actions might affect users PC systems. At network level, a simple way to prevent a malware infected PC sending out the information is to prevent all the direct TCP/IP connection from clients. However allowing HTTP protocol is really leaking hole which might be exploited by HTTP malware. In [12], to overcome this issue, they monitored DNS queries to determine C&C server, but actually, many botnets use hacked URL as C&C server. Therefore, the detection method might be insufficient.

Some of approaches use lexical features or keywords extracted from URL and web contents as in [13–16]. However, many other types of malicious web pages are disguised by domain names or URLs like normal website and can harm users PC systems. In this case, lexical or keywords features might be compromised. Bartlett et al. [17] proposed an approach to identify low-rate periodic network traffic and changes in regular communication of autoware. Their research also focuses on many types of autoware and monitor TCP flows to detect, but, in this paper, the target does not just focus only on detecting general types of autoware but also on particular URLs where autoware request to. In addition, our method just collects and processes with basic features of HTTP Traffic at application layer. This will help reduce process cost compared with method used TCP packets features since the number of packets to be processed increases.

3. Features Extraction and Terminology

In this paper, classification and detection method is based on autoware communication behaviour. For that target, by observation of HTTP traffic, autoware communication is analyzed, from which beneficial features are extracted in order to classify and detect various types of autoware. In this section, background related contents and also core terminologies are presented.

3.1. Features Extraction. HTTP traffic from a client consisted of many requests from that client to outside. At application layer, a request includes basic information: IP address of client, full URL, and request method. Full URL's parts contain webpage/server URL and parameter path, as shown in Figure 2. At network level, numerous features are extracted which are made from basic client requests information as follows:

(i) Client IP: source IP address of machine in network which generated requests.

(ii) Request method: main methods of HTTP requests, POST/GET.

(iii) Request date time: date and time when a client sends request.

(iv) Webpage/server URL (shorten as URL): URL requested by a client IP but without parameters' part,

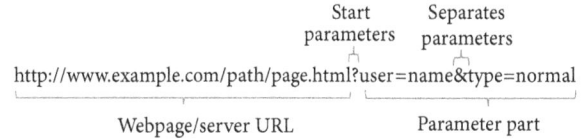

FIGURE 2: Main parts of URL.

as shown in Figure 2. Some normal web servers are hacked and some of their resource paths are exploited as C&C servers. Additionally, parameter parts are easily changed based on the specification of requests content, but actually the functionality of that webpage/server URL, such as C&C server or advertise content update, is the same in each request. Therefore, nonparameter URL is used instead of domain or full URL (will be parameter part), and this matter will help the classification of autoware access behaviour become more detailed and accurate.

(v) Unique URL: set of unique URLs requested by a client.

(vi) Request interval: break time between two consecutive requests to the same URLs.

(vii) Request count: number of requests to URL from a client in a period of observation data.

(viii) Access time: a period of time in seconds during which a client accessed to URL from the fist request to the end request.

3.2. Access Graph. Access graph presents communication behaviour of a client to a specific URL in a duration of time. It is formed on request interval which are extracted from HTTP traffic. Assuming that $R = \{r_1, r_2, \ldots, r_N\}$ is set of requests from a client to a webpage/server and all r_i have the same webpage/server URL, as described in Figure 2, then access graph G is a sequence which included $N - 1$ items, $G = \{g_1, g_2, \ldots, g_{N-1}\}$, where g_i is a pair of (t_i, d_i), where t_i is timing of request r_{i+1} and d_i is request interval between r_i and r_{i+1}. An access graph is shown as in Figure 3, in which, X-axis is timing of request (except the first request) and Y-axis shows the request interval value in second. An installed or infected autoware client will establish a different access graph for each URL which it sends requests to. For that, this graph can present the behaviour in communication between an autoware to its webpage or server URLs.

3.3. Autoware Communication Behaviour. For keeping communication, update or receive command, all kinds of HTTP-based autoware have common characteristics that they generate repetitively legal traffic and requests to their servers/domains. However, in detail, there are some sophisticated differences in the way of communication behaviour of autoware to their sites.

(i) Malicious HTTP-based bots always follow the PULL style where they connect to their command and control server periodically in order to get the commands and updates. The number of requests from malicious bots are not high as normal autoware (e.g., updater

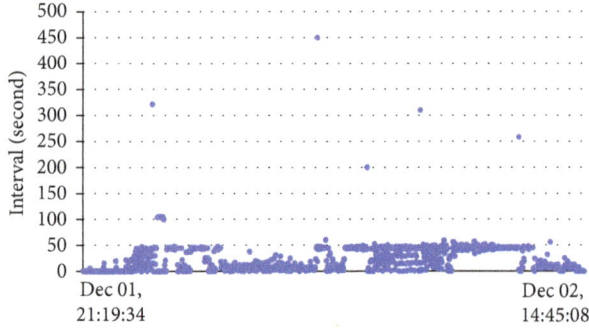

FIGURE 3: An access graph of a client request to URL.

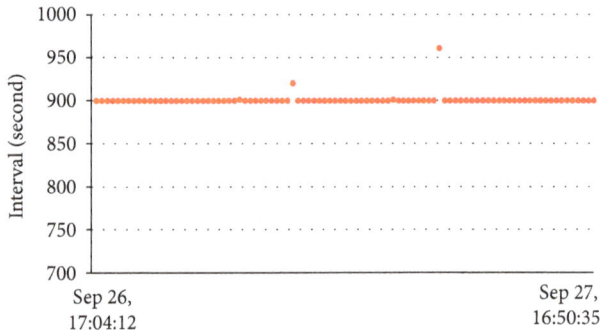

FIGURE 4: An access graph of HTTP malicious bot.

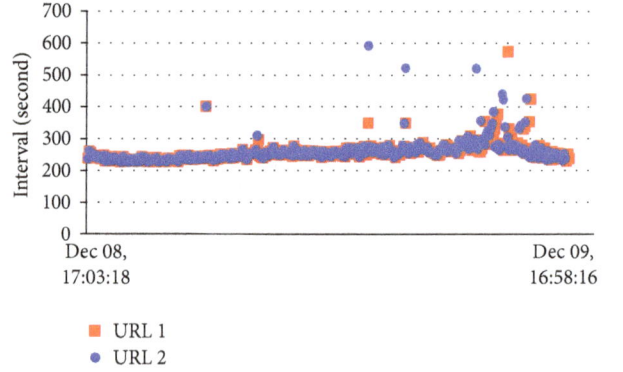

FIGURE 5: Access graphs from an autoware of a client IP to two different URLs are similar, and the access times of URLs are the equal since both of them are requested from Dec 08, 17:03:18, to Dec 09, 16:58:16.

recent years, many sites (e.g., shopping online site or social media webpage) append advertisement path to their sites and use JavaScript or Flash as autoaware part to automatically collect the advertising content as adware or spyware. Therefore parts of users access sites can generate HTTP traffic which act as autoware communication.

3.4. Access Graph Distance. As analysis in Section 3.3, even URLs are different; if they are requested by the same autoware then the access graphs look similar, as can be seen in Figure 5. This part proposes a distance to measure the similarity of autoware access behaviour in communication to URLs from a client. The calculation method is based on Modified Hausdorff (MH) distance which is presented in [19].

Assume that there are two access graphs $A = (a_1, \ldots, a_N)$ and $B = (b_1, \ldots, b_M)$. Define that the distance between two points a_i and b_j is calculated as Euclidean distance $d(a_i, b_j) = \|a_i - b_j\|$. From that, distance between point a_i and graph B is defined as $d(a_i, B) = \min_{b_j \in B} \|a_i - b_j\|$. Generalized Hausdorff distance of A and B in [19, 20] is defined as follows:

$$d(A, B) = \frac{1}{N} \sum_{a_i \in A} d(a_i, B). \tag{1}$$

Based on (1), distance between access graphs A and B, which follow by MH distance (MHD), is formed as follows:

$$\mathrm{MHD}(A, B) = \max(d(A, B), d(B, A)). \tag{2}$$

The smaller the MH distance between A and B is, the more A and B are similar to each other.

3.5. Suspicious Score. As described in Section 3.3, malicious bots connect to their command and control server (C&C server) periodically in order to get the commands and updates; therefore, almost there is no large variation in the access graph from malicious bot to its C&C, as can be seen in Figure 4. Based on this analysis, a score is proposed to measure the variation of a access graph, from which it shows suspicious of communication between client to its URL.

and downloader) which just generate requests with a long interval than unusual malicious bots [10, 11, 18]. Because interval in communication between a malicious bot to their C&C server is stable, there is almost no variation in their access graph as can be seen in Figure 4 showing the access graph of a bot communication.

(ii) Malicious bots often connect to one control domain and to a specific server resource. Difference with that, unwanted HTTP applications, or greyware, such as annoying adware or spyware, often report back to or request new information from many external resources [17]. Therefore, they keep communicating to their numerous advertising sites or URLs to update pop-up or advertisement and commercial content areas. Autoware will behave the same communication pattern to its URLs if they are requested at the same or approximately equivalent timing so access graph of URLs from a specified autoware is looked similar. In addition, many URLs are requested with the same timing by a specified autoware, so the access duration to these URLs is approximately equal. It means that the first and the last requests timing to these URLs are the same with others. In Figure 5, a sample of two similar access graphs presents the communication from one autoware to two different URLs, and the first and the last requests moment of them are equal.

(iii) On the contrary with autoware, there are no interval or periodic patterns in users' web access; however, in

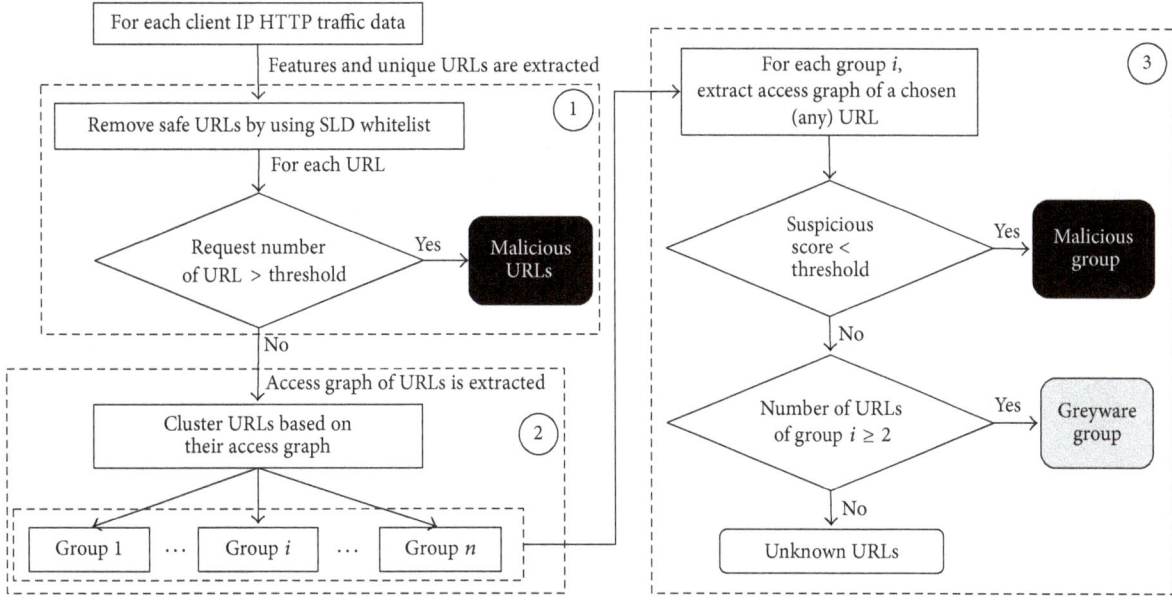

FIGURE 6: Proposed method diagram in classification and detection of HTTP automated software. Labels of 1, 2, and 3 are preprocesing, clustering, and detection/classification phase, respectively.

Assuming that the access graph of URL S is specified and denoted as $X = (x_1, \ldots, x_N)$, a suspicious score will be defined as *coefficient of variation* of X as follows:

$$\text{Suspicious Score}(X) = \frac{\sigma}{\mu} \qquad (3)$$

in which σ and μ are standard deviation and mean of X, respectively. The smaller suspicious score shows that URL is more suspicious.

4. Proposed Method

Based on the autoware communication behaviour which is described in Section 3 and the observation of access graphs in Section 3.1, a classification and detection method, including three phrases, is proposed as in Figure 6; details are as follows.

4.1. Preprocessing Phase. This preprocessing phase is objective to eliminate unnecessary processed data. For each client IP, the one-day HTTP traffic features are extracted and preprocessed; in order to process this phase two methods are applied:

(i) The first one is to filter URLs requests from client IP through a whitelist of second level domain names (SLDN). This filter method is described in [13]; according to that, the tokens in the URLs of phishing websites are less consistent with their content when compared with those of legal websites. An example is illustrated in Figure 7. In this example, the legitimate website contains the brand names *apple* in the SLDN. Even though the phishing website also contains the brand name *apple* in the URL, it is not in the SLDN. Therefore, a domain name which contains a second

Legitimate URL

https://**secure1.store.apple.com**/au/shop/sign_in

Second level domain name

Phishing URL   Phishing position   Second level domain name

*http://**secure1.store.apple.com**.australia.peeie.projektenet.de/apache/include/jquery/i18n/cgisys/WebObjects/iTunesConnect.html (*http://phishtank.com)

FIGURE 7: Phishing websites are less consistent with their content when compared with those of legitimate websites.

level domain name which is defined in SLDN whitelist is marked as benign.

(ii) The second method is based on the number of requests to URL from a client IP. Based on the observations number of requests from autoware to URL, it can be seen that suspicious autoware has access many times to URL in a duration of time. Therefore, if the number of requests to URL is too small, it seems not to be requested by an autoware.

Also in this phase, URLs which are requested with extremely fast speed in a duration time will pose a malicious autoware communication; access speed is defined as follows:

$$\text{Access Speed}(\text{URL}_i) = \frac{\text{Request Count}(\text{URL}_i)}{\text{Access Time}(\text{URL}_i)}. \qquad (4)$$

In that access time and request count features are described in Section 3.1.

4.2. Clustering Phase. After preprocessing phase, in this phase, remaining URLs will be clustered into number of

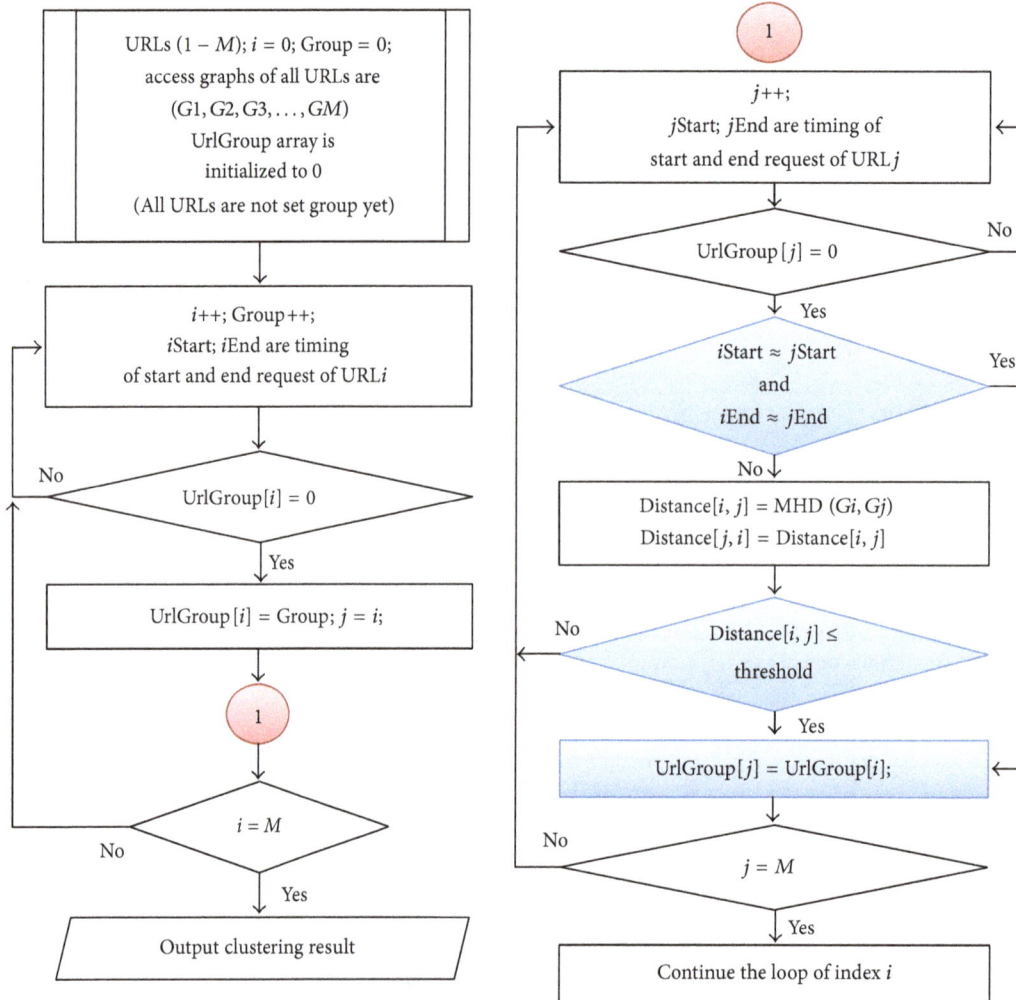

FIGURE 8: Autoware communication clustering algorithm.

groups based on their characteristics which are presented in Section 3.3. Accordingly, two URLs are of the same group (requested by the same autoware from a client) if they match one of following conditions:

(i) The first and the last request timing to two URLs are approximately the same.

(ii) Based on the similarity of its access graph, MH distance between two access graphs of URLs is calculated; if this distance is small enough, they will be recognized as in the same group.

An algorithm is suggested to decide a group for any two URLs. In order to optimize the consumption processing time of method, the steps of algorithm are proposed in Figure 8. By using a group label array, from this algorithm, distance between all pair of access graphs need not to be calculated. If URL is labeled to a group, it will not need to check group again with other URLs.

4.3. Detection and Classification Phase. The third phase is detection and classification. For each group, a URL (any in

the group) is chosen and its access graph is extracted. Then the suspicious score of this URL is calculated; in order to detect whether it is malicious or not a threshold is proposed as 0.04. If the suspicious score is less than or equal to the threshold it is detected as malicious. Finally, remaining groups will be detected by examining the number of unique URLs in group. As analyzed in Section 3, difference with malicious bots, greyware commonly access to various URLs instead of only one server or URL. Therefore, a group having number of unique URLs which are not less than 2 will be marked as greyware groups.

5. Big Data Proposed Framework and Experiment Results

5.1. Big Data Proposed Framework. In this paper, based on above proposed method, big data application is suggested to classify and detect autoware communication. Data for experiment are collected from web proxy of a certain network which served about 2000 clients. Collected data are divided by day saved into logs' file as raw data. Big data application is

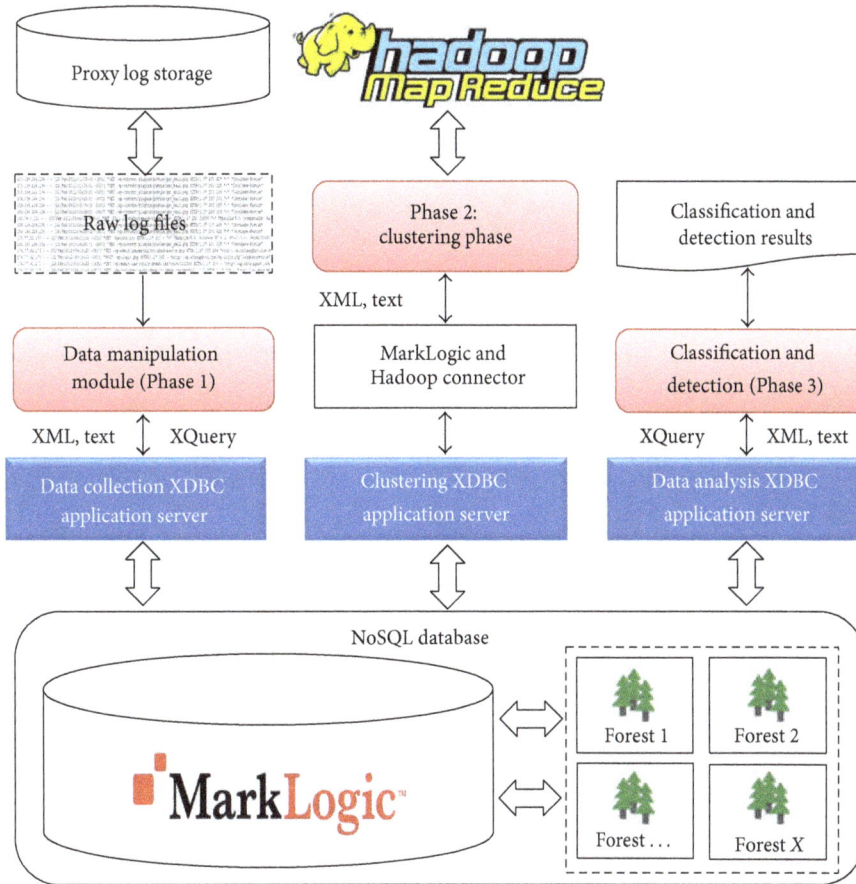

FIGURE 9: Big data based framework proposal.

composed by combination of MarkLogic database and MapReduce of Hadoop.

As described in [2, 21], MarkLogic is an enterprise NoSQL (Not only Structured Query Language) database which supports a very flexible and convenient XQuery when working with structured and also unstructured data. In addition, it also has had ACID transactions (ACID stands for Atomicity, Consistency, Isolation, and Durability). In a transactional application ACID's properties are necessary so that reads and writes are durably logged to disk and strongly isolated from other transactions. Without this feature, users run the risk of encountering data corruption, stale reads, and inconsistent data. In this framework, XML and text data format are suggested to use because of easily transforming from raw data log file into database.

Hadoop is a great tool to help database application developers and organizations to store and analyze massive amounts of structured and unstructured data from disparate data sources, of which data are too massive to manage effectively with traditional relational databases. Hadoop has become popular because it is designed to cheaply store data in the Hadoop Distributed File System (HDFS) and run large-scale MapReduce jobs for batch analysis. MapReduce is a processing framework that uses a divide-and-conquer paradigm that takes a huge task and breaks it into small parts (Map) and then aggregates the resulting outputs from each

part (Reduce). Any large task that can be broken into smaller pieces is a candidate for use with Hadoop [2].

The combination between MarkLogic database and MapReduce of Hadoop in this framework is described in Figure 9, whereby a cluster of MarkLogic is set, and due to optimizing performance in query to database, three XDBC application servers, Data Collection, Clustering, and Data Analysis, are configured along with a number of forests. There are three modules working independently for each phase in Figure 6; details are expressed as follows:

(i) Phase 1 is processed as a part in Data Manipulation Module which will read raw log files, convert to XML and text format, and do the preprocessing before being stored into MarkLogic database via Data Collection Application Server.

(ii) Core functions of heavy Phase 2, Clustering Phase, are implemented according to algorithm in Figure 8 and deployed in the middle part between MarkLogic database and MapReduce of Hadoop. This module will archive results from Phase 1, and URLs are clustered in MapReduce by the distributed processing paradigm. Finally, results of Phase 2 will be returned to MarkLogic database through CLUSTERING XDBC application server. The data exchange between MarkLogic and MapReduce of Hadoop will be

TABLE 1: Experimental data statistic.

Item	Statistic	Unit	Note
Number of logs	95	PC	Log equals HTTP traffic in a day of IP
Total of requests	13,905,165	Request	All requests of 95 logs
Max requests	479,751	Request	
Min requests	22,305	Request	Requests from log
Average requests	146,370	Request	
Max access time	24	Hour	
Min access time	6	Hour	From the first request to the last request
Average requests	20	Hour	

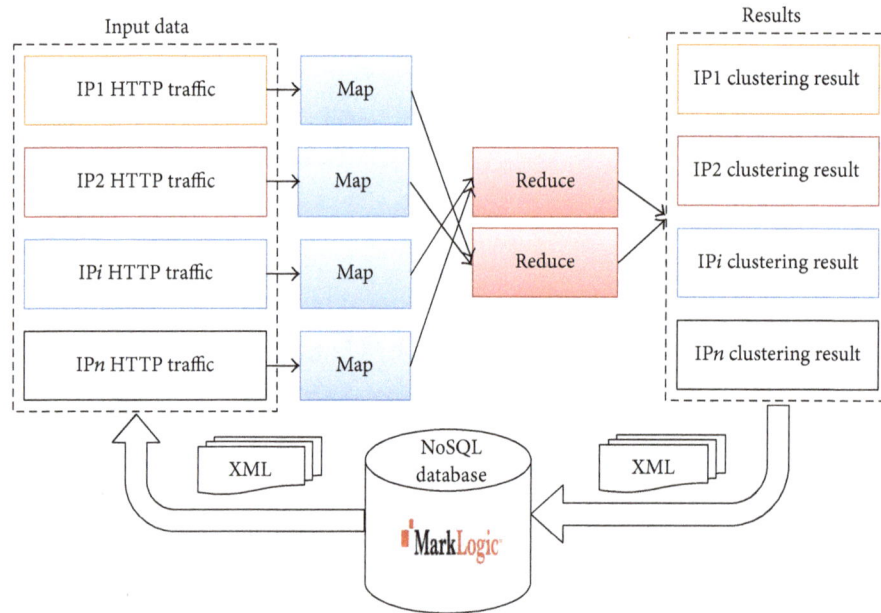

FIGURE 10: Process flow of clustering phase.

undertaken by a connector. Detailed process flow of this phase is described in Figure 10.

(iii) Classification and Detection Module is implemented for Phase 3, Detection and Classification Phase. It will process the result which is archived from Phase 2 and work with database through Data Analysis Application Server and after that give out processed results.

5.2. Experimental Analysis and Results. Experiment environment is shown in Figure 11; in that free developer licenses of MarkLogic verion 8.0.1 and Hadoop 2.6.0 are used [22]. From this experiment model, HTTP traffic from a university network is captured through a proxy server in separated files which are divided by date and stored in a proxy storage. These logs' raw data files will import to system through Data Manipulation Module as in Figure 9. Denoted log is HTTP traffic of IP in one day, which will be stored in its own directory in MarkLogic; 95 logs' data of clients are extracted, analyzed, and classified through the proposed method. Experiment data is detailed and summarized in Table 1. In that there are two Zeus bots [23] which are installed into a client with

difference interval in communication to C&C. All output results are manually checked with the support of VirusTotal online system [24] and McAfee Web Gateway which is installed in experiment network [25].

After preprocessing phase of proposed method described in Figure 6, a set of unique URLs (for logs of each IP) is established with 5621 URLs. In that, there are 14 URLs requested by numerous IPs which are generated with extreme speed over a threshold which is set as 0.8 in this experiment. In Table 2, details of 14 malicious URLs detected by preprocessing phase are summarized. The request per second (access speed) is determined by request count and access time via (4). Based on the characters of malicious autoware which is infected into client IP, the access speed and also communication behaviour to these URLs are determined. For example, as can be seen in Table 2, just in only 0.6 hours, URL2 is requested 80,903 times so it owns highest access speed at 32.98 requests per second. Vise versa, with URL12, it is requested with lowest speed at 0.82 requests per second, 71,004 times in 24 hours; however it is still higher than access speed to other URLs in experimental data. By manually checking the support of [24, 25], all these 14 URLs from domains/web servers contain unwanted software

TABLE 2: Malicious URLs detected in Phase 1 (preprocessing phase).

| Number | Malicious URL | Requests | | Access time (h) | Requests per second |
		Count	Percent		
1	URL1	237,291	1.71%	2.04	32.38
2	URL2	80,903	0.58%	0.68	32.98
3	URL3	80,032	0.58%	24.00	0.93
4	URL4	303,633	2.18%	10.56	7.98
5	URL5	81,256	0.58%	24.00	0.94
6	URL6	149,966	1.08%	12.53	3.32
7	URL7	496,781	3.57%	4.40	31.39
8	URL8	364,809	2.62%	11.69	8.67
9	URL9	80,761	0.58%	24.00	0.93
10	URL10	297,938	2.14%	16.65	4.97
11	URL11	80,423	0.58%	24.00	0.93
12	URL12	71,004	0.51%	24.00	0.82
13	URL13	80,549	0.58%	24.00	0.93
14	URL14	81,040	0.58%	24.00	0.94
	Total	*2,486,386*	*17.88%*		

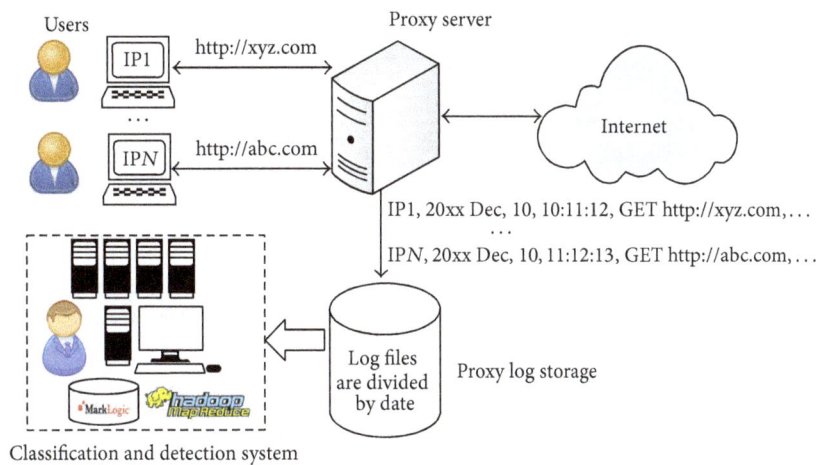

FIGURE 11: Experiment environment.

and are marked as malicious by many network security companies and software. These 14 URLs are requested 2,486,386 times, and they derive 17.88% of 13,905,165 total requests in experimental data.

Remaining 5607 URLs are classified in 673 groups in which 393 groups which contain 2 URLs above are detected as greyware. MapReduce just needed about 30 seconds to process all these URLs of 95 logs. As results summarized in Table 3, beside 14 malicious URLs which are detected in Phase 1 (preprocessing phase), 5 URLs requested are detected as malicious in Phase 3 (classification and detection phase), 2 of them are matched with C&C servers communicated by installed Zeus bots and other 3 URLs are detected from experimental captured data. All the detected greyware communication groups are confirmed when they come from shopping sites, social media, and adverting companies. Remaining 275 URLs are unclustered; system can not detect these URLs.

TABLE 3: Experimental results.

| Phase | Malicious URLs | Greyware | | Unknown URLs |
		Group	URLs	
Phase 1	14			
Phase 3	5	393	5327	275

These constitute a false negative of 4.9% and the accuracy rate reaches 95.1%.

6. Conclusion and Future Work

In this paper, a new method is proposed to detect and classify autoware communication based on its behaviour via analysis of HTTP traffic. The major advantage of the proposed method is that it just used minor features in HTTP traffic and does not use any signature or content-based technique. In addition, big data application framework also is proposed by combination of two leading technologies, which are the power of distributed processing of MapReduce of Hadoop and the convenient in working with unstructured data through XDBC servers of NoSQL database MarkLogic. Experiment results are promised and methods are working well in private network environment.

There are some reasons contributing undetected rate. First, even autoware commonly communicates with sites by the same behaviour, some rare cases of autowares' requests are different. Second, some types of autoware have less activities in network since they just send out little requests. In other situations, users' Internet accessed traffic also might be auto-mated communication since their access sites automatically refresh its contents via HTML script such as JavaScript or Flash. In these cases, clustering and detection of these URLs access graphs are become tougher. Based on this result, with the objective of reducing the undetected rate, some new features need to be considered in the future work. For that matter, data size sent in each request is regarded since this feature from malicious bot communication to its C&C server is almost steady whist variation of adware's data size in each request depends on the content which they get. In addition, unclustered URLs are also considered to be classified by checking the matching between domain name part of them and clustered group which is in clustering phase.

Competing Interests

The authors declare that there are no competing interests regarding the publication of this paper.

References

[1] MapReduce Tutorial, *Apache Hadoop*, 2008, https://hadoop.apache.org/docs/current/hadoop-mapreduce-client/hadoop-mapreduce-client-core/MapReduceTutorial.html.

[2] MarkLogic database, "What is Marklogic," 2015, http://www.marklogic.com/what-is-marklogic/.

[3] MarkLogic 8 Product Documentation, https://docs.marklogic.com/.

[4] M. A. Rajab, L. Ballard, N. Lutz, P. Mavrommatis, and N. Provos, "CAMP: content-agnostic malware protection," in *Proceedings of the Network and Distributed Systems Security Symposium (NDSS '13)*, Internet Society, 2013.

[5] A. Averbuch, M. Kiperberg, and N. J. Zaidenberg, "An efficient VM-based software protection," in *Proceedings of the 5th International Conference on Network and System Security (NSS '11)*, pp. 121–128, IEEE, Milan, Italy, September 2011.

[6] P. Royal, M. Halpin, D. Dagon, R. Edmonds, and W. Lee, "PolyUnpack: automating the hidden-code extraction of unpack-executing malware," in *Proceedings of the 22nd Annual Computer Security Applications Conference (ACSAC '06)*, pp. 289–298, IEEE, Miami Beach, Fla, USA, December 2006.

[7] J. Oberheide, E. Cooke, and F. Jahanian, "Cloudav: N-version antivirus in the network cloud," in *Proceedings of the 17th Conference on Security Symposium*, pp. 91–106, USENIX Association, 2008.

[8] D. Ashley, *An Algorithm for HTTP Bot Detection*, University of Texas at Austin—Information Security Office, Austin, Tex, USA, 2011.

[9] W. Lu, M. Tavallaee, and A. A. Ghorbani, "Automatic discovery of botnet communities on large-scale communication networks," in *Proceedings of the 4th International Symposium on Information, Computer, and Communications Security (ASIACCS '09)*, pp. 1–10, ACM, Sydney, Australia, March 2009.

[10] M. Eslahi, H. Hashim, and N. M. Tahir, "An efficient false alarm reduction approach in HTTP-based botnet detection," in *Proceedings of the IEEE Symposium on Computers & Informatics (ISCI '13)*, pp. 201–205, Langkawi, Malaysia, April 2013.

[11] B. AsSadhan and J. M. F. Moura, "An efficient method to detect periodic behavior in botnet traffic by analyzing control plane traffic," *Journal of Advanced Research*, vol. 5, no. 4, pp. 435–448, 2014.

[12] S. Shin, Z. Xu, and G. Gu, "EFFORT: a new host-network cooperated framework for efficient and effective bot malware detection," *Computer Networks*, vol. 57, no. 13, pp. 2628–2642, 2013.

[13] Y.-S. Chen, H.-S. Liu, Y.-H. Yu, and P.-C. Wang, "Detect phishing by checking content consistency," in *Proceedings of the 15th IEEE International Conference on Information Reuse and Integration (IRI '14)*, pp. 109–119, Redwood City, Calif, USA, August 2014.

[14] A. Blum, B. Wardman, T. Solorio, and G. Warner, "Lexical feature based phishing URL detection using online learning," in *Proceedings of the 3rd ACM Workshop on Artificial Intelligence and Security (AISec '10)*, pp. 54–60, 2010.

[15] J. Ma, L. K. Saul, S. Savage, and G. M. Voelker, "Beyond blacklists: learning to detect malicious web sites from suspicious URLs," in *Proceedings of the 15th ACM SIGKDD International Conference on Knowledge Discovery and Data Mining (KDD '09)*, pp. 1245–1254, ACM, Paris, France, July 2009.

[16] T.-C. Chen, S. Dick, and J. Miller, "Detecting visually similar web pages: application to phishing detection," *ACM Transactions on Internet Technology*, vol. 10, no. 2, article 5, pp. 5:1–5:38, 2010.

[17] G. Bartlett, J. Heidemann, and C. Papadopoulos, "Low-rate, flow-level periodicity detection," in *Proceedings of the IEEE Conference on Computer Communications Workshops (INFOCOM WKSHPS '11)*, pp. 804–809, April 2011.

[18] M. C. Tran and Y. Nakamura, "In-host communication pattern observed for suspicious HTTP-based auto-ware detection," *International Journal of Computer and Communication Engineering*, vol. 4, no. 6, pp. 379–389, 2015.

[19] M.-P. Dubuisson and A. K. Jain, "A modified Hausdorff distance for object matching," in *Proceedings of the 12th IAPR International Conference on Pattern Recognition, Conference A: Computer Vision & Image Processing*, vol. 1, pp. 566–568, IEEE, Jerusalem, Israel, 1994.

[20] D. P. Huttenlocher, G. A. Klanderman, and W. J. Rucklidge, "Comparing images using the Hausdorff distance," *IEEE Transactions on Pattern Analysis and Machine Intelligence*, vol. 15, no. 9, pp. 850–863, 1993.

[21] C. Brooks, *Enterprise NoSQL for Dummies*, John Wiley & Sons, Hoboken, NJ, USA, 2014.

[22] MarkLogic Developer License, *Enterprise NoSQL Power for Developers*, 2008, https://developer.marklogic.com/free-developer.

16

Novel Chaos Secure Communication System based on Walsh Code

Gang Zhang, Niting Cui, and Tianqi Zhang

School of Communication and Information Engineering, Chongqing University of Posts and Telecommunications, Chongqing 400065, China

Correspondence should be addressed to Niting Cui; 1276707796@qq.com

Academic Editor: George S. Tombras

A multiuser communication scheme which is a hybrid of Walsh code with DCSK and CDSK is proposed to improve low data transmission rate of Differential Chaos Shift Keying (DCSK), poor bit error ratio (BER) performance of Correlation Delay Shift Keying (CDSK), and disadvantage of orthogonality in traditional multiuser DCSK. It not only overcomes the disadvantages of DCSK and CDSK, but also has better performance than CDSK and higher transmission data rate than DCSK. It has been proved that the novel multiuser CDSK-DCSK has better properties than traditional Multiple Input Multiple Output-Differential Chaos Shift Keying (MIMO-DCSK) and Modified-Differential Chaos Shift Keying (M-DCSK). Also the multiuser interference is greatly suppressed due to the orthogonality of Walsh code.

1. Introduction

Spread spectrum technology is used in communication system to bear low data rate information by using spread spectrum sequence due to its broad bandwidth characteristics. The technology has advantages such as high security, anti-interference, and antimultipath fading and being easy to be realized in code division multiple access (CDMA) [1]. In recent years, chaotic spread spectrum communication system has been deeply researched in spread spectrum technology [2]. It is different from the traditional spread spectrum technology since the carriers are high speed chaotic signals which are generated by different chaotic maps. Chaotic signal has the advantages of high bandwidth, being nonperiodic, being difficult to predict, and good autocorrelation and cross correlation features [3].

According to the way of demodulation, chaotic communication system is divided into two types: the coherent and the noncoherent demodulation [4]. Coherent demodulation needs the receiver to reconstruct chaotic signal, which means that the security and noise immunity are better than noncoherent demodulation. But it is difficult to realize chaotic synchronization. The security in noncoherent demodulation is worse than in coherent demodulation, but it is easy to implement and the cost of hardware is much lower. In the existing noncoherent systems, reference signal is used for dispreading in most of the receivers [5–10]. Differential Chaos Shift Keying (DCSK) has disadvantages of low data rate [11]. In Correlation Delay Shift Keying (CDSK), the data rate is 2 times that of DCSK [12], but the BER performance is worse than DCSK [13]. Much attention has been attracted since the concept of multiple access DCSK has been proposed [14]. In [15], different interval between transmit signal and carrier is used to distinguish different users. But the orthogonality is poor when using the smaller spread factor.

A combination of multiuser CDSK-DCSK with Walsh coded scheme is proposed in this paper. It can not only overcome the disadvantages of DCSK and CDSK, which means that the BER is better than CDSK and the data rate is higher than DCSK, but also suppress multiuser interference well, due to the orthogonality of Walsh code.

2. Novel Multiuser CDSK-DCSK System

The novel multiuser CDSK-DCSK scheme is shown in Figure 1, where the system has U users totally, and the uth user is discussed for special purpose.

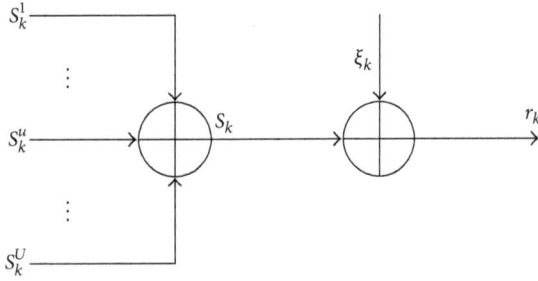

FIGURE 1: U users through Additive White Gaussian Noise (AWGN) channel.

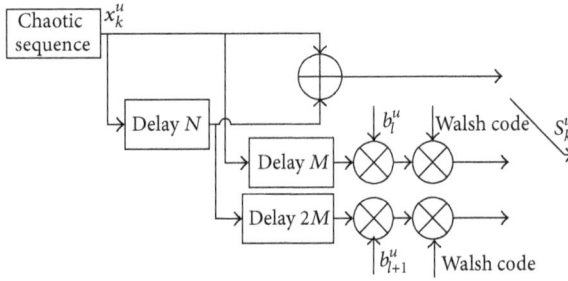

FIGURE 2: Transmitting end of user u.

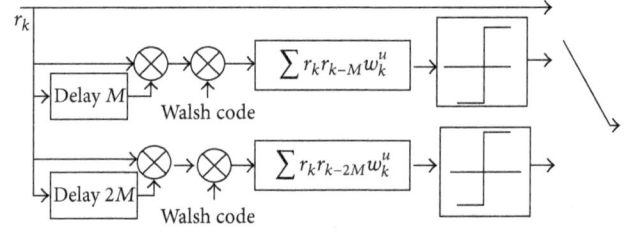

FIGURE 3: Receiving end of user u.

Chaotic signal is generated by the logistic map firstly. Then, chaotic sequence x_k is generated after the symbolic function mapping as follows:

$$y_{k+1} = 1 - 2y_k^2, \quad y_k \in (-1, 1),$$

$$x_k^u = \text{sgn}(y_k),$$

$$k = 0, 1, 2, \ldots, \tag{1}$$

where y_k is chaotic signal, sgn() is symbolic function, and x_k^u is chaotic sequence of the uth user.

The transmitter is illustrated in Figure 2. A pair of bit (b_l^u, b_{l+1}^u) is modulated and transmitted in a frame, where $b_l^u, b_{l+1}^u \in \{+1, -1\}$. In the lth frame, x_k^u and x_{k-N}^u are transmitted in the first slot, where N is the time delay of x_k^u and $N > M$. In the second and third slots, b_l^u and b_{l+1}^u are multiplied with chaotic sequence and the assigned Walsh code, respectively. Transmitting signal of uth user is shown in

$$s_k^u$$

$$= \begin{cases} x_k^u + x_{k-N}^u, & 3lM + 1 < k < (3l+1)M, \\ b_l^u x_{k-M}^u w_{1,k}^u, & (3l+1)M + 1 < k < (3l+2)M, \\ b_{l+1}^u x_{k-N-2M}^u w_{2,k}^u, & (3l+2)M + 1 < k < (3l+3)M, \end{cases} \tag{2}$$

where M is the spread factor. $w_{1,k}^u$ and $w_{2,k}^u$ are Walsh code of uth user.

From Figure 1 and (2), the total transmitting signal s_k is easy to be obtained. The received signal after transmission is shown as follows:

$$r_k = s_k + \xi_k = \sum_{u=1}^{U} s_k^u + \xi_k$$

$$= \begin{cases} \sum_{u=1}^{U} (x_k^u + x_{k-N}^u) + \xi_k, & 3lM + 1 < k < (3l+1)M, \\ \sum_{u=1}^{U} (b_l^u x_{k-M}^u w_{1,k}^u) + \xi_k, & (3l+1)M + 1 < k < (3l+2)M, \\ \sum_{u=1}^{U} (b_{l+1}^u x_{k-N-2M}^u w_{2,k}^u) + \xi_k, & (3l+2)M + 1 < k < (3l+3)M, \end{cases} \tag{3}$$

where ξ_k is assumed as Additive White Gaussian Noise (AWGN) and U is the total number of users.

Figure 3 shows the receiver's structure. Walsh code is multiplied with r_k and r_{k-M} or r_{k-2M}. The Walsh code in the uth receiver must agree with the one of the uth transmitter. After correlation demodulation the original signal is obtained and is shown in

$$c_l^u = \sum_{k=1}^{M} r_k r_{k-M} w_{1,k}^u$$

$$= \sum_{k=1}^{M} \left((s_k + \xi_k)(s_{k-M} + \xi_{k-M}) w_{1,k}^u \right)$$

$$= \sum_{i=1}^{U} \sum_{j=1}^{U} \sum_{k=1}^{M} b_l^j x_{k-M}^i x_{k-M}^j w_{1,k}^j w_{1,k}^u + \sum_{i=1}^{U} \sum_{k=1}^{M} x_{k-M}^i \xi_k w_{1,k}^u$$

$$+ \sum_{i=1}^{U} \sum_{k=1}^{M} x_{k-N-M}^i \xi_k w_{1,k}^u$$

$$+ \sum_{j=1}^{U} \sum_{k=1}^{M} b_l^j x_{k-M}^j w_{1,k}^j w_{1,k}^u \xi_{k-M}$$

$$+ \sum_{i=1}^{U} \sum_{j=1}^{U} \sum_{k=1}^{M} b_l^j x_{k-M}^j x_{k-N-M}^i w_{1,k}^j w_{1,k}^u$$

$$+ \sum_{k=1}^{M} \xi_k \xi_{k-M} w_{1,k}^u. \tag{4}$$

In order to simplify the output, c_l^u can be divided into three parts A, B, and C as follows:

$$A = \sum_{i=1}^{U}\sum_{j=1}^{U}\sum_{k=1}^{M} b_l^j x_{k-M}^i x_{k-M}^j w_{1,k}^j w_{1,k}^u$$

$$= \sum_{i=1;i\neq j}^{U}\sum_{j=1}^{U}\left(\sum_{k=1}^{M} b_l^j x_{k-M}^i x_{k-M}^j w_{1,k}^j w_{1,k}^u\right) \qquad (5)$$

$$+ \sum_{j=1;j\neq u}^{U}\left(\sum_{k=1}^{M} b_l^j x_{k-M}^j x_{k-M}^j w_{1,k}^j w_{1,k}^u\right) + b_l^u M,$$

$$B = \sum_{i=1}^{U}\sum_{k=1}^{M} x_{k-M}^i \xi_k w_{1,k}^u + \sum_{i=1}^{U}\sum_{k=1}^{M} x_{k-N-M}^i \xi_k w_{1,k}^u$$

$$+ \sum_{j=1}^{U}\sum_{k=1}^{M} b_l^j x_{k-M}^j w_{1,k}^j \xi_{k-M} w_{1,k}^u \qquad (6)$$

$$+ \sum_{i=1}^{U}\sum_{j=1}^{U}\sum_{k=1}^{M} b_l^j x_{k-M}^j w_{1,k}^j x_{k-N-M}^i w_{1,k}^u,$$

$$C = \sum_{k=1}^{M} \xi_k \xi_{k-M} w_{1,k}^u. \qquad (7)$$

The chaotic sequence has the following properties [16]:

(1) Chaotic sequence generated by the same map but with different initial value is noncorrelated.

(2) The chaotic sequence is the same as impulse function after normalized autocorrelation.

Besides, due to the orthogonal property of Walsh code such as $\sum_{k=1}^{M} w_k^j w_k^u = 0$, $w_k^u w_k^u = 1$, and $x_k^u x_k^u = 1$, $b_l^u M$ in (5) is the only useful signal and the rest of (5), (6), and (7) are interference. The first item is the cross correlation of chaotic sequence and the second item equals 0. b_l^u can be demodulated according to the following rules:

$$b_l^u = f\left(c_l^u\right) = \begin{cases} -1, & c_l^u < 0, \\ +1, & c_l^u \geq 0. \end{cases} \qquad (8)$$

Similarly, the decision rules for b_{l+1}^u are

$$b_{l+1}^u = f\left(c_{l+1}^u\right) = \begin{cases} -1, & c_{l+1}^u < 0, \\ +1, & c_{l+1}^u \geq 0. \end{cases} \qquad (9)$$

3. Performance Analysis

By central limit theorem, the correlation output approximately obeys the normal distribution. The mean and variance of c_l^u are required to get the system's BER. Features of chaotic sequence and Walsh code are presented in [17].

(1) For different chaotic sequences x_i ($i = 0, 1, 2, \ldots$) and x_j ($j = 0, 1, 2, \ldots$) generated by the same map, $E[x_i x_j] = E[x_i]E[x_j] = 0$ and $\mathrm{var}[x_i x_j] = \mathrm{var}[x_i]\mathrm{var}[x_j] = 1$, when $i \neq j$.

(2) For different Walsh codes w_k^p and w_k^q, where $p, q \in [1, U]$ and $k = (1, 2, \ldots, M)$, when $p \neq q$, $\mathrm{var}[w_k^p w_k^q] = \mathrm{var}[w_k^p] = \mathrm{var}[w_k^q] = 1$.

(3) Correlation among chaotic sequences, AWGN and Walsh codes, is 0.

(4) For (4), $\mathrm{cov}[A, B] = \mathrm{cov}[A, C] = \mathrm{cov}[B, C] = 0$.

Suppose the uth user's first bit in lth frame is "+1":

$$E\left\{c_l^u \mid b_l^u = +1\right\} = EA + EB + EC = M, \qquad (10)$$

where E represents the mean of c_l^u. Consider

$$\mathrm{var}(A) = \mathrm{var}\left(\sum_{i=1;i\neq j}^{U}\sum_{j=1}^{U}\sum_{k=1}^{M} x_{k-M}^i x_{k-M}^j w_{1,k}^j w_{1,k}^u + M\right)$$

$$= \sum_{i=1;i\neq j}^{U}\sum_{j=1}^{U}\sum_{k=1}^{M} \mathrm{var}\left(x_{k-M}^i x_{k-M}^j\right)\mathrm{var}\left(w_{1,k}^j w_{1,k}^u\right)$$

$$= M(U-1)^2$$

$$\mathrm{var}(B)$$

$$= \sum_{i=1}^{U}\sum_{k=1}^{M} \mathrm{var}\left(x_{k-M}^i \xi_k w_{1,k}^u\right)$$

$$+ \sum_{i=1}^{U}\sum_{k=1}^{M} \mathrm{var}\left(x_{k-N-M}^i \xi_k w_{1,k}^u\right)$$

$$+ \sum_{j=1}^{U}\sum_{k=1}^{M} \mathrm{var}\left(x_{k-M}^j w_{1,k}^j \xi_{k-M} w_{1,k}^u\right)$$

$$+ \sum_{i=1}^{U}\sum_{j=1}^{U}\sum_{k=1}^{M} \mathrm{var}\left(x_{k-M}^j w_{1,k}^j x_{k-N-M}^i w_{1,k}^u\right) \qquad (11)$$

$$= \sum_{i=1}^{U}\sum_{k=1}^{M} \mathrm{var}\left(x_{k-M}^i\right)\mathrm{var}\left(\xi_k\right)\mathrm{var}\left(w_{1,k}^u\right)$$

$$+ \sum_{i=1}^{U}\sum_{k=1}^{M} \mathrm{var}\left(x_{k-N-M}^i\right)\mathrm{var}\left(\xi_k\right)\mathrm{var}\left(w_{1,k}^u\right)$$

$$+ \sum_{j=1}^{U}\sum_{k=1}^{M} \mathrm{var}\left(x_{k-M}^j\right)\mathrm{var}\left(\xi_{k-M}\right)\mathrm{var}\left(w_{1,k}^u\right)$$

$$+ \sum_{i=1}^{U}\sum_{j=1}^{U}\sum_{k=1}^{M} \mathrm{var}\left(x_{k-M}^j\right)\mathrm{var}\left(x_{k-N-M}^i\right)\mathrm{var}\left(w_{1,k}^u\right)$$

$$= 2MUN_0,$$

$$DC = \sum_{k=1}^{M} D\left(\xi_k\right) D\left(\xi_{k-M}\right) D\left(w_{1,k}^u\right) = \frac{1}{4}MN_0^2,$$

where D represents variance and $N_0/2$ is noise power density.

So the variance of c_l^u is

$$
\begin{aligned}
D\left\{c_l^u \mid b_l^u = +1\right\} \\
= D[A] + D[B] + D[C] \\
+ 2\left\{\text{cov}[A, B] + \text{cov}[A, C] + \text{cov}[B, C]\right\} \\
= M\left(U^2 - U\right) + 2MUN_0 + \frac{1}{4}MN_0^2,
\end{aligned}
\tag{12}
$$

where cov represents covariance.

Similarly, when $b_l^u = -1$, the mean and variance of c_l^u are

$$
E\left\{c_l^u \mid b_l^u = -1\right\} = -M,
$$

$$
D\left\{c_l^u \mid b_l^u = -1\right\} = M\left(U^2 - U\right) + 2MUN_0 + \frac{1}{4}MN_0^2.
\tag{13}
$$

The system's bit error ratio (BER) is

$$
\begin{aligned}
\text{BER} &= \frac{1}{2}P\left(c_l^u > 0 \mid b_l^u = -1\right) + \frac{1}{2}P\left(c_l^u < 0 \mid b_l^u = +1\right) \\
&= \frac{1}{2}\,\text{erfc}\left(\frac{E\left(c_l^u \mid b_l^u = +1\right)}{\sqrt{2D\left(c_l^u \mid b_l^u = +1\right)}}\right) = \frac{1}{2} \\
&\cdot \text{erfc}\frac{M}{\sqrt{2M(U-1)^2 + 4MUN_0 + (1/2)MN_0^2}} = \frac{1}{2} \\
&\cdot \text{erfc}\left(\frac{2(U-1)^2}{M} + 12U\left(\frac{E_b}{N_0}\right)^{-1}\right. \\
&\left. + \frac{9M}{2}\left(\frac{E_b}{N_0}\right)^{-2}\right)^{-1/2},
\end{aligned}
\tag{14}
$$

where erfc is the error function, $\text{erfc}(\varphi) = (2/\sqrt{\pi})\int_{\varphi}^{\infty}\exp(-x^2)dx$, and E_b is bit energy, $E_b = 3M\,\text{var}[x_i]$.

From (14), with a certain value of U and E_b/N_o, there exists M_{opt} to realize the best system performance. Suppose $y = 2(U-1)^2/M + 12U(E_b/N_0)^{-1} + (9M/2)(E_b/N_0)^{-2}$. It is easy to obtain M_{opt} after differentiating y:

$$
y' = -\frac{2(U-1)^2}{M^2} + \frac{9}{2}\left(\frac{E_b}{N_0}\right)^{-2}.
\tag{15}
$$

Suppose that $y' = 0$; the equation of M_{opt} is as follows:

$$
M_{\text{opt}} = \frac{2(U-1)}{3}\frac{E_b}{N_0}.
\tag{16}
$$

By (16), for certain E_b/N_o, under different U, M_{opt} is different. For example, suppose $E_b/N_0 = 10$ dB; when $U = 3$ and $U = 5$, M_{opt} is 13.33 and 26.66, respectively.

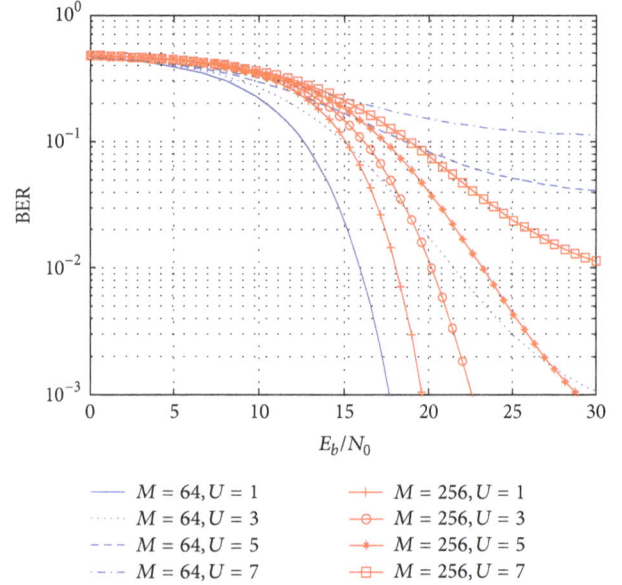

FIGURE 4: BER performance versus E_b/N_0 for different spreading factor M.

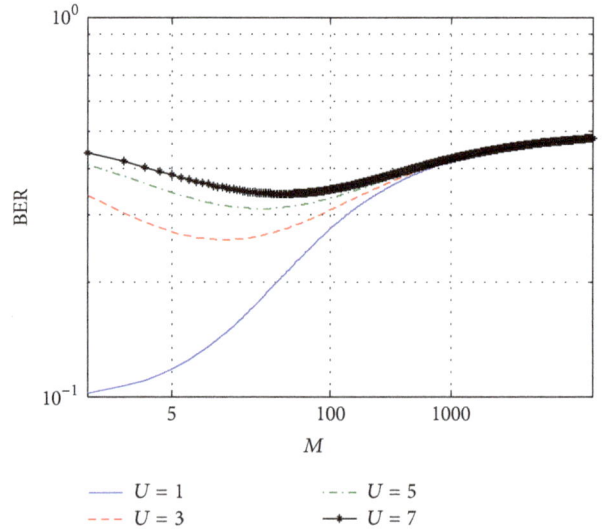

FIGURE 5: BER performance versus spreading factor M under $E_b/N_0 = 10$ dB.

4. Simulation Comparisons

Figure 4 shows that the smaller the value of M, the stronger the influence on BER by E_b/N_0. It is obvious that the intervals between different curves under the same E_b/N_0 in $M = 64$ are significantly larger than that of $M = 256$. With M increasing, BER gets smaller and the system's performance gets better.

Figure 5 shows that selecting an appropriate M has great impact on the system's performance. On one hand, there exists an optimum M to achieve the best BER. If M is increased continuously, the system's performance gets worse. On the other hand, the transmission efficiency is too low if M is too large.

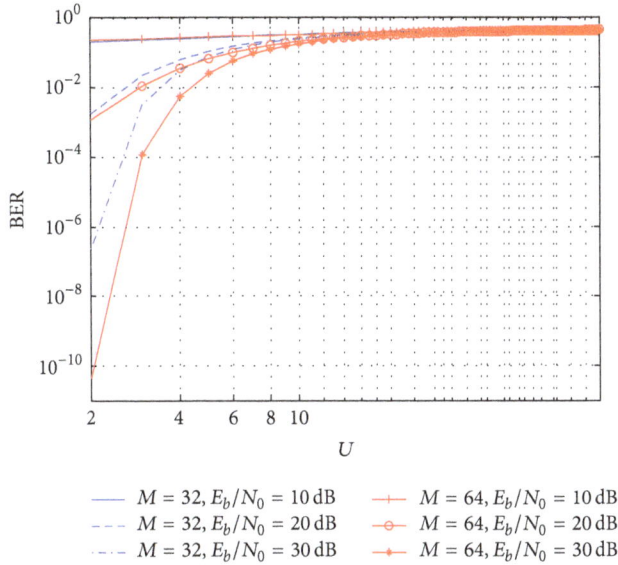

FIGURE 6: BER performance versus different total number of users.

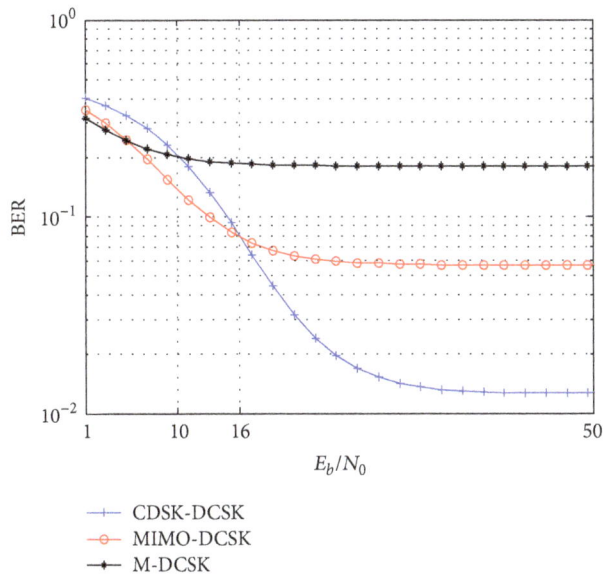

FIGURE 7: Simulated BERs versus E_b/N_0 for multiuser CDSK-DCSK, MIMO-DCSK, and M-DCSK with $M = 5$ and $U = 2$.

Figure 6 displays that multiuser interference increases with the users' total number increasing, so the system's performance gets worse. Under different E_b/N_0, the BER gradually tends to be constant and unrelated with M.

It can be seen from Figure 7 that the performance of multiuser CDSK-DCSK is slightly inferior to MIMO-DCSK [11] when E_b/N_0 is low. But the proposed system's data rate is 2 times that of MIMO-DCSK. When $E_b/N_0 > 16$ dB, the proposed system's performance is much better than that of MIMO-DCSK. Compared with M-DCSK proposed in [9], in low E_b/N_0, there is not much difference between multiuser CDSK-DCSK and M-DCSK. But when $E_b/N_0 > 10$ dB, the proposed system's BER is improved one-order magnitude.

5. Conclusion

The excellent autocorrelation and cross correlation characteristics of chaotic signal are used in traditional multiuser DCSK system [15] to distinguish different users. When the spread factor is small, the orthogonality between different chaotic signals is poor. A hybrid CDSK-DCSK combined with Walsh code system is proposed in this paper to realize the multiuser transmission. The signal is transmitted in pairs, so transmitting rate is 2 times the traditional ones. Interference between different users can be reduced due to the application of Walsh code. Also, the simulation results show that, under the same circumstances, the performance of multiuser CDSK-DCSK system is much better than that of M-DCSK and MIMO-DCSK, especially when E_b/N_0 is larger than 16 dB and 10 dB, respectively.

Conflict of Interests

The authors declare that there is no conflict of interests regarding the publication of this paper.

Acknowledgments

This research is supported by the National Natural Science Foundation of China (Grants nos. 61371164, 61071196, and 61102131), the Program for New Century Excellent Talents in University (Grant no. NCET-10-0927), the Project of Key Laboratory of Signal and Information Processing of Chongqing (Grant no. CSTC2009CA2003), the Chongqing Distinguished Youth Foundation (Grant no. CSTC2011jjjq40002), the Natural Science Foundation of Chongqing (Grants nos. CSTC2010BB2398, CSTC2010BB2409, CSTC2010BB2411, and CSTC2012JJA40008), and the Research Project of Chongqing Educational Commission (Grants KJ120525 and KJ130524).

References

[1] G. Kolumban, M. P. Kennedy, Z. Jako, and G. Kis, "Chaotic communications with correlator receivers: theory and performance limits," *Proceedings of the IEEE*, vol. 90, no. 5, pp. 711–732, 2002.

[2] G. Kolumban, M. P. Kennedy, and L. O. Chua, "The role of synchronization in digital communications using chaos. II. Chaotic modulation and chaotic synchronization," *IEEE Transactions on Circuits and Systems. I. Fundamental Theory and Applications*, vol. 45, no. 11, pp. 1129–1140, 1998.

[3] L. F. He, G. Zhang, and T. Q. Zhang, "A secure image transmission scheme based on improved DCSK," *Telecommunications Science*, no. 6, pp. 94–99, 2013.

[4] Z. G. Chen and W. K. Xu, "Performance analysis of CS-DCSK over Nakagami-m fading channels," *Journal of Chongqing University of Posts and Telecommunications*, vol. 24, pp. 395–399, 2012.

[5] H. Yang and G.-P. Jiang, "Reference-modulated DCSK: a novel chaotic communication scheme," *IEEE Transactions on Circuits and Systems II: Express Briefs*, vol. 60, no. 4, pp. 232–236, 2013.

[6] G. Kolumban, Z. Jako, and M. P. Kennedy, "Enhanced versions of DCSK and FM-DCSK data transmission systems," in *Proceedings of the IEEE International Symposium on Circuits and*

Systems (ISCAS '99), vol. 4, pp. 475–478, Orlando, Fla, USA, 1999.

[7] J.-Y. Duan, G.-P. Jiang, and H. Yang, "A new chaotic communication scheme: differential correlation delay shift Keying," in *Proceedings of the International Conference on Communications, Circuits and Systems (ICCCAS '13)*, pp. 446–449, IEEE, Chengdu, China, November 2013.

[8] W. N. Tam, F. C. M. Lau, and C. K. Tse, "Generalized correlation-delay-shift-keying scheme for noncoherent chaos-based communication systems," *IEEE Transactions on Circuits and Systems I: Regular Papers*, vol. 53, no. 3, pp. 712–721, 2006.

[9] H. Yang and G.-P. Jiang, "High-efficiency differential-chaos-shift-keying scheme for chaos-based noncoherent communication," *IEEE Transactions on Circuits and Systems II: Express Briefs*, vol. 59, no. 5, pp. 312–316, 2012.

[10] G. Kaddoum and F. Gagnon, "Design of a high-data-rate differential chaos-shift keying system," *IEEE Transactions on Circuits and Systems II: Express Briefs*, vol. 59, no. 7, pp. 448–452, 2012.

[11] S. P. Li, L. Wang, and Y. L. Xia, "System devising and simulation of FM-DCSK in non-flat fading channel," *Journal of Chongqing University of Posts and Telecommunications*, vol. 17, pp. 282–286, 2005.

[12] M. Sushchik, L. S. Tsimring, and A. R. Volkovskii, "Performance analysis of correlation-based communication schemes utilizing chaos," *IEEE Transactions on Circuits and Systems I: Fundamental Theory and Applications*, vol. 47, no. 12, pp. 1684–1691, 2000.

[13] Q. Ding and J. N. Wang, "Design of frequency-modulated correlation delay shift keying chaotic communication system," *IET Communications*, vol. 5, no. 7, pp. 901–905, 2011.

[14] M. P. Kennedy, G. Kolumban, G. Kis, and Z. Jako, "Recent advances in communicating with chaos," in *Proceedings of the IEEE International Symposium on Circuits and Systems (ISCAS '98)*, vol. 4, pp. 461–464, IEEE, Monterey, Calif, USA, May-June 1998.

[15] F. C. M. Lau, M. M. Yip, C. K. Tse, and S. F. Hau, "A multiple-access technique for differential chaos-shift keying," *IEEE Transactions on Circuits and Systems I: Fundamental Theory and Applications*, vol. 49, no. 1, pp. 96–104, 2002.

[16] S. Mandal and S. Banerjee, "Analysis and CMOS implementation of a chaos-based communication system," *IEEE Transactions on Circuits and Systems I: Regular Papers*, vol. 51, no. 9, pp. 1708–1722, 2004.

[17] Z.-B. Zhou, T. Zhou, and J.-X. Wang, "Performance analysis of an improved multiple-access differential chaos shift keying," *Journal of Xidian University*, vol. 36, no. 4, pp. 730–735, 2009.

17

Parallel Resampling of OFDM Signals for Fluctuating Doppler Shifts in Underwater Acoustic Communication

Shingo Yoshizawa ⓘ,[1] **Takashi Saito,**[2] **Yusaku Mabuchi,**[2] **Tomoya Tsukui,**[3] **and Shinichi Sawada**[3]

[1]*Kitami Institute of Technology, Kitami, Japan*
[2]*Mitsubishi Electric TOKKI Systems Corporation, Kamakura, Japan*
[3]*IHI Corporation, Yokohama, Japan*

Correspondence should be addressed to Shingo Yoshizawa; yosizawa@mail.kitami-it.ac.jp

Academic Editor: Jit S. Mandeep

Reliable underwater acoustic communication is demanded for autonomous underwater vehicles (AUVs) and remotely operated underwater vehicles (ROVs). Orthogonal frequency-division multiplexing (OFDM) is robust with multipath interference; however, it is sensitive to Doppler. Doppler compensation is given by two-step processing of resampling and residual carrier frequency offset (CFO) compensation. This paper describes the improvement of a resampling technique. The conventional method assumes a constant Doppler shift during a communication frame. It cannot cope with Doppler fluctuation, where relative speeds between transmitter and receiver units are fluctuating. We propose a parallel resampling technique that a resampling range is extended by measured Doppler standard deviation. The effectiveness of parallel resampling has been confirmed in the communication experiment. The proposed method shows better performance in bit error rates (BERs) and frame error rates (FERs) compared with the conventional method.

1. Introduction

Autonomous underwater vehicles (AUVs) and remotely operated underwater vehicles (ROVs) are of wide interest for marine survey and offshore engineering. Reliable underwater communication is essential for command control and image/video data transmission. In recent underwater acoustic communication (UAC), orthogonal frequency-division multiplexing (OFDM) is widely used in terms of high-frequency utilization [1]. OFDM inserts a cyclic prefix (CP) or a guard interval (GI) into block data and makes use of frequency domain equalization (FDE) as multipath compensation. We have presented the countermeasure against strong multipath, where a multipath spread exceeds a CP length in [2].

Although OFDM is robust with multipath interference, it is sensitive to Doppler. Due to the low velocity of acoustic waves, the wideband modulation of OFDM is considerably affected by Doppler and results in intercarrier interference (ICI). Doppler compensation is mandatory for moving platforms such as AUVs and ROVs. Doppler compensation is given by two steps of resampling and residual carrier frequency offset (CFO) correction [3, 4]. This paper describes the improvement of a resampling technique.

Doppler is expressed by frequency scaling in narrow band signals. This model is often used in electromagnetic wave communication. In UAC, the wideband model of time scaling (expansion or compression) is assumed when the signal bandwidth is close to carrier frequency. Resampling manipulates a received signal to compensate the time-scale change. The resampling ratio is determined according to the Doppler estimation result observed as the change of signal time length or frequency.

If a Doppler shift is simple and known, Doppler compensation would be perfectly performed by the incorporation of resampling and CFO correction. However, nonuniform Doppler shifts must be considered in actual communication. Nonuniform Doppler shifts are classified into Doppler spread and Doppler fluctuation as shown in Figure 1.

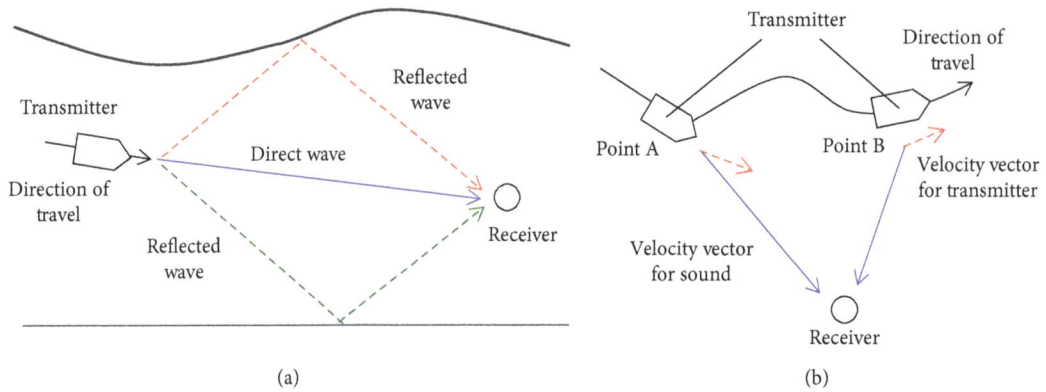

FIGURE 1: Types of nonuniform Doppler shifts. (a) Doppler spread and (b) Doppler fluctuation.

Doppler spread is caused by the acoustic propagation that each path has an individual Doppler scale factor. In Figure 1(a), the transmitter unit has the same direction as the direct wave. The reflected waves have the different directions, whose Doppler scales are slightly shifted from that of the direct wave. These Doppler scales are measured as frequency spread in the receiver unit. The countermeasures of Doppler spread have been presented in recent papers such as multiple resampling [5], Bayesian resampling [6], frequency domain oversampling [7], and Doppler-resilient orthogonal signal-division multiplexing (D-OSDM) [8].

Doppler fluctuation is caused by the irregular motions of transmitter or receiver units. The case that the transmitter unit is moving in meandering trajectory is depicted in Figure 1(b). A Doppler shift becomes positive in Point A and negative in Point B according to the relation between sound and transmitter velocity vectors. While the Doppler shift in Figure 1(a) is regarded to be time invariant, the Doppler shift in Figure 1(b) rapidly fluctuates with plus or minus as time goes on. There are few papers to tackle Doppler fluctuation because of difficulty in acoustic propagation modeling. The authors in [9] have presented linear multiscale compensation by smoothing estimated Doppler scales. It assumes that a Doppler shift is linearly changing. This paper discusses more severe Doppler fluctuation where a Doppler shift changes with plus or minus. The conventional method assumes a time invariant (or linearly changing) Doppler shift during a communication frame. It cannot cope with severe Doppler fluctuation, where a relative speed between transmitter and receiver units is fluctuating within a communication frame.

This paper presents a parallel resampling technique that a resampling range is extended by measured Doppler dispersion. If an OFDM block is affected by Doppler fluctuation, using Doppler shift average is insufficient for finding an appropriate resampling ratio. We apply statistical analysis that uses both mean and variance in Doppler estimation to cope with complicated Doppler shift variations. The parallel resampling receiver takes multiple resampling blocks whose resampling ratios are settled according to Doppler mean and deviation. The best decoded data are selected by checking data errors. The parallel resampling stricture is applicable to typical OFDM

systems. The effectiveness of parallel resampling has been tested by our sea trial, where a transmitter unit moves with various speeds and directions.

A part of contents in this paper has been presented in the conference paper as our previous work [10]. The following two points have been improved in this paper. One is the frequency allocation of a continuous wave (CW) whose signal is given by a sinusoidal function. The previous work requires frequency spacing between CW and OFDM bands to avoid their own interference, which results in low-frequency utilization. This work has improved that CW is allocated as one of the OFDM subcarriers. The signal interference can be avoided by making use of frequency orthogonality of OFDM. The second is the dynamic control of branches in the parallel resampling receiver. The parallel numbers of resampling units are controlled on frame-by-frame basis according to the measured Doppler shift mean and deviation. It can reduce computational complexity of resampling compared with the fixed branches in our previous work.

This paper is organized as follows. Section 2 explains fundamentals of Doppler and resampling and reports examples of Doppler tolerance in OFDM. Section 3 investigates the influence of Doppler fluctuation. Section 4 proposes a parallel resampling technique. Section 5 reports the experimental results in the Doppler test. Section 6 summarizes our work.

2. Resampling

The velocity of acoustic waves (about 1500 m/s underwater) is much lower than that of electromagnetic waves. The effect of Doppler gives a great impact on acoustic communication. The relative Doppler shift Δ is defined as a ratio of source relative velocity to propagation wave velocity. For a single-frequency component f, the Doppler effect is expressed by a frequency scaling:

$$f' = f(1 + \Delta). \tag{1}$$

In wideband signals, each frequency component is affected by a different amount. This Doppler effect is modeled by a time scaling (expansion or compression) of the signal waveform:

$$r(t) = s((1 + \Delta)t), \qquad (2)$$

where $s(t)$ and $r(t)$ are the source and received signals, repectively. When we use a discrete time-sampled source signal $s[nT_s]$, where n is an integer number and T_s is a sampling period, the received signal is expressed as

$$r[nT_s] = s[(1 + \Delta)nT_s]. \qquad (3)$$

The rate conversion of $1/(1 + \Delta)$, i.e., resampling, eliminates the effect of Doppler by the following equation:

$$r\left[\frac{1}{1 + \Delta}nT_s\right] = s[nT_s]. \qquad (4)$$

Doppler estimation for resampling is accomplished by inserting a known sequence to a communication frame. A popular approach is to detect the times of arrival of preamble and postamble [11] shown in Figure 2. A Doppler shift can be measured as the change of signal length by $(1 + \Delta)LT_s$, where LT_s is the interval between preamble and postamble.

The tolerance for Doppler estimation error can be investigated from a simple OFDM model in Figure 3. The relative Doppler shift is added to a transmit signal. The OFDM parameters of simulation are enumerated in Table 1. These parameters are just one of the examples because adequate OFDM parameters depend on acoustic propagation conditions. The similar parameters have been adopted for our experiment in Section 5.

The results of error vector magnitude (EVM) and bit error rate (BER) are plotted in Figure 4. The horizontal axis shows a Doppler shift corresponding to the residual error of Doppler estimation. A bit error occurs when EVM exceeds 40% when comparing the results in Figures 4(a) and 4(b). Since the center frequency is 45 kHz, a Doppler shift of 4.5 Hz is equivalent to 1/10000 in a relative Doppler shift. It indicates that a resolution of utmost 1/10000 is required within an allowable resampling ratio error. Frequency domain oversampling [7] can extend the error tolerance; however the tolerance is less than 5 Hz. We assume the Doppler deviation of 10 Hz in Doppler fluctuation, which is tested in Section 3. Only applying frequency domain oversampling cannot cope with this large Doppler fluctuation.

3. Influence of Doppler Fluctuation

The influence of Doppler fluctuation is investigated by applying various Doppler shift patterns into an OFDM block. First, we explain our Doppler fluctuation model illustrated in Figure 5, comparing two types of Doppler shift models. The constant Doppler shift model gives the same direction in the sound and transmitter velocity vectors, where a constant Doppler shift of Δ_C is observed. In the Doppler fluctuation model, the direction of a transmitter unit swings during sending one OFDM frame. The Doppler shift changes Δ_C, $\Delta_C + \Delta$, and $\Delta_C - \Delta$ due to the relation between sound and transmitter velocity vectors.

Since the constant component Δ_C can be eliminated by a typical Doppler compensation technique, we consider the

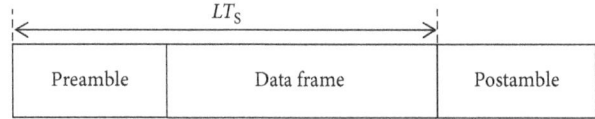

FIGURE 2: Frame format for Doppler estimation.

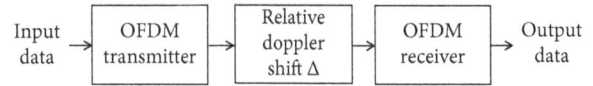

FIGURE 3: OFDM model with Doppler shift.

TABLE 1: OFDM parameters.

Modulation	QPSK-OFDM
Sampling frequency (kHz)	200
Center frequency (kHz)	45
Frequency band (kHz)	30 to 60
FFT size	1024
FFT length (ms)	34.1
GI length (ms)	4.3
Number of data subcarriers	512
Number of pilot subcarriers	512
OFDM frame length (ms)	38.4
Frequency domain oversampling factor	1, 2
FEC (coding rate)	Convolutional coding (3/4)
Channel model	AWGN
CNR	30 dB

Doppler shifts of 0, Δ, and $-\Delta$ by applying $\Delta_C = 0$. Figure 6 shows the OFDM block with Doppler shift patterns. One OFDM block with N samples is partitioned into three sections of A, B, and C. We make various Doppler shift patterns by shuffling the Doppler shifts in three sections so that all the shifts are allocated in a whole frame. Although further complex shifts should be considered in the actual environment, the proposed parallel resampling technique can cope with arbitrary unknown Doppler shifts (to be described later). We represent arbitrary shifts by shuffling the three Doppler shifts (expressing the swing of a transmitter unit as one of the examples) for simplification. Our model that a Doppler shift fluctuates with plus or minus in a whole frame can be observed in actual communication, which will be reported in Section 5.

The resampling procedure is illustrated in Figure 7. We evaluate various resampling ratios of $1/(1 - \Delta)$, $1/(1 - (\Delta/2))$, 1, $1/(1 + (\Delta/2))$, and $1/(1 + \Delta)$ to observe which ratio is the best choice. The magnitude of Doppler shift is set by $\Delta \cdot f_C = 10$ Hz, assuming the speed fluctuations of ± 1.2 km/h in moving platforms. $f_C (=45$ kHz) denotes the center frequency of OFDM. The other simulation parameters are the same in Table 1. The OFDM block sample N becomes 7,740 including the guard interval. The frequency domain oversampling factor of 2 is applied to extend the Doppler tolerance.

The results of EVMs and BERs are shown in Tables 2 and 3. The OFDM block length is the same for all cases. The highlighted cell shows the best performance for each case.

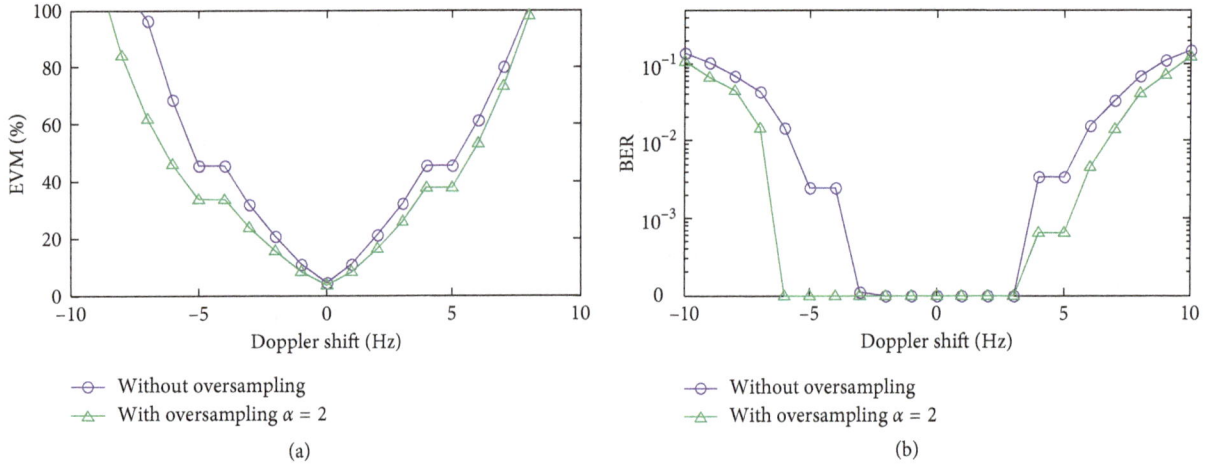

FIGURE 4: Tolerance for Doppler estimation errors. (a) EVM and (b) BER.

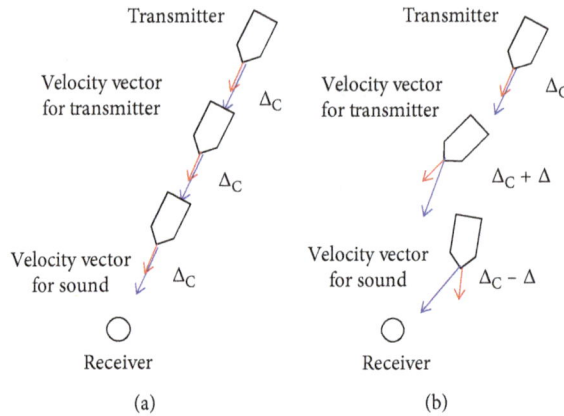

FIGURE 5: Doppler shift models. (a) Constant Doppler shift and (b) Doppler fluctuation.

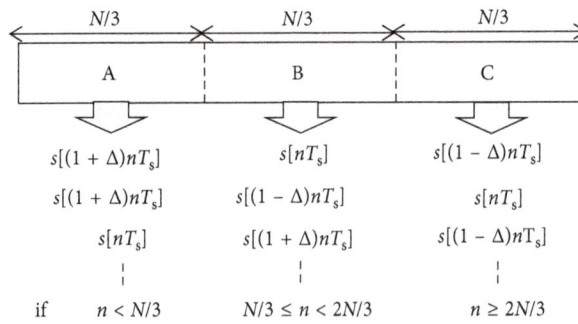

FIGURE 6: OFDM block with Doppler shift patterns.

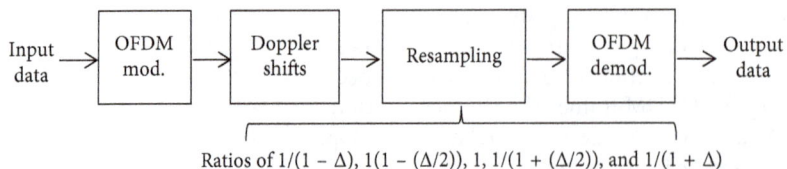

FIGURE 7: Resampling procedure.

TABLE 2: EVM results for various Doppler shift patterns and resampling ratios.

	Doppler shift patterns			Resampling ratio				
Case	A	B	C	$1/(1-\Delta)$	$1/(1-(\Delta/2))$	1	$1/(1+(\Delta/2))$	$1/(1+\Delta)$
(1)	Δ	0	$-\Delta$	76.9	36.1	36.3	90.0	128.8
(2)	Δ	$-\Delta$	0	131.8	83.1	35.1	32.6	80.3
(3)	0	Δ	$-\Delta$	137.5	160.8	47.4	29.9	60.0
(4)	0	$-\Delta$	Δ	154.3	122.3	68.0	57.2	103.1
(5)	$-\Delta$	Δ	0	197.2	149.1	119.6	62.1	29.9
(6)	$-\Delta$	0	Δ	188.8	163.6	71.4	21.8	32.2

TABLE 3: BER results Doppler shift patterns and resampling ratios.

	Doppler shift patterns			Resampling ratio				
Case	A	B	C	$1/(1-\Delta)$	$1/(1-(\Delta/2))$	1	$1/(1+(\Delta/2))$	$1/(1+\Delta)$
(1)	Δ	0	$-\Delta$	0.06	0.002	0.002	0.02	0.1
(2)	Δ	$-\Delta$	0	0.2	0.05	0.002	0	0.03
(3)	0	Δ	$-\Delta$	0.3	0.08	0	0	0.01
(4)	0	$-\Delta$	Δ	0.3	0.1	0.02	0.008	0.05
(5)	$-\Delta$	Δ	0	0.4	0.3	0.1	0.02	0
(6)	$-\Delta$	0	Δ	0.3	0.2	0.03	0	0

The average Doppler shift becomes zero for all Doppler shift patterns because the OFDM block length does not change. It indicates that resampling is not needed in the conventional approach. However, the appropriate resampling ratios having the lowest EVM and BER depend on Doppler shift patterns. If these Doppler shift patterns are known, the appropriate resampling ratios can be determined according to the results in Tables 2 and 3. Actual Doppler shift patterns are much complex and cannot be prospected.

The conventional approach takes the Doppler average in determining a resampling ratio. It induces the mismatch between determined and appropriate resampling ratios in the presence of large Doppler fluctuation. We take another approach to grasp the Doppler fluctuation, inserting a continuous wave (CW) into the OFDM frame to observe short-term Doppler shifts. The block diagram of CW-aided Doppler estimation is illustrated in Figure 8. The CW signal of $\sin(2\pi f_{CW}t)$ and the OFDM transmitted signal are multiplexed in frequency. The CW signal is allocated as one of the OFDM subcarriers, and frequency spacing between CW and OFDM is not required. The Doppler-affected CW signal is extracted by the bandpass filter. The short-term Doppler shifts can be measured by detecting phase offsets in IQ demodulation, where the observed Doppler shifts are plotted in Figure 8.

The statistical results based on short-term Doppler shifts are summarized in Table 4. The CW signal of 30 kHz is added into the OFDM block in Figure 6. When we look into the results of average, standard deviation, maximum, and minimum values, the standard deviation gives useful information. These standard deviation values are close to the Doppler fluctuation of 10 Hz. The phenomena that the average values do not become zero would be caused by sudden changes in Doppler shifts. Doppler dispersion provides useful information to grasp large Doppler fluctuation.

4. Parallel Resampling

As reported in Section 3, it is difficult to find an optimal resampling ratio even for the simple Doppler shift patterns given by three sections if their Doppler shift patterns are unknown. Doppler shift patterns should be assumed as almost unknown for actual Doppler fluctuation caused by the irregular motions of transmitter or receiver units. We consider statistical information in short-term Doppler shifts and present a parallel resampling technique that a resampling range is extended by measured Doppler dispersion.

The conceptions of single and parallel resampling techniques are compared in Figure 9. The single resampling technique, regarded as a conventional method, determines a resampling ratio according to the Doppler mean. As long as the Doppler shift is time invariant, the single resampling can provide adequate performance for Doppler compensation. However, it cannot cope with Doppler fluctuation especially for the case that Doppler shift patterns are unknown. The parallel resampling technique uses both Doppler mean and standard deviation and extends the resampling range. The optimal resampling ratio is detected by checking all outputs in the extended resampling units (called as branches); i.e., the exhaustive search is adopted. The details of the exhaustive search are explained by the following receiver structure.

The block diagram of the parallel resampling receiver is illustrated in Figure 10. The CW and OFDM signals are multiplexed across the communication frame in the transmitter side (Figure 10(a)). As shown in Figure 8, we have improved the frequency allocation of CW and OFDM that the CW is regarded as one of the OFDM subcarriers. Their frequency bands are 30 kHz and 30 kHz to 60 kHz, respectively. Our previous work required frequency spacing, where CW and OFDM are allocated in 45 kHz and 50 kHz to 60 kHz [10]. In the receiver side (Figure 10(b)), the CW signal is separated by the bandpass filter. The Doppler

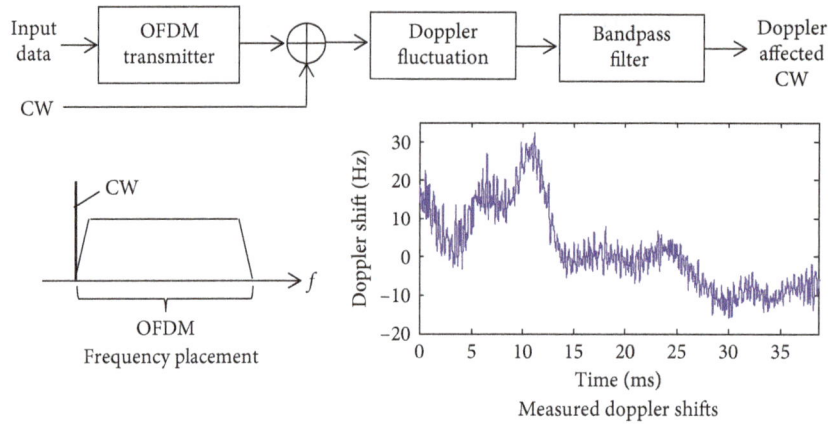

FIGURE 8: Block diagram of CW-aided Doppler estimation.

TABLE 4: Statistical results based on short-term Doppler shifts.

Case	Doppler shift patterns			Estimated Doppler shift (Hz)			
	A	B	C	Ave.	Std.	Max.	Min.
(1)	Δ	0	−Δ	1.8	10.5	32.4	−15.7
(2)	Δ	−Δ	0	4.9	12.5	44.1	−15.9
(3)	0	Δ	−Δ	2.9	9.9	29.8	−15.4
(4)	0	−Δ	Δ	3.3	13.1	49.6	−15.9
(5)	−Δ	Δ	0	7.0	12.9	48.9	−20.7
(6)	−Δ	0	Δ	4.4	10.2	43.0	−20.7

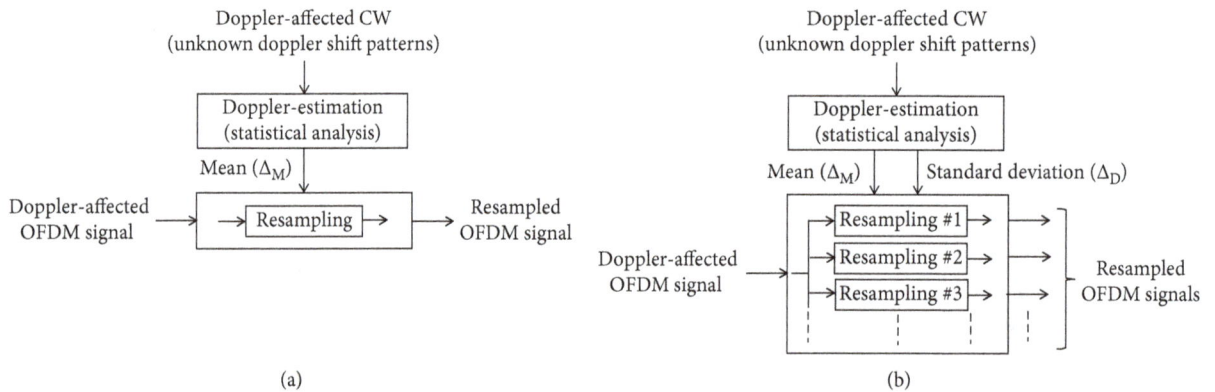

(a)   (b)

FIGURE 9: Conceptions of single and parallel resampling techniques. (a) Single resampling and (b) parallel resampling.

estimation unit calculates the mean and standard deviation values (Δ_M and Δ_D) from the measured short-term Doppler shifts. According to the values of Δ_M and Δ_D, the resampling range is extended as $1/(1 - \Delta_M - \Delta_D), \ldots, 1/(1 + \Delta_M), \ldots, 1/(1 + \Delta_M + \Delta_D)$.

The receiver takes multiple branches for resampling and OFDM demodulation blocks, where every branch has a different resampling ratio. As mentioned in Section 3, the appropriate resampling ratio depends on Doppler shift patterns. For unknown shift patterns, it is very difficult to find the appropriate resampling ratio by only evaluating the received signals before OFDM demodulation. We apply the exhaustive search to cope with this problem. The parallel resampling receiver performs data checking for the outputted bit data after OFDM demodulation and selects the best branch having no data error. In the transmitter side, cyclic redundancy check (CRC) codes are inserted in bit data before forward error correcting (FEC) coding. In the receiver side, all branches are checked whether their outputted data have an error or not. The best branch having no error is selected, which corresponds to selecting the appropriate resampling ratio. If all branches have data errors, the final data are generated by merging all decoded data of branches in the bit level. The overhead of CRC codes might be counted. The overhead would be very small as far using CRC-16 or CRC-8.

The procedure of OFDM demodulation is shown in Figure 10(c), which is adopted in typical OFDM systems.

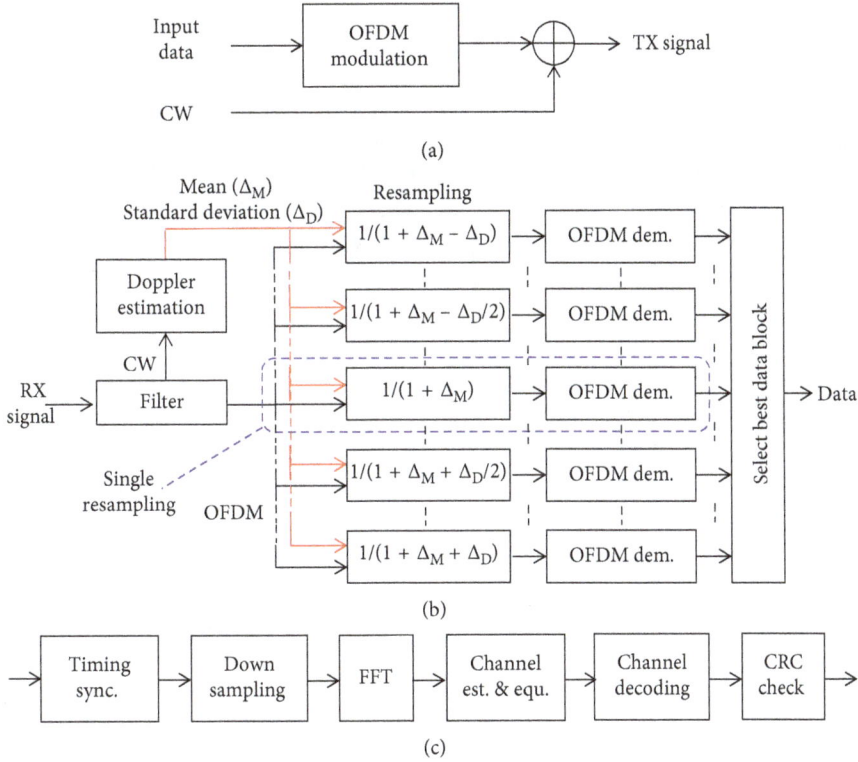

FIGURE 10: Block diagram of parallel resampling receiver. (a) Transmitter, (b) receiver, and (c) OFDM demodulation.

The single resampling is highlighted by the dashed lines in Figure 10(b), where the resampling ratio is given by $1/(1 + \Delta_M)$ according to the Doppler mean value Δ_M. The performance comparison of single and parallel resampling techniques is reported in Section 5.

We mention examples of resampling range and number of branches in parallel resampling. Table 5 shows the examples of resampling ratios and number of branches (N_P). We express the resampling ratio by an integer format as

$$\frac{1}{1 + \Delta_M + \Delta_D} = \frac{R}{R + \Delta_M + \Delta_D}, \tag{5}$$

where R relates the resolution of resampling as explained in Section 2. When we set $R = 10000$, and $\Delta_M = 0$, the resampling ratio becomes $10000 : 10000 + \Delta_D$ as an integer ratio. Note that the parameter of R depends on the Doppler tolerance. We set it according to the Doppler test results in Section 2. For the Doppler standard deviation value of 5 Hz (with zero mean), the parallel resampling receiver takes three branches whose ratios are $10000 : 9999$, $10000 : 10000$, and $10000 : 10001$. It indicates that parallel resampling has a trade-off between improving communication performance and increasing computational complexity.

The number of branches has been fixed as $N_P = 11$ based on the empirical data in our acoustic communication tests in our previous work [10]. This paper applied the dynamic control of branches (we call it variable branches) in order to reduce computational complexity. The procedure for the variable branches is explained as follows:

TABLE 5: Examples of resampling ratios and number of branches.

Doppler std. ($\Delta_D f_C$) (Hz)	Resampling range (10000:m)	Number of branches (N_P)
5	(9999, 10000, 10001)	3
10	(9998, 9999,..., 10002)	5
20	(9996, 9997,..., 10004)	9

(a) Cut out a received signal for a certain frame

(b) Calculate Δ_M and Δ_D from the extracted CW

(c) Determine N_P according to Δ_D

Since the resampling ratio is adjusted by the unit of 1/10000, the target signal for resampling should hold at least 10000 samples. For example, the numbers of samples per frame amount from 10,000 to 100,000 for 200 kHz sampling frequency and 1024 subcarriers. It is advantageous to perform both resampling and statistical analysis of Doppler mean and standard deviation for the entire frame. The calculation of standard deviation for every frame is required for the variable branches. Its computational cost is $O(3N)$, which is much smaller than resampling processing itself given by $O(20N^2)$ (N denotes the number of samples per frame).

When the magnitude of Doppler fluctuation changes as time goes, the complexity reduction is available by changing the number of branches for each frame.

5. Experimental Results

We conducted the communication experiment on November 2017 in Mombetsu Port, Hokkaido, Japan. The

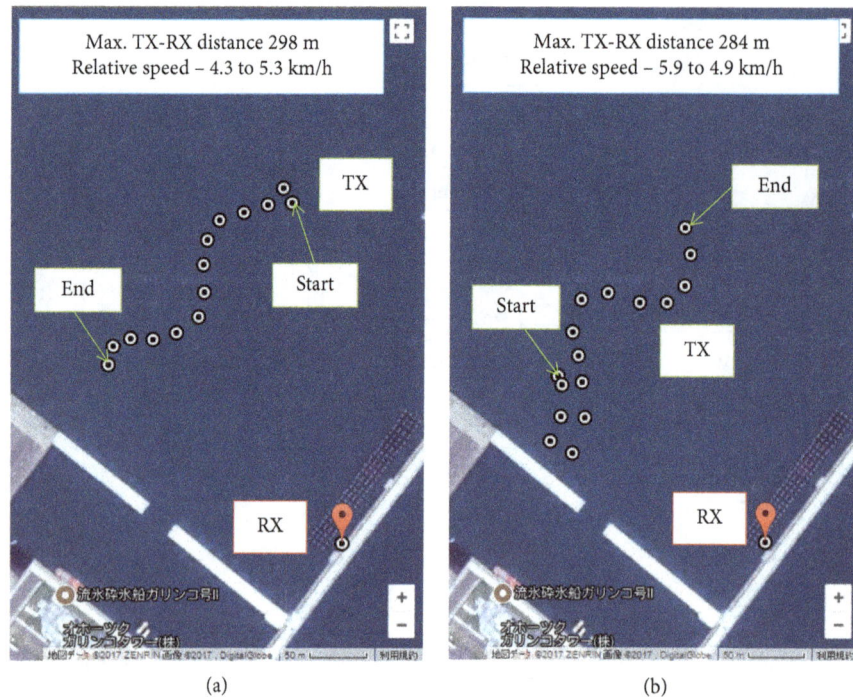

FIGURE 11: Locations of transmitter and receiver units on Google Maps [12]. (a) Test 1 and (b) Test 2.

locations of transmitter (TX) and receiver (RX) units are plotted on aerial photograph in Figure 11. The sea depth was about 10 m. The receiver was fixed at 2 m below the sea surface. The TX unit was located at 1 m below the surface and moved by a ship with meandering trajectory. The average ship' speeds were about 3.7 km/h (2 knot) and 5.6 km/h (3 knot) in Test 1 and Test 2. The relative Doppler speeds between TX and RX units are not the same as the ship speeds because the directions of the ship do not agree with those of sound propagation.

The measured delay profile at the RX unit is plotted in Figure 12. The cluster of large delay waves is observed around 40 ms, inducing strong multipath interference. These delay waves are caused by the reflection of sea surface, bottom, and two side walls near the RX unit. Table 6 enumerates the experimental parameters. Due to the severe conditions on Doppler and multipath, the transmit data rate was set to 3 kbps. We have compared communication performance in single and parallel resampling techniques.

The results of carrier-to-noise ratio (CNR) and relative Doppler speed are reported in Figures 13 and 14 for Test 1 and Test 2. The horizontal axis denotes the frame number. We can observe Doppler fluctuation that the relative speed changes with plus or minus because the transmitter unit moves in the meandering trajectory as well as Figure 1(b). The CNR values are not uniform, which would be caused by the wave disturbance by a ship propeller and the ship direction.

The results of BER and number of branches (N_p) are plotted in Figures 15 and 16. We compared BER results in baseline (without resampling), single resampling, and parallel resampling (with fixed and variable branches). The results of BER = 0 are plotted on the line of BER = 10^{-4}.

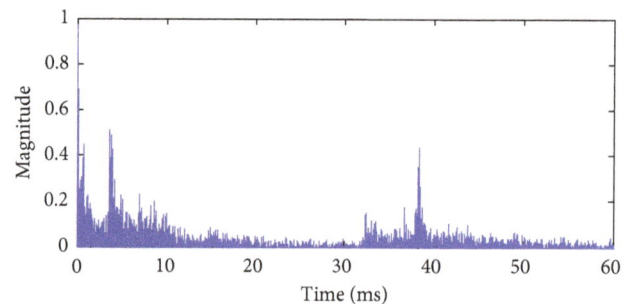

FIGURE 12: Delay profile at RX unit.

TABLE 6: Experimental parameters.

Modulation	QPSK-OFDM
Sampling frequency (kHz)	200
Center frequency (kHz)	45
Frequency band (kHz)	30 to 60
FFT size	1024
FFT length (ms)	34.1
GI length (ms)	4.3
Number of data subcarriers	512
Number of pilot subcarriers	512
Number of OFDM blocks	4
OFDM frame length (ms)	154
Frequency domain oversampling factor	2
FEC (coding rate)	Convolutional coding (1/2)
Transmit data rate (kbps)	3

The baseline degrades communication performance in the presence of Doppler, where most of frames have a larger BER close to 0.5. The single resampling can decrease BERs in

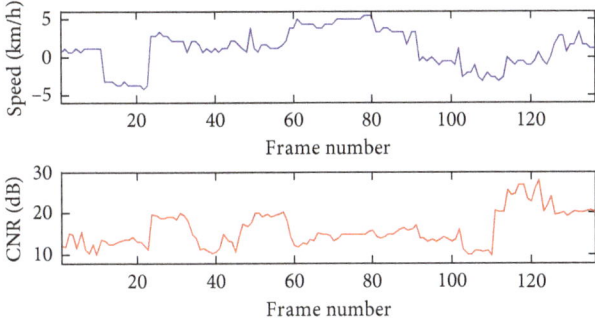

FIGURE 13: Results of CNR and relative Doppler speed in Test 1.

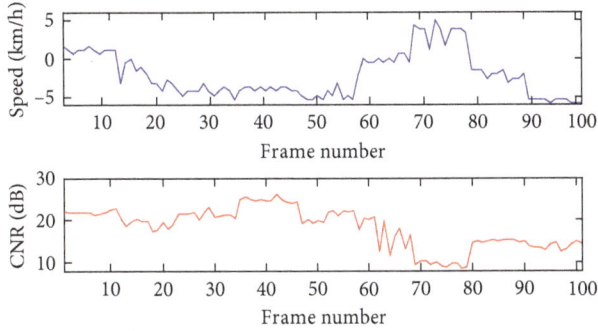

FIGURE 14: Results of CNR and relative Doppler speed in Test 2.

many frames, however, remains as bit errors in some frames. The parallel resampling has better BER performance than the baseline and single resampling. All communication frames are successfully demodulated in Test 1. Comparing fixed branches with variable branches in parallel resampling, their BER results are the same for all frames. The transitions of number of branches in the variable branches are also plotted in the graph. The number of branches changes for each frame, which indicates that the complexity reduction is available without sacrificing communication performance.

The measured Doppler shifts for the frame numbers of 33, 49, and 102 in Test 1 are shown in Figure 17. We can observe that the Doppler shift fluctuates with plus or minus within an OFDM frame as mentioned in Section 3. The BERs for the frame numbers of 33, 49, and 102 are 0.17, 0.45, and 0.37 in single resampling and all zeros in parallel resampling. Their results indicate that parallel resampling can improve communication performance under actual Doppler fluctuation.

Table 7 shows the summary of experimental results. The overall results of frame error rate (FER) are 0.096 and 0.24 for the single resampling in Test 1 and Test 2. The parallel resampling achieves FERs of 0 and 0.10. The parallel resampling with the variable branches saves computational cost by 28% to 33% compared with the fixed branches.

The parallel resampling has improved communication performance for large Doppler fluctuation; however, it relies on the exhaustive search with low computation efficiency. The further complexity reduction should be considered as a future issue. One of the key ideas is use of compressing sensing (CS). CS-based channel estimation in OFDM has

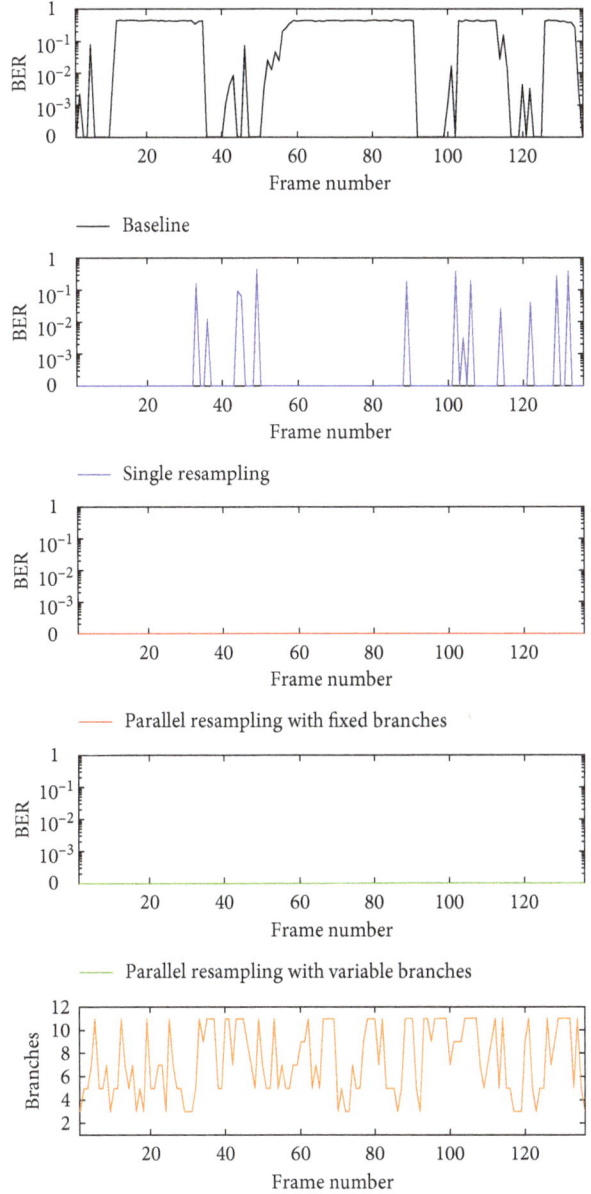

FIGURE 15: Results of BER and number of branches in Test 1.

been discussed in [13]. In UAC, a doubly spread acoustic channel (dispersion in time and frequency) is estimated by orthogonal matching pursuit (OMP) algorithm in [14, 15], which makes use of the sparse structure of channel impulse response. If Doppler fluctuation could be accurately estimated by the similar approach, an optimal resampling ratio is detected without the exhaustive search. Since the computational cost of OMP algorithm is much higher than that of Doppler mean and standard deviation, we would have to compare the overall complexity including Doppler estimation, resampling, and demodulation.

6. Conclusion

This paper improved a resampling technique for large Doppler fluctuation in underwater acoustic communication.

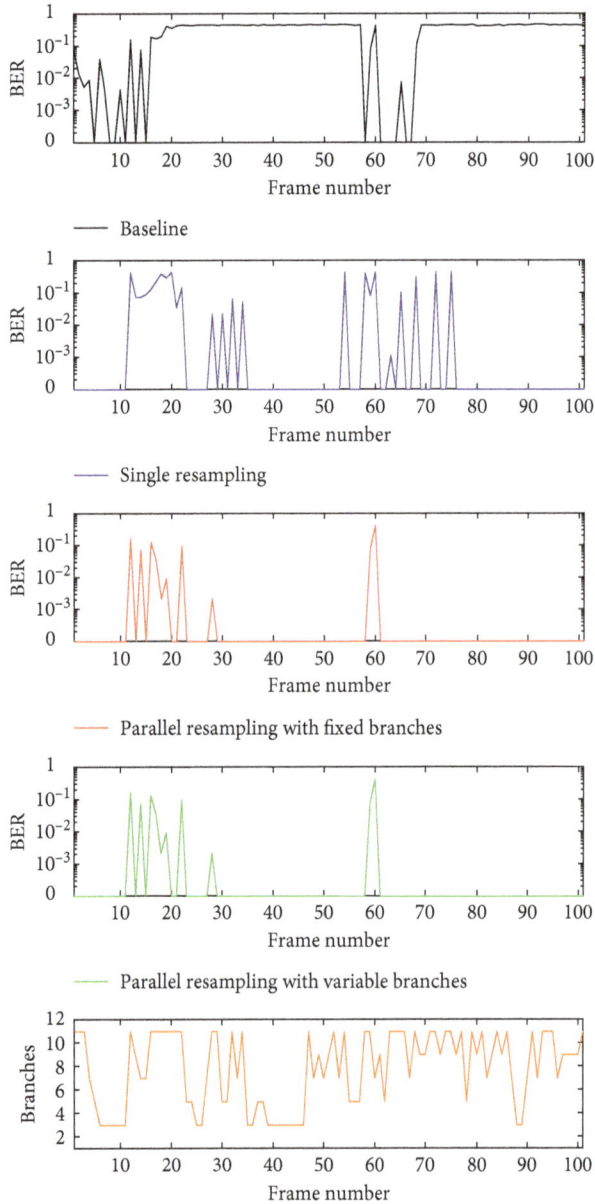

FIGURE 16: Results of BER and number of branches in Test 2.

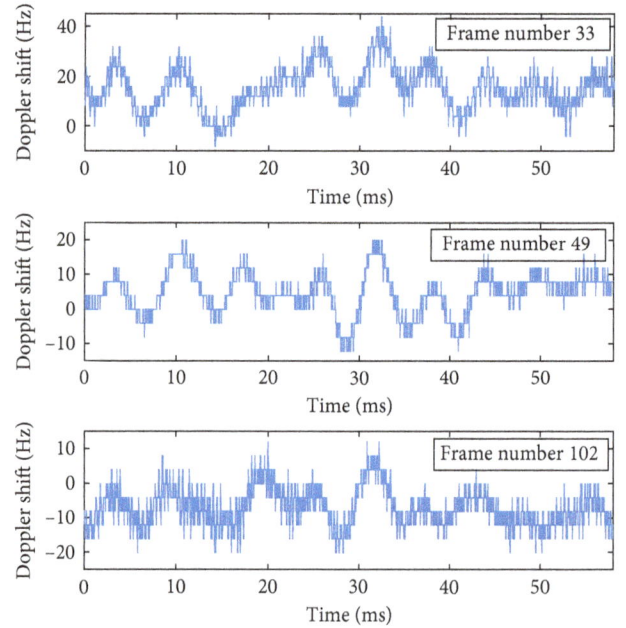

FIGURE 17: Measured Doppler shifts for frame numbers of 33, 49, and 102 in Test 1.

TABLE 7: Summary of experimental results.

	Single resampling		Parallel resampling with fixed branches [10]		Parallel resampling with variable branches (this work)	
	Test 1	Test 2	Test 1	Test 2	Test 1	Test 2
FER	0.096	0.24	0	0.10	0	0.10
Number of branches (N_P)	1	1	11	11	7.4	7.9

The influence of Doppler fluctuation has been investigated by giving unknown Doppler shift patterns. According to the observation that Doppler dispersion gives useful information to grasp Doppler fluctuation, we have presented the parallel resampling technique that the resampling range is extended according to the measured Doppler deviation. The effectiveness of parallel resampling has been confirmed by evaluating communication performance comparing with single resampling.

The complexity reduction has been done by applying the variable branches in the parallel resampling receiver. It saves computational cost by 30 % compared with the fixed branches. However, its computation cost remains rather high compared with the single resampling. The further complexity reduction will be studied in our future work.

Conflicts of Interest

The authors declare that there are no conflicts of interest regarding the publication of this paper.

Acknowledgments

The authors would like to thank Dr. Hiroshi Tanimoto for their support and assistance with this project. This work was supported by the City of Mombetsu, Mombetsu Okhotsk Tower, and JSPS KAKENHI (Grant Number 16K18099).

References

[1] M. Stojanovic, "Low complexity OFDM detector for underwater acoustic channels," *IEEE Oceans*, pp. 18–21, 2006.

[2] S. Yoshizawa, H. Tanimoto, and T. Saito, "Data selective rake reception for underwater acoustic communications in strong

multipath interference," *Journal of Electrical and Computer Engineering*, vol. 2017, Article ID 5793507, 9 pages, 2017.

[3] B. S. Sharif, J. Neasham, O. R. Hinton, and A. E. Adams, "A computationally efficient Doppler compensation system for underwater acoustic communications," *IEEE Journal of Oceanic Engineering*, vol. 25, no. 1, pp. 52–61, 2000.

[4] B. Li, S. Zhou, M. Stojanovic, L. L. Freitag, and P. Willett, "Multicarrier communication over underwater acoustic channels with nonuniform Doppler shifts," *IEEE Journal of Oceanic Engineering*, vol. 33, no. 2, pp. 198–209, 2008.

[5] K. Tu, T. M. Duman, M. Stojanovic, and J. G. Proakis, "Multiple-resampling receiver design for OFDM over Doppler-distorted underwater acoustic channels," *IEEE Journal of Oceanic Engineering*, vol. 38, no. 2, pp. 333–346, 2013.

[6] S. Beygi and U. Mitra, "Optimal Bayesian resampling for OFDM signaling over multi-scale multi-lag channels," *IEEE Signal Processing Letters*, vol. 20, no. 11, pp. 1118–1121, 2013.

[7] Z. Wang, S. Zhou, G. B. Giannakis, C. R. Berger, and J. Huang, "Frequency-domain oversampling for zero-padded OFDM in underwater acoustic communications," *IEEE Journal of Oceanic Engineering*, vol. 37, no. 1, pp. 14–24, 2012.

[8] T. Ebihara and G. Leus, "Doppler-resilient orthogonal signal-division multiplexing for underwater acoustic communication," *IEEE Journal of Oceanic Engineering*, vol. 41, no. 2, pp. 408–427, 2016.

[9] A. E. Abdelkareem, B. S. Sharif, C. C. Tsimenidis, and J. A. Neasham, "Compensation of linear multiscale Doppler for OFDM-based underwater acoustic communication systems," *Journal of Electrical and Computer Engineering*, vol. 2012, Article ID 139416, 16 pages, 2012.

[10] S. Yoshizawa, H. Tanimoto, T. Saito et al., "Parallel resampling receiver for underwater acoustic communication with non-uniform Doppler shifts," *IEEE Oceans–Anchorage*, p. 4, 2017.

[11] C.-H. Hwang, K.-M. Kim, S.-Y. Chun, and S.-K. Lee, "Doppler estimation based on frequency average and remodulation for underwater acoustic communication," *International Journal of Distributed Sensor Networks*, vol. 2015, Article ID 746919, 2015.

[12] https://www.google.com/permissions/geoguidelines.html.

[13] Y. Zhang, R. Venkatesan, O. A. Dobre, and L. Cheng, "Novel compressed sensing-based channel estimation algorithm and near-optimal pilot placement scheme," *IEEE Transaction on Wireless Communication*, vol. 15, no. 4, pp. 2590–2603, 2016.

[14] W. Li and J. C. Preisig, "Estimation of rapidly time-varying sparse channels," *IEEE Journal of Oceanic Engineering*, vol. 32, no. 4, pp. 927–939, 2007.

[15] F. Qu, X. Nie, and W. Xu, "A two-stage approach for the estimation of doubly spread acoustic channels," *IEEE Journal of Oceanic Engineering*, vol. 40, no. 1, pp. 131–143, 2015.

18

An Optimization Synchronization Algorithm for TDDM Signal

Fang Liu and Yongxin Feng

School of Information Science and Engineering, Shenyang Ligong University, Shenyang 110159, China

Correspondence should be addressed to Yongxin Feng; onceowned_1019@163.com

Academic Editor: Jit S. Mandeep

The time division data modulation (TDDM) mechanism is recommended to improve the communications quality and enhance the antijamming capability of the spread spectrum communication system, which will be used in the next generation global navigation satellite (GNSS) systems. According to the principle and the characteristics of TDDM signal, an optimization synchronization algorithm is proposed. In the new algorithm, the synchronization accuracy and environmental adaptability have been improved with the special local sequence structure, the multicorrelation processing, and the proportion threshold mechanism. Thus, the inversion estimation formula was established. The simulation results demonstrate that the new algorithm can eliminate the illegibility threat in the synchronization process and can adapt to a lower SNR. In addition, this algorithm is better than the traditional algorithms in terms of synchronization accuracy and adaptability.

1. Introduction

To improve the communications quality and the antijamming capability of the spread spectrum communication system, the time division data modulation (TDDM) [1] mechanism is recommended based on the traditional modulation methods. The new mechanism will be used in the next generation global navigation satellite systems [2, 3], for example, the GPS-III [4, 5] system. The difference of the TDDM and the traditional modulation mechanism is that nondata element is permitted in the new signal, and the spread spectrum processing depends on the pseudonoise (PN) code odd or even scheduling. The typical TDDM principle is that the even sequences sign is the same as the PN code sign, whereas the odd sequence sign is uncertain. The uncertain factor will cause some illegibility problems in the signal synchronization processing, such as inconspicuous correlation peak and higher data inverse error.

If the traditional algorithms are applied in TDDM signal synchronization, the correlation peak will decrease to half of the autocorrelation function when the data inversion position is in the middle of the received signal, and the correlation result is the same as the cross-correlation function when the data sign is successive negative. Thus, the approximate noncorrelation characteristic appears in

this TDDM signal receiving. Currently, several PN code receiving [6, 7] and detection [8] technologies have been proposed, which can provide some referenced methods for TDDM signals synchronization. And the synchronization circuit system [9] based on field programmable gate array has become one of the development directions for GPS and even for a GNSS embedded real-time software receiver. Furthermore, large numbers of synchronization algorithms are proposed to improve the navigation signals receiving capability. For example, in [10], an efficient differential coherent accumulation algorithm for weak GPS signal bit synchronization was presented to reduce the computational load by approximately sixfold. In [11], the authors provided a detailed analysis of the significance of the cell-correlation phenomenon in MF correlates for the two widely used signal families in GNSS, namely, BPSK and BOC; next, the theoretical analysis was validated by Monte Carlo simulations. In [12], the code acquisition architecture for GNSS was presented to reduce the buffer resource required without reducing the numbers of code, frequency, and satellite bins. In [13], the authors proposed to reconstruct the correlation results of DBZP to improve the detection performance of the previously published double-block zero-padding (DBZP) method for weak GNSS long PN code signals.

In summary, from the perspective of algorithm generality, the TDDM signal synchronization algorithm is usually divided into two categories, namely, the filled zero (FZ) algorithm [14] and the positive-negative (PN) algorithm [15]. The principle of the FZ algorithm is that the odd bits of the local code are replaced by zero, and then the correlation arithmetic is executed by using the preprocessed receiving signal and the local signal. This algorithm reduces the uncertainty factor and enhances the estimation inversion precision by sacrificing the correlation peak in the synchronization processing. The principle of the PN algorithm is that the local code is processed by the positive and negative TDDM mechanism, and next, the correlation arithmetic is executed by using the preprocessed receiving signal and the positive local sequence. This algorithm imports the inversion judgment mechanism and combines the correlation peaks of the two channels to estimate the inversion position. The FZ algorithm and PN algorithm can be applied to receive TDDM signal and to estimate the inversion position, but the estimation precision of these algorithms is expected to improve in a complicated environment.

2. The Optimization Synchronization Algorithm with Illegibility Elimination

To describe the new algorithm, the digital intermediate frequency received signal $S'(n)$ is modeled after frequency downconversion, ADC, and band-pass filtering. The processed signal is expressed as

$$S'(n) = D(n) C_T(n) \sin(\omega n + \omega_0 n) + N_0(n), \qquad (1)$$

where $D(n)$ is the binary data, $C_T(n)$ is the TD codes, ω is the intermediate frequency due to the downconversion, ω_0 is the Doppler frequency, and $N_0(n)$ is the band-limited mixed noise signals. L is the number of the received signal.

To eliminate the data inversion illegibility threat, we structure the special local sequences. The local sequence is produced using the same principle with the PN code of the received signal, and it is denoted as $C_i(n)$. The phase of the local code is expressed as variable i, for which the initialization value is set as p. Furthermore, the filling zero sequence is produced and expressed as $C_{0i}(n)$, in which the odd number sequence is zero and the even number sequence is the same as $C_i(n)$; these processes are expressed as

$$C_{0i}(n) = \{a_i, a_{1+i}, a_{2+i}, \ldots, a_{L+i-1}\}$$
$$C_i(n) = \{b_i, b_{1+i}, b_{2+i}, \ldots, b_{L+i-1}\} \qquad (2)$$

with

$$a_n = \begin{cases} 0, & n = \text{odd number} \\ b_n, & n = \text{even number.} \end{cases} \qquad (3)$$

The preprocessing received signal is executed by the in-phase channel and quadrature channel frequency compensation arithmetic to overcome the influence of the uncertain frequency. The processing is expressed as

$$\begin{aligned} S'_I(n) &= S'(n) \sin(\omega n) \\ &= [D(n) C_T(n) \sin(\omega n + \Delta n) + N_0(n)] \sin(\omega n) \\ &= \frac{1}{2} D(n) C_T(n) \cos(\Delta n) \\ &\quad - \frac{1}{2} D(n) C_T(n) \cos(2\omega n + \Delta n) \\ &\quad + N_0(n) \sin(\omega n), \end{aligned} \qquad (4)$$
$$\begin{aligned} S'_Q(n) &= S'(n) \cos(\omega n) \\ &= [D(n) C_T(n) \sin(\omega n + \Delta n) + N_0(n)] \cos(\omega n) \\ &= \frac{1}{2} D(n) C_T(n) \sin(2\omega n + \Delta n) \\ &\quad + \frac{1}{2} D(n) C_T(n) \sin(\Delta n) + N_0(n) \cos(\omega n). \end{aligned}$$

The two-channel compensated signals are processed using low-pass filtering, which are given by

$$\begin{aligned} S_I(n) &= \frac{1}{2} D(n) C_T(n) \cos(\Delta n), \\ S_Q(n) &= \frac{1}{2} D(n) C_T(n) \sin(\Delta n). \end{aligned} \qquad (5)$$

The local code phase i is moved from p to $p + L - 1$, and next, the filling zero sequence $C_{0i}(n)$ is, respectively, multiplied by the received signal $S_I(n)$ and $S_Q(n)$. These quantities are given by

$$\begin{aligned} X_{I0}(n) &= S_I(n) \cdot C_{0i}(n) \\ &= \frac{1}{2} D(n) C_T(n) C_{0i}(n) \cos(\Delta n), \\ X_{Q0}(n) &= S_Q(n) \cdot C_{0i}(n) \\ &= \frac{1}{2} D(n) C_T(n) C_{0i}(n) \sin(\Delta n). \end{aligned} \qquad (6)$$

Combining the product results of the two channels, the square sum is calculated to inhibit the Doppler influence and then to achieve the correlation purpose. The processing is expressed as

$$\begin{aligned} X_i &= \sum \left(X_{I0}(n) \right)^2 + \left(X_{Q0}(n) \right)^2 \\ &= \sum \left[\left(\frac{1}{2} D(n) C_T(n) C_{0i}(n) \cos(\Delta n) \right)^2 \right. \\ &\quad \left. + \left(\frac{1}{2} D(n) C_T(n) C_{0i}(n) \sin(\Delta n) \right)^2 \right] \end{aligned}$$

$$= \sum \left[\frac{1}{4} \left(D(n) C_T(n) C_{0i}(n) \right)^2 \right.$$

$$\left. \cdot \left(\cos(\Delta n)^2 + \sin(\Delta n)^2 \right) \right]$$

$$= \sum \left[\frac{1}{4} \left(D(n) C_T(n) C_{0i}(n) \right)^2 \right]. \tag{7}$$

From (7), we can see that the influences of Doppler and wave have been restrained. Furthermore, the correlation results X_i are obtained using the change of variable i, so the correlation sequence is expressed as

$$X(n) = \left\{ X_p, X_{p+1}, X_{p+2}, \ldots, X_{p+L-1} \right\}. \tag{8}$$

Although the existing algorithms can automatically adjust the threshold, they cannot improve the correlation peak values. Thus, we structure the proportion peak to improve the new algorithm adaptability. We define the proportion peak V as

$$V = \frac{\max[X(n)]}{\sum(X(n))}. \tag{9}$$

Thus, the peak judgment operation is performed after the proportion peak processing. If the proportion peak cannot exceed the threshold G, then reacquisition will be operated after the local PN codes and adjusted Doppler shift, or else the value i of X_i is calculated using the proportion peak V. Let the calculated i be P_M; namely, $i = P_M$. Thus, the calculated i is imported to formula (2), and the results are expressed as

$$C_{0P_M}(n) = \left\{ a_{P_M}, a_{1+P_M}, a_{2+P_M}, \ldots, a_{L+P_M-1} \right\},$$
$$C_{P_M}(n) = \left\{ b_{P_M}, b_{1+P_M}, b_{2+P_M}, \ldots, b_{L+P_M-1} \right\}. \tag{10}$$

The received signal $S_I(n)$ is multiplied by the new local signals $C_{0P_M}(n)$ and $C_{P_M}(n)$, whose results are given by (11) and (12). Thus, the accumulative totals are given by (13) and (14):

$$X'_{I0}(n) = S_I(n) \cdot C_{0P_M}(n)$$
$$= \frac{1}{2} D(n) C_T(n) C_{0P_M}(n) \cos(\Delta n), \tag{11}$$

$$X'_{IZ}(n) = S_I(n) \cdot C_{P_M}(n)$$
$$= \frac{1}{2} D(n) C_T(n) C_{P_M}(n) \cos(\Delta n), \tag{12}$$

$$R_{I0} = \sum \left(X'_{I0}(n) \right)$$
$$= \frac{1}{2} \sum \left(D(n) C_T(n) C_{0P_M}(n) \cos(\Delta n) \right), \tag{13}$$

$$R_{IZ} = \sum \left(X'_{IZ}(n) \right)$$
$$= \frac{1}{2} \sum \left(D(n) C_T(n) C_{P_M}(n) \cos(\Delta n) \right). \tag{14}$$

In this condition, let A be the correlation result of the even number sequence when the data sign is negative, and let α be the influence coefficient, which is given by (15). At the same time, let B be the correlation result of the even number sequence when the data sign is positive, and let β be the influence coefficient, which is given by (16):

$$A = \frac{1}{2\beta} \sum \left(D(n) C_T(n) C_{0P_M}(n) \cos(\Delta n) \right), \tag{15}$$

when the data sign is negative,

$$B = \frac{1}{2\alpha} \sum \left(D(n) C_T(n) C_{0P_M}(n) \cos(\Delta n) \right), \tag{16}$$

when the data sign is positive.

In this paper, the theoretical correlation result between the local sequence and the received signal is defined as

$$R'_{I0} = \sum \left(D(n) C_T(n) C_{0P_M}(n) \right). \tag{17}$$

In mixed noise condition, the relationship between the correlation result and A, B, α, and β can been given by

$$\alpha \cdot B + \beta \cdot A = R_{I0}, \tag{18}$$

$$\alpha \cdot B - \beta \cdot A = R_{IZ} - R_{I0}. \tag{19}$$

In view of these characteristics and combining (18) and (19), we can calculate the sequence number of the same sign as

$$B = \frac{R_{IZ}}{2\alpha}. \tag{20}$$

Furthermore, the relationship between the real correlation and the theoretical correlation result is given by (21), and then the influence coefficient α is calculated and given by (22):

$$R_{I0} = \alpha \cdot R'_{I0}, \tag{21}$$

$$\alpha = \frac{R_{I0}}{R'_{I0}}. \tag{22}$$

We can rectify formula (20) as (23), and then we obtain the formula (24), according to the relationship of the theoretical correlation result and the processing number. Thus, formula (23) is renewed as (25):

$$B = \frac{R_{IZ}}{2} \cdot \frac{R'_{I0}}{R_{I0}} = \frac{R_{IZ} \cdot R'_{I0}}{2R_{I0}}, \tag{23}$$

$$R'_{I0} = \frac{L}{2}, \tag{24}$$

$$B = \frac{R_{IZ} \cdot L}{4R_{I0}}. \tag{25}$$

In view of the same relation of the odd number and the even number, the sum of the positive sign and the negative sign is equal to the processing number L. Finally, the inversion estimation formula is established to calculate

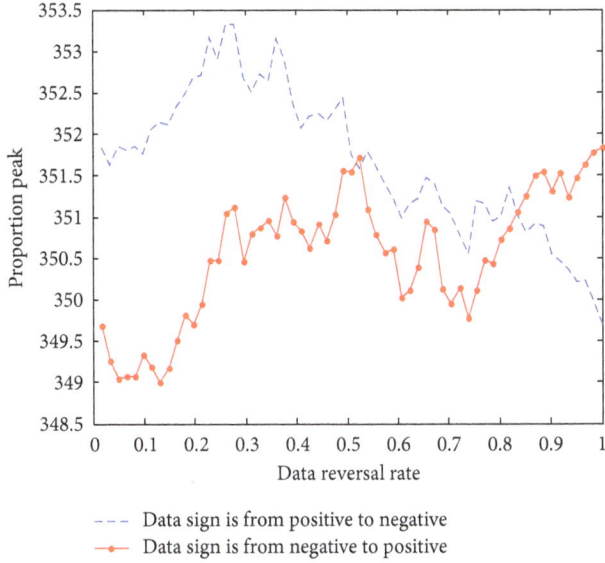

FIGURE 1: The proportion peak change with changing data reversal rate.

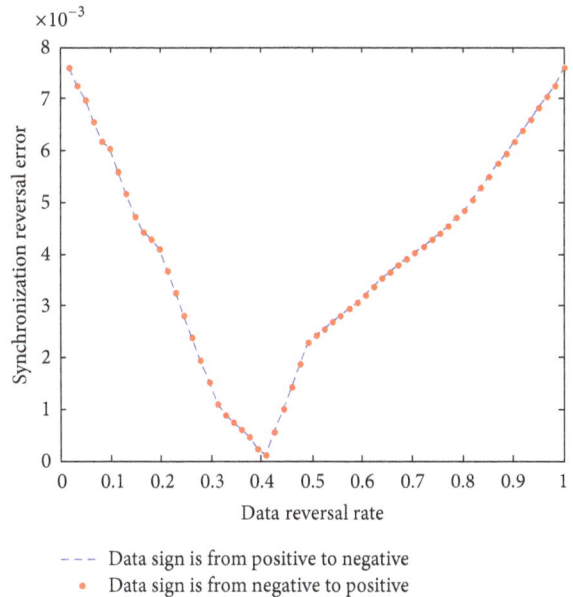

FIGURE 2: The reversal position change with changing data reversal rate.

the reversal position combining the above formulas, and the estimation reversal position is expressed as

$$\rho = \frac{L - 2B}{L} = 1 - \frac{R_{IZ} \cdot L}{2R_{I0} \cdot L} = 1 - \frac{R_{IZ}}{2R_{I0}}. \qquad (26)$$

3. Simulation and Analysis

3.1. The Capability Analysis. To furnish the judgment criterion for the synchronization algorithms, the thresholds G for the new algorithm and the existing algorithms are ascertained. These algorithms are analyzed using the following parameters: 5 MHz PN code frequency, 10 MHz intermediate wave frequency, 40 MHz sampling frequency, and 50 bps data rate. In this simulation condition, three algorithms' thresholds are as high as 10.

The ratio of the actual inversion position to the total length is defined as the data reversal rate. And with changing data reversal rate, the synchronization proportion peak change is shown in Figure 1. The results show that the new algorithm proportion peak can exceed 10 and can achieve the threshold required when the data sign is from positive to negative or is from negative to positive. With changing data reversal rate, the changes of the estimated reversal position and the theoretical reversal position are shown in Figure 2, the results indicate that the estimated value is approximately equal to the theoretical value. Furthermore, there may be some error between the estimated inversion position and the actual inversion position, so the estimated reversal error is tested. With changing actual data reversal rate, the estimated reversal error change is shown in Figure 3, in which x-axis is the ratio of the actual inversion position to the total length and y-axis is the estimated error between the estimated inversion position and the actual inversion position. The results show that the error is highest when the actual data

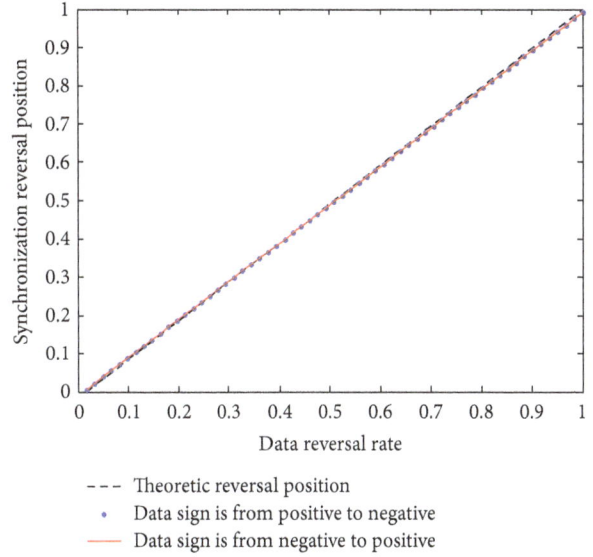

FIGURE 3: The estimated reversal error change with changing data reversal rate.

reversal position is in the middle of the received signal; however, all errors can be less than the 10^{-4} level.

In the conditions that the data sign is from positive to negative or is from negative to positive, the relationship of the synchronization proportion peak and SNR is shown in Figure 4, revealing that the proportion peak decreases gradually with decreasing SNR. The proportion peak cannot achieve the threshold required when the SNR is less than -37 dB; thus, the new algorithm's adaptability to the SNR environment is more than -37 dB.

Furthermore, when the input parameter of the actual reversal rate is constant, which is 0.32, the new algorithm

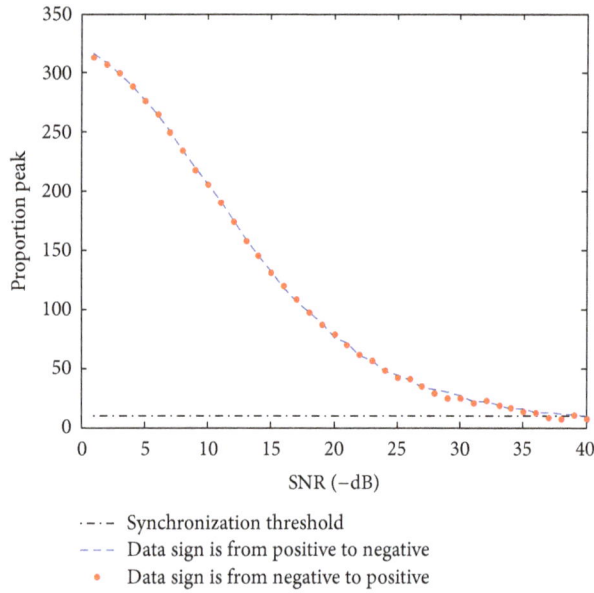

- · - Synchronization threshold
- - - Data sign is from positive to negative
- • Data sign is from negative to positive

FIGURE 4: The relationship of the synchronization proportion peak and the SNR.

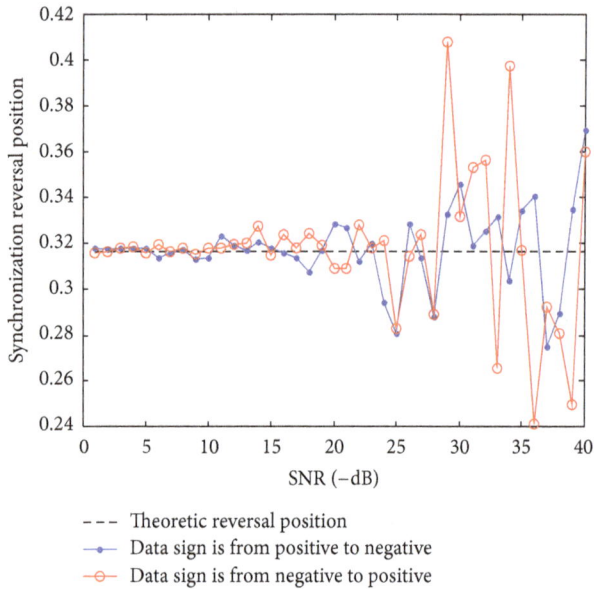

- - - Data sign is from positive to negative
- • Data sign is from negative to positive

FIGURE 6: The estimated reversal error change with changing SNR.

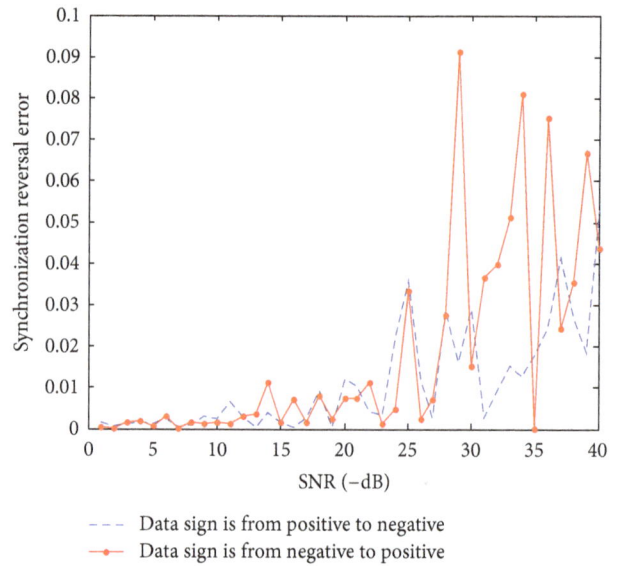

- - - Theoretic reversal position
- • Data sign is from positive to negative
- ○ Data sign is from negative to positive

FIGURE 5: The estimated reversal position change with changing SNR.

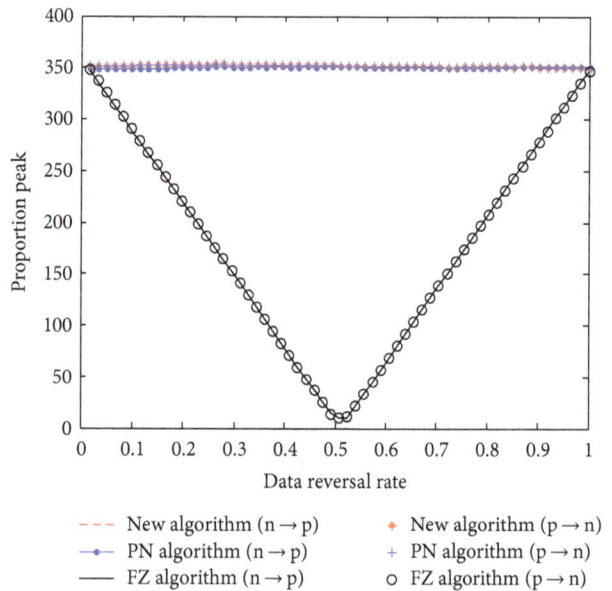

- - - New algorithm (n → p)	◆ New algorithm (p → n)
• PN algorithm (n → p)	+ PN algorithm (p → n)
— FZ algorithm (n → p)	○ FZ algorithm (p → n)

FIGURE 7: The proportion peak comparison with changing data reversal rate.

is tested. With changing SNR, the changes of the estimated reversal position and the theoretical reversal position are shown in Figure 5, revealing that the estimated value increases gradually with decreasing SNR. Next, the absolute value of the estimated reversal error is shown in Figure 6, the result of which indicates that the error increases clearly when SNR is less than −25 dB.

3.2. The Comparison Analysis of the Algorithms. To verify the superiority of the new algorithm, it is compared with other algorithms, namely, the PN algorithm and the FZ algorithm.

The parameter n → p represents that the data sign is from negative to positive, and p → n represents that the data sign is from positive to negative. With changing data reversal rate, the proportion peak comparison is shown in Figure 7. The result shows that the FZ algorithm's proportion peak is the smallest when the data reversal position is in the middle of the received signal, whereas the proportion peaks of the new algorithm and the PN algorithm are both higher. With changing data reversal rate, the estimated reversal errors comparison is shown in Figure 8. The result shows that the FZ algorithm's estimated reversal error is the highest, and the new algorithm's estimated reversal error is the lowest,

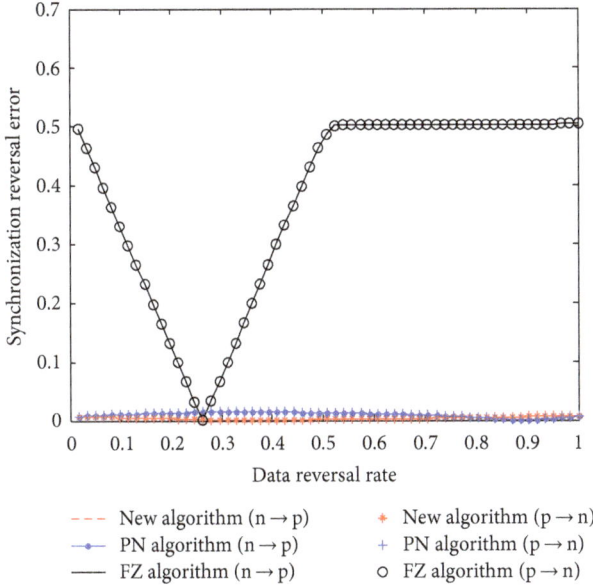

FIGURE 8: The estimated reversal errors comparison with changing data reversal rate.

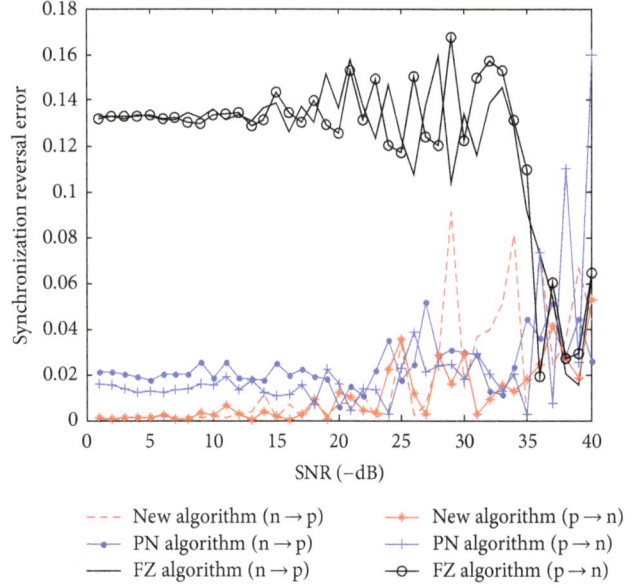

FIGURE 10: The estimated reversal errors compared with changing SNR.

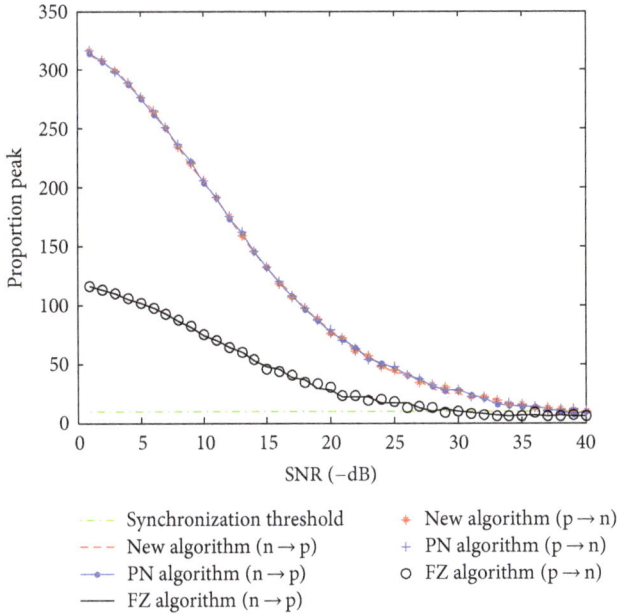

FIGURE 9: The proportion peak comparison with changing SNR.

demonstrating that the new algorithm's synchronization performance is the best.

With changing SNR, the proportion peak comparison is shown in Figure 9. The results show that the FZ algorithm's proportion peak is the smallest under the same SNR condition, but there are no significant differences between the new algorithm and the PN algorithm. And the FZ algorithm can achieve the threshold required when the SNR is less than −29 dB, but the new algorithm can achieve the threshold required when the SNR is less than −37 dB. Furthermore, the estimated reversal errors comparison is

shown in Figure 10, revealing that the estimated reversal errors of three algorithms increase gradually with decreasing SNR. The errors of the FZ algorithm, the PN algorithm, and the new algorithm obviously increase when the SNR is less than −15 dB, −23 dB, and −25 dB, respectively. Thus, the results show that the new algorithm's estimated reversal error is the smallest, and the FZ algorithm's error is the largest, demonstrating that the new algorithm's applicability is the best.

4. Conclusions

In this paper, the principle and characteristics of TDDM signals were studied, and the effective algorithms were studied, including the filled zero (FZ) algorithm and the positive-negative (PN) algorithm. Considering the deficiency problems in accuracy and environment adaptability for traditional algorithms, we proposed the optimization synchronization algorithm with illegibility elimination. We eliminated the data inversion illegibility threat by structuring special local sequence and establishing inversion estimation formula. In addition, we improved the correlation judgment mechanism by multicorrelation processing and proportion threshold processing. We demonstrated the new algorithm's correlation precision and SNR applicability by simulation analysis. In addition, we demonstrated that the new algorithm is better than the traditional algorithms in synchronization accuracy and adaptability.

Competing Interests

The authors declare that they have no competing interests.

Acknowledgments

This work was supported by the National Natural Science Foundation of China (no. 61501309) and China Postdoctoral Science Foundation (no. 2015M580231).

References

[1] J. K. Holmes and S. Raghavan, "A summary of the new GPS IIR-M and IIF modernization signals," in *Proceedings of the IEEE 60th Vehicular Technology Conference (VTC '04)*, pp. 4116–4126, September 2004.

[2] J. J. H. Wang, "Antennas for global navigation satellite system (GNSS)," *Proceedings of the IEEE*, vol. 100, no. 7, pp. 2349–2355, 2012.

[3] F. Vejrazka, "Galileo and the other satellite navigation systems," in *Proceedings of the 17th International Radioelektronika Conference*, pp. 1–4, IEEE, Brno, Czech Republic, 2007.

[4] O. Luba, L. Boyd, A. Gower, and J. Crum, "GPS III system operations concepts," *IEEE Aerospace and Electronic Systems Magazine*, vol. 20, no. 1, pp. 10–18, 2005.

[5] P. G. Mattos and F. Pisoni, "GPS-III L1C signal reception demonstrated on QZSS," in *Proceedings of the IEEE/ION Position, Location and Navigation Symposium (PLANS '12)*, pp. 1162–1168, IEEE, Myrtle Beach, SC, USA, April 2012.

[6] F. Martín, S. D'Addio, A. Camps, and M. Martín-Neira, "Modeling and analysis of GNSS-R waveforms sample-to-sample correlation," *IEEE Journal of Selected Topics in Applied Earth Observations and Remote Sensing*, vol. 7, no. 5, pp. 1545–1559, 2014.

[7] S. U. Qaisar and A. G. Dempster, "Assessment of the GPS L2C code structure for efficient signal acquisition," *IEEE Transactions on Aerospace and Electronic Systems*, vol. 48, no. 3, pp. 1889–1902, 2012.

[8] M. L. Psiaki, B. W. O'Hanlon, J. A. Bhatti, D. P. Shepard, and T. E. Humphreys, "GPS spoofing detection via dual-receiver correlation of military signals," *IEEE Transactions on Aerospace and Electronic Systems*, vol. 49, no. 4, pp. 2250–2267, 2013.

[9] X. Yu, Y. Sun, J. Liu, and J. Miao, "Design and realization of synchronization circuit for GPS software receiver based on FPGA," *Journal of Systems Engineering and Electronics*, vol. 21, no. 1, pp. 20–26, 2010.

[10] X. Li and W. Guo, "Efficient differential coherent accumulation algorithm for weak GPS signal bit synchronization," *IEEE Communications Letters*, vol. 17, no. 5, pp. 936–939, 2013.

[11] T. H. Ta, N. C. Shivaramaiah, A. G. Dempster, and L. L. Presti, "Significance of cell-correlation phenomenon in GNSS matched filter acquisition engines," *IEEE Transactions on Aerospace and Electronic Systems*, vol. 48, no. 2, pp. 1264–1286, 2012.

[12] C.-W. Chen, S.-H. Chen, H.-W. Tsao, and W.-L. Mao, "Memory-based two-dimensional-parallel differential matched filter correlator for global navigation satellite system code acquisition," *IET Radar, Sonar & Navigation*, vol. 8, no. 5, pp. 525–535, 2014.

[13] H. Li, M. Lu, and Z. Feng, "Partial-correlation-result reconstruction technique for weak global navigation satellite system long pseudo-noise-code acquisition," *IET Radar, Sonar & Navigation*, vol. 5, no. 7, pp. 731–740, 2011.

[14] C.-J. Li, M.-Q. Lu, Z.-M. Feng, and Q. Zhang, "Study on GPS L2C acquisition algorithm and performance analysis," *Journal of Electronics & Information Technology*, vol. 32, no. 2, pp. 296–300, 2010.

[15] F. Liu and Y. Feng, "A long code acquisition algorithm on resolve time-frequency uncertainty problem," *Acta Aeronautica et Astronautica Sinica*, vol. 34, no. 8, pp. 1924–1933, 2013.

19

Energy Efficient Partial Permutation Encryption on Network Coded MANETs

Ali Khan,[1] Qifu Tyler Sun,[1] Zahid Mahmood,[1] and Ata Ullah Ghafoor[2]

[1]*School of Computer and Communication Engineering, University of Science and Technology Beijing, Beijing, China*
[2]*Department of Computer Science, National University of Modern Languages, Islamabad, Pakistan*

Correspondence should be addressed to Qifu Tyler Sun; qfsun@ustb.edu.cn

Academic Editor: Xiong Li

Mobile Ad Hoc Networks (MANETs) are composed of a large number of devices that act as dynamic nodes with limited processing capabilities that can share data among each other. Energy efficient security is the major issue in MANETs where data encryption and decryption operations should be optimized to consume less energy. In this regard, we have focused on network coding which is a lightweight mechanism that can also be used for data confidentiality. In this paper, we have further reduced the cost of network coding mechanism by reducing the size of data used for permutation. The basic idea is that source permutes only global encoding vectors (GEVs) without permuting the whole message symbols which significantly reduces the complexity and transmission cost over the network. We have also proposed an algorithm for key generation and random permutation confusion key calculation. The proposed scheme achieves better performance in throughput, encryption time, and energy consumption as compared to previous schemes.

1. Introduction

Mobile Ad Hoc Networks (MANETs) have dynamic topology, which means, on requirement, devices act as nodes and establish a network for communication (Figure 1). All nodes are battery powered [1]. As MANETs do not need any infrastructure, nodes move freely in arbitrary direction, so nodes easily enter and leave the network at any moment. There is a possibility when a node cannot send information to another node directly within its communication range. To overcome this issue some intermediate nodes are used for routing the information from one node to another by multiple hops [2].

MANETs were thought as one of the most innovative and challenging wireless networking paradigms [3] at its evolution. Potentials of MANETs made ad hoc networking a key area for building Forth-Generation (4G) wireless networks and hence MANETs gained thrust and resulted in remarkable innovation for mobile network paradigm [4]. With advances in research, MANET becomes essential communication technology in military tactical environment to help in military deployment among soldiers, vehicles, and operational command centers [5]. MANET applications can also be used in law enforcement, other security sensitive environments, emergency relief scenarios, public meetings, and virtual class rooms.

Network encoding is used to transmit the maximum flow of data in a message by using encoding mechanism. It achieves transmission and reduction in communication overhead and better throughput instead of just storing and forwarding as in traditional routing. The traditional routing where the routers typically store and forward the information cannot be overlaid. Network coding, as introduced by the pioneers Ahlswede et al. [6], shows how information is encoded at intermediate or source nodes for efficient transmissions. The basic concept of network coding can be easily understood by butterfly network as explored in Figure 2. In this butterfly network topology a source (l) wants to transmit information to two nodes acting as destinations m and n. Each edge is represented as error-free channel which has the ability to deliver a single bit per channel use. The source sends two bits a and b, but instead of routing one and blocking the other, node x transmits their XOR. Node m receives a and $a \oplus b$. Since $a \oplus (a \oplus b) = b$, node m can recover

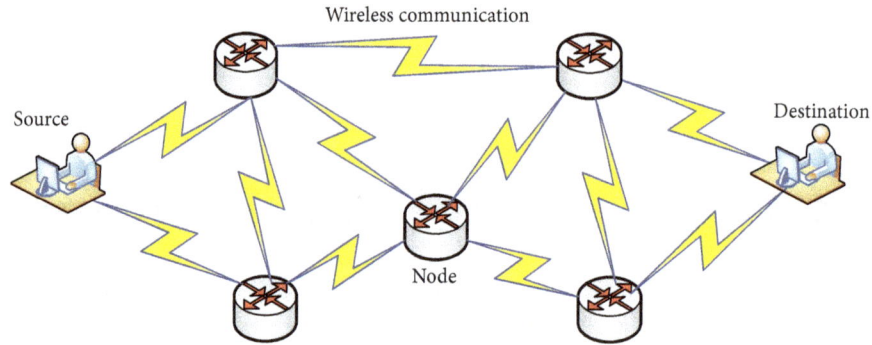

FIGURE 1: MANET topology.

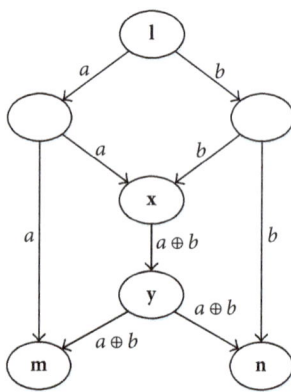

FIGURE 2: Network coding (butterfly network).

idea has been considered in a scheme in [15], which randomly permutes the whole message of length n with $n!$ possibilities and uses AES for encryption. Instead of permuting the whole message of length n, we permuted only the components in GEVs of length h which gives much lower computational complexity as $h \ll n$. It reduces the complexity of the scheme and energy consumption and is thus a more efficient encryption scheme. To further enhance the security of the message, we proposed an algorithm for dynamic random permutation key generation of the key.

The paper is organized as follows: Section 2 discusses the related work; Section 3 explores the proposed partial permutation based encryption scheme on network coded MANETs. Section 4 evaluates the performance and Section 5 concludes the work.

2. Related Work

The transmission energy reduction in network coding applications has received a lot of attention. We have explored different schemes in this context to identify the network coding mechanisms that claim to achieve efficient energy consumption. In this regard, Wu et al. [16] explores the solution to finding minimum energy of multicast tree expressed as linear program that is solved in polynomial time encoding at intermediate nodes. This is in contrast to the fact that the same problem is NP-complete in the case of traditional routing as investigated by Čagalj et al. [17]. Fragouli et al. [18] investigated efficient broadcasting problem in MANET and proposed probabilistic algorithms. Energy saving is done by lowering number of transmissions, as illustrated in Figure 3, where nodes are allowed to do the encoding of packets. Suppose there are six nodes where every node communicates only with both sides of its neighbors. Every node has to broadcast a message to other nodes.

It shows that without using network coding (NC), every message will need four broadcasts but with NC transmissions per message has decreased to three. So in this way 25% energy is saved in transmission only. The same problem is considered by Li et al. [19], where authors proposed deterministic approach based on PDP (Partial Dominant Pruning). The algorithm depends on two hop neighbors and opportunistic listening to encode packets.

both a and b. By using coding operation at bottleneck node and then decoding at sink nodes, the multicast throughput has been enhanced to 2 bits, beyond what can be done in conventional routing. Network coding has advantages like enhancing better throughput [7], network robustness [8], and reducing network congestion [9]. Network coding has applications in many areas such as Ad Hoc Network [10], delay tolerant networks [11], P2P network [12], wireless sensor network [13], and content distribution network [14].

The main problem in the conventional routing approach is highlighted when a message is transmitted through a number of intermediate nodes. At one point as depicted in the butterfly network, the intermediate node x is the bottleneck as it receives multiple packets but has a single channel to forward the packets. It results in network congestion. Secondly nodes also receive redundant data from different other nodes which can be removed by using network coding.

This paper presents an encryption scheme for MANETs which fully takes advantage of the security property of network coding. As global encoding vectors (GEVs) are essential for decoding so reordering the components in a GEV randomly makes significant confusion for the eavesdropper to get meaningful information. In the paper, we studied energy efficient encryption by merely permuting the components of GEVs, without manipulating the entire messages during encryption. We refer to the proposed scheme as partial permutation based encryption scheme. A similar

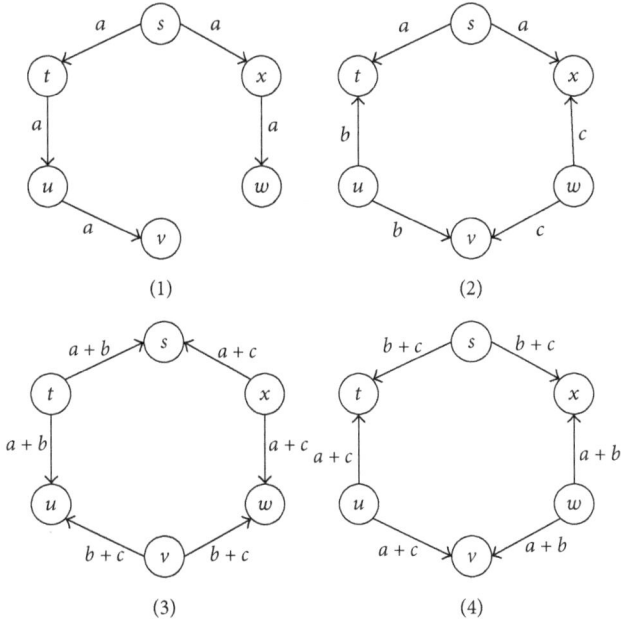

FIGURE 3: Reduction of transmission in MANETs.

Other than minimizing energy consumption during transmission in MANETs, network coding also has security properties as follows. Bhattad and Narayanan [20] introduced weak security, which means that a system is secured if adversary cannot get any meaningful information from the adversarial attack. Authors showed that random linear network coding is weakly secured with high probability when coding is applied on a large finite field. Lima et al. [21], after considering the threats of the intermediate nodes, developed a security criterion to access intrinsic security provided by network coding. Authors observed that security is directly dependent on the network topology as well by deriving the relationship between security level and field size. Based on this weak security research model, Wang et al. [22] designed a polynomial time deterministic code to secure linear network coding. They showed that optimal throughput between a single source and a paired destination for multiple streams is achieved by using this algorithm.

Many secure NC based cryptographic schemes have been proposed. A signature based scheme proposed by Yu et al. [23] detects and filters out the polluted messages. It used homomorphic signature function by source to delegate signing authorities to the forwarders that means intermediate nodes can generate signature for their output messages without contacting the source. SPOC (Secure Practical Network Coding), an end-to-end lightweight security scheme, is proposed by Vilela et al. [24], in which source encrypts GEVs of every message after performing random linear coding with an additional set of GEVs for network coding. Receiver recovers source messages by using decode-decrypt-decode steps. Fan et al. [25] proposed another similar scheme using HEF (Homomorphic Encryption Function) as introduced by Benaloh [26]. In this proposed scheme, the

coding coefficients used HEF approach for encryption. Linear combinations can be performed directly on these encrypted coefficients because of the homomorphic nature of HEF. This result does not need additional coding coefficients by SPOC.

Wei et al. [27] proposed an efficient encryption scheme that used permutation function that randomizes the message vectors to make confusion for the adversary. Zhang et al. [15] introduced a permutation coding scheme, called P-Coding, which is a lightweight encryption above network coding in MANETs. Their scheme significantly reduced energy consumption because of minimizing the security cost. Their scheme also exploits intrinsic property of security in network coding by using simple permutation encryption. In P-Coding scheme GEVs and message symbols are permuted together. Its complexity is significantly large as compared to the proposed scheme which only permutes components of GEVs. As data needs to be protected at every node in MANETs, energy efficiency can be achieved by the more efficient encryption and decryption processes in the proposed scheme. The conventional approach of encrypting the information is to use a symmetric key algorithm. We proposed a dynamic random permutation key that uses symmetric parameters.

Researchers have given considerable attention towards the energy efficient schemes. Energy efficient scheme, which is also termed as green computing, is one of the most important areas of research these days. Dutta and Culler [28] proposed a mechanism to reduce the energy utilization of mobile wireless nodes by reducing idle listening time of the nodes. Energy efficiency has also been the focal point of routing protocols. An OSLR based routing algorithm is proposed by Tan et al. [29] in which during node selection process different trust levels are used. Venkanna et al. [30] proposed a route splitting algorithm in which a solution to battery faults is provided by the persistent performance adapting the nodes. Another exclusive method for power consumption is proposed by Takeuchi et al. [31] which provides high performance in the dynamic environment based on creating assurance network. The energy consumption has made critical requirement to adopt effective green computing in wireless communication. Our work contributes to green computing from two facets: it saves energy in wireless environment during secure data transmission and introduces energy efficient partial permutation based encryption.

3. Preliminaries

We adopt the similar system model discussed by Zhang et al. in [15]. Let π be a sequence containing each element of set $(1, \ldots, h)$ once and only once as a permutation with length h. Let $\pi(i)$ be the ith element of π. The product of two permutations π_1 and π_2, defined by $\pi_1 \times \pi_2$ or $\pi_1\pi_2$, is calculated using $\pi_1\pi_2(i) = \pi_1(\pi_2(i))$. Denote by π^{-1} the inverse of π; that is, $\pi^{-1}\pi(i) = i$.

Definition 1. Consider a sequence $\mathbf{a} = (a_1, a_2, \ldots, a_h)$ over a finite field $\mathbb{F}_q$ and a permutation k on $(1, \ldots, h)$. The

Permutation Encryption Function (PEF) using key k on $\mathbf{a}$ is defined in

$$E_k(\mathbf{a}) = [a_{k(1)}, a_{k(2)}, \dots, a_{k(h)}]. \quad (1)$$

In the same way, we define permutation decryption function on a ciphertext c using key k as $D_k(c)$, satisfying $D_k(E_k(\mathbf{a})) = \mathbf{a}$.

There is a Key Distribution Center (KDC), responsible for establishing symmetric key so that source and destinations share a PEF key at the initial stage of the scheme. For effectiveness of PEF, the encryption key should be generated randomly. In the existing scheme [15], the generated session key sharing process uses AES encryption key management technique which is not resource efficient in MANETs. Any node, in a MANET consisting of N nodes, can act as a source. MANET can be modeled as an acyclic directed graph represented by $G(V, E)$ where V denotes the set of nodes and E the set of links of unit capacity, that is, transmitting one packet per link use. Let $\Gamma^-(v)$ be the set of terminating links at v whereas let $\Gamma^+(v)$ be the set of originating links from v. We assume that each link $e \in E$ has the capacity of one packet per unit time and $\mathbf{y}(e)$ is the packet carried on it. Here a packet is defined as a row vector of n elements from finite field $\mathbb{F}_q$. A unique source s sends a series of packets $\mathbf{x}_i, \dots, \mathbf{x}_h$ to a set of sinks $T \subset V$. Every source packet $\mathbf{x}_i$ is a row vector of length n over a finite field $\mathbb{F}_q$ and can be divided into two parts $[\mathbf{a}_i \ \mathbf{m}_i]$: the header vector $\mathbf{a}_i$ consists of h elements and the message vector consists of $n - h$ elements. Initially, the header vector $\mathbf{a}_i$ in the packet $\mathbf{x}_i$ is just the unit vector $\mathbf{u}_i$ of length h.

The vector matrix of source packet is defined as $\mathbf{X} = (\mathbf{x}_i^T, \dots, \mathbf{x}_h^T)^T$, where $\mathbf{X}$ has all packets of the source as its rows. Let $\Gamma^-(s)$ denote the set consisting of h imaginary links $\tilde{e}_1, \dots, \tilde{e}_h$ with $\mathbf{y}(\tilde{e}_i) = \mathbf{x}_i$. For any $e \in \Gamma^+(v)$, $v \notin T$, linearly combining the incoming packets of v, and $y(e)$ is calculated as illustrated in

$$\mathbf{y}(e) = \sum_{e' \in \Gamma^-(v)} \beta_{e'}(e) \, y(e') = \beta(e) \left[\mathbf{y}^T(e')\right]_{e' \in \Gamma^-(v)}^T. \quad (2)$$

In this equation, $\beta_{e'}$ are from $\mathbb{F}_q$ and $\beta(e) = [\beta_{e'}]_{e' \in \Gamma^-(v)}$ which is called LEV (Local Encoding Vector) of the link e. By induction, $\mathbf{y}(e)$ is represented as the linear combination of source packets.

$$\mathbf{y}(e) = \sum_{i=1}^{n} g_i(e) \, x_i = \mathbf{g}(e) \mathbf{X}. \quad (3)$$

Equation (3) elucidates that $\mathbf{g}(e) = [g_1(e), \dots, g_m(e)]$ that can be recursively calculated using (2). This is named as GEV (global encoding vector) of link e.

Assume that h packets $\mathbf{y}(e_1), \dots, \mathbf{y}(e_h)$ are received by a sink node v from links $e_1, \dots, e_h$ incoming to v. Then by (3) we have the packets $\mathbf{Y}$ received by v as follows:

$$\mathbf{Y} = \begin{bmatrix} y(e_1) \\ \vdots \\ y(e_m) \end{bmatrix} = \begin{bmatrix} g(e_1) \\ \vdots \\ g(e_m) \end{bmatrix} \mathbf{X} = \mathbf{G}\mathbf{X}, \quad (4)$$

TABLE 1: Notations for PPE.

Symbol	Explanation
$\mathbf{x}_i$	Source packets
$\mathbf{X}$	Vector matrix of source packets
h	Number of messages, permutation length
$\mathbf{y}(e)$	Coded packet carried on link e
$\beta(e)$	LEV of link e
$\mathbf{g}(e)$	GEV of link e
$\mathbf{G}$	Global Encoding Matrix
$\mathbf{a}$	Sequence of GEVs
k	PEF key
c	Ciphertext
D	Data generations
$f(h)$	Function used for confusion key
k'	Confusion key for each data generation

where $\mathbf{G}$ is referred to as global encoding matrix (GEM) of v. When $\mathbf{G}$ is invertible, then the source packets $\mathbf{X}$ can be reconstructed by using $\mathbf{X} = \mathbf{G}^{-1}\mathbf{Y}$.

4. Partial Permutation Based Encryption (PPE) Scheme

The main idea of the proposed scheme, called PPE, is that only GEVs are permuted instead of the whole packets at the source. This makes sufficient confusion for an adversary to locate GEVs in order to get meaningful information. As we are permuting only GEVs, this significantly reduces the complexity and making the scheme more efficient. On the other hand, as the header length may not be long enough for permutation to achieve sufficient security, we additionally propose using random encryption key on message to further increase the security of the proposed scheme against adversarial attacks. A list of notations to be used to introduce the PPE scheme is provided in Table 1.

Figure 4 briefly illustrates the PPE scheme. Initially, the source has h original messages $\mathbf{m}_1, \mathbf{m}_2, \dots, \mathbf{m}_h$ of length $n - h$, each of which is padded with a unit vector $\mathbf{u}_j$ as the header. Then, it performs linear combinations on the packets to generate h linearly independent packets. Subsequently, the PPE scheme will conduct encryption on the h packets. First, GEVs are permuted based on some permutation key k and then the message is encrypted using dynamic random key encryption based on a key k', whose generation will be discussed in detail in the following subsections.

A typical MANET scenario involves a source node, intermediate nodes, and sink nodes. Figure 5 depicts the data transmission in a MANET based on the proposed PPE scheme and network coding and the PPE scheme is only performed at the source nodes for encryption and at the sink nodes for decryption, and the intermediate nodes perform recoding of the message packets.

4.1. Dynamic Random Key Generating Algorithm. In a scenario where source needs to transmit huge data volume

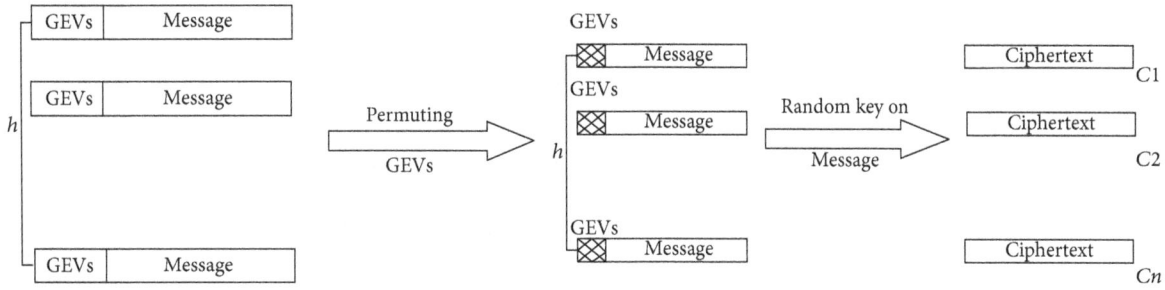

FIGURE 4: PPE scheme on coded message.

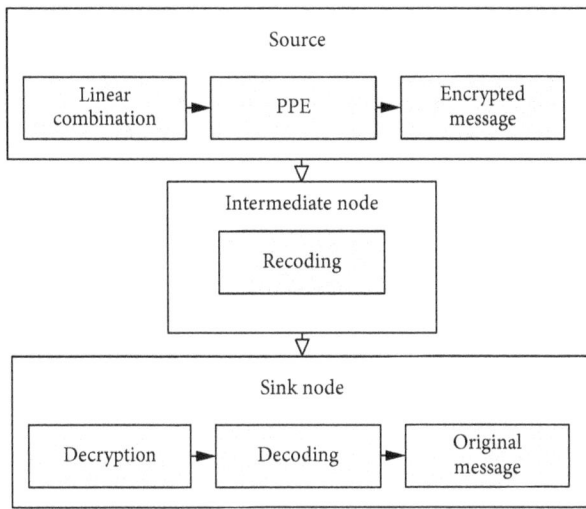

FIGURE 5: Stages of data transmission in a MANET based on the PPE scheme and network coding.

```
(1)   Array Key[] /* size m */
(2)   Function Key_Gen (integer m)
(3)   Initialization(m)
(4)     For i ← 1 to m − 1
(5)       ψ ← rand() /* between i to m */
(6)       Key[i] ← perm(ψ)
(7)     End for
(8)   Function Initialization (integer q)
(9)   For a ← 1 to q
(10)  Key[a] ← a
(11)  End for
```

ALGORITHM 1: Key generation.

Step (1)–(3). Key size m is declared. Function *key_Gen* gets the key size as its argument. Initialization function is used to call key size m.

Step (4)–(7). A random value between i and m is stored in ψ. The permutation function takes this random value as a seed to generate a value as $Key[i]$. The loop ends on $m − 1$.

Step (8)–(11). Initialization function has values from a to q where a is stored as $key[a]$ and loop ends at q.

In Algorithm 2, we use symmetric encryption for secured transmission of the secret key from source node to sink. We propose partial permutation data generation key. In this scenario, instead of using the same generation data stream, some data elements remain at their original position. In the proposed scheme, source randomly selects the length of data packet and makes a generation. On the basis of shared random value for partial permutation and confusion key, both ends have the ability to generate common session key on a distributed manner. In this scenario, we assume that secret parameter has been distributed by a trusted authority between source and sink as illustrated in Figure 6. To guard against replay attacks, we have used dynamic random key generation. Dynamic keys are the cryptographic keys that are not the same for the whole network lifetime. Instead, it is established either periodically or on demand. This helps increase the network survivability by revoking keys of the compromised nodes in the process of rekeying.

from one node to other nodes, source should first divide data into generations and use perturbing key on each data generation. If a single perturbing key is used in the course of transmission then there is a chance that this key would be disclosed, which will result in compromising security of the whole data volume. This is known as single generation failure. Assume the following steps are performed by source on ith data generation D_{Gi}, and key k_i is generated.

(a) Randomly choose h positions among data generation D_{Gi} which are known as perturbing key.

(b) Corresponding to data generation D_{Gi}, key k_i is calculated according to Algorithm 2.

(c) Encrypt D_{Gi} based on k_i and the encrypted data generation is sent from source node to all participant nodes that can update key.

Algorithm 1 explains the perturbing function for GEVs and is also used to generate dynamic random key.

(1) *Set s* = rand, *h* = 0, $\gamma = D/\omega$, *d* = (0, *h*! − 1)

(2) $D = \{D_{G_1}, D_{G_2}, D_{G_3}, \ldots, D_{G_\gamma}\}$

(3) $k(i) = \{D_{k_1}, D_{k_2}, D_{k_3}, \ldots, D_{k_\gamma}\}$

(4) *Loop from i* = 1 *to* γ

       (i) $g(i) = d\%/(i + 1)$;

       (ii) $d = d/(i + 1)$

       (iii) $g = \omega i/h$

(5) $CK[i] = Confusion\_Key\,(g, d, i)$

(6) $Cipher(Ci) = Encr(D_{G_i}, CK[i])$

(7) *Send Ci*

(8) *EndLoop*

Function Confusion_Key (int r_n, int *d*, int *index*) *BEGIN*

(9) $b(index) = D - d - (f(d/h)/(index + 1))$

(10) *For each* (*index* $\in [1, h]$)

(11) *do* $\theta(s - 1 + index) \leftrightarrow \theta(s - 1 + b(index))$

(12) $(k'(index)) \leftarrow \theta(k(index))$

(13) *Return* $k'(index) \leftarrow (k'(index) \parallel r_n)$

(14) *END*

ALGORITHM 2: Random permutation based keying.

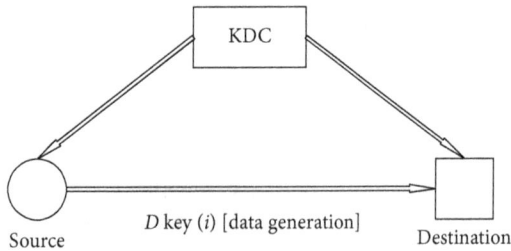

FIGURE 6: Dynamic key generation by a Key Distribution Center (KDC).

Dynamic Source Routing (DSR) protocol is simple and efficient routing mechanism designed typically for multihop MANETs. It makes the network completely self-configuring and self-organizing without requiring any existing infrastructure. As DSR is source initiated link state routing the nodes dynamically discover an efficient route to send the packet by multiple network hopping till the destination which makes it be the loop-free packet routing. The DSR protocol consists of two mechanisms: one is route discovery and the other is route maintenance. Based on this scenario, we adopt dynamic key for each data generation (D_G) to handle replay attack in MANET. Based on the sharing parameters at both ends, source generates dynamic key for each generation. Suppose we have key (ki) then the $D_{key(i)} = D_{key(1)}, D_{key(2)}, \ldots, D_{key(\gamma)}$ where (D_γ) represents the number of dynamic keys distributed for each generation and $D_{key(i)}$ is dynamic key.

As discussed in [15], traditional cryptographic technique like AES for end-to-end encryption cannot be used due to the limited resource capabilities of MANETs. This work contributes to generate lightweight encryption key as shown in Algorithm 2, which introduces a random number and updating key for each data generation to enhance security and reduce computation and communication overheads from source to destination in MANETs.

Steps (1)–(3). Set random number *n* used to generate the value of *h* for permutation. In this step, γ represents the number of generations and *D* is the total data length that is divided into generations like D_{G_1} and D_{G_2} where ω is chunk size to produce a single generation. The value of *h* can vary from 4 bits to 32 bits where $h \ll \omega$ and it is used for permutation. Moreover, $K(i)$ represents the set of keys where D_{k_1} represents key for the first generation. Other parameters are initialization factor $K(h)$, packet length (n), and random division of perturb-vectors ($d = (0, h! − 1)$).

Steps (4)–(8). Loop for generation from 1 to γ where $g(i)$ is the index of generation and calculated as chunk over permutation length *h* is used in Steps (5) and (6) for calculating confusion key and ciphertext which is generated as a function of data generation and confusion key. After the loop ends we will get ciphertext.

Steps (9)-(10). Confusion key function $b(index)$ is calculated by subtracting $f(d/h)/(index + 1)$ from random division of perturbed-vectors *d* and again from the data length *D*, where indexing is from 1 to the permutation length *h*.

Steps (11)–(14). Confusion key $k'(index)$ is generated for an end-to-end communication which consists of a perturbing key and a random number. The perturbing function θ is calculated by subtracting the sum of 1 and indexed confusion function, which will continue till the end of the permutation length.

5. Performance Analysis

We take into account typical cryptanalysis on permutation cipher which is a case of transposition cipher and evaluate how effectively a permutation encryption can be broken. This is based on nonuniform occurrences of *n-letter* combinations named as *n-gram*. Taking an example, the frequency of bigram "*TH*" in English is much higher as compared to bigram "*QZ*." The ability of guessing permutation π is accessed by using *n-grams* frequency statistics: first large cipher texts are decrypted by using inverse of permutation π and then evaluated on the basis of how close the statistics of *n-grams* decrypted messages are as compared to statistics of underlying languages. Then we find the permutation of other letters that have better ability by searching "*ps*" neighborhood with good ability until key *k* is found. Although this is rather effective in transposition ciphers breaking, we contend that for the case of our permutation encryption it does not work fine. (1) First we permuted GEVs; then we introduced dynamic random encryption key on message. This double encryption makes it much stronger against replay attack. So this means that, as compared with transposition cipher, the proposed scheme requires a lot of time to access the ability of permutation. (2) A small change of permutation, for example, change of just two positions, will give different

TABLE 2: Simulation parameters.

Simulation time	150 sec
Terrain area	$2000 \times 2000 \, \text{m}^2$
Number of nodes	250
Node placement strategy	Random
Propagation model	Two-ray model
Mobility model	Random
Routing protocol	AODV
Traffic type	Constant bit rate (CBR)
Pause time	0 s

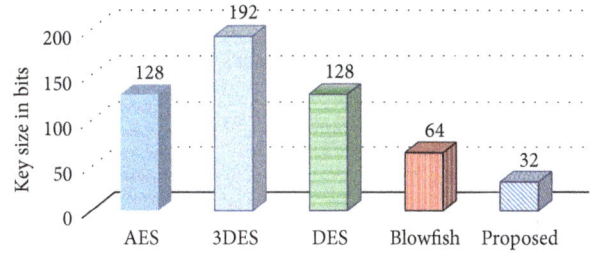

FIGURE 7: Key size of the storage analysis.

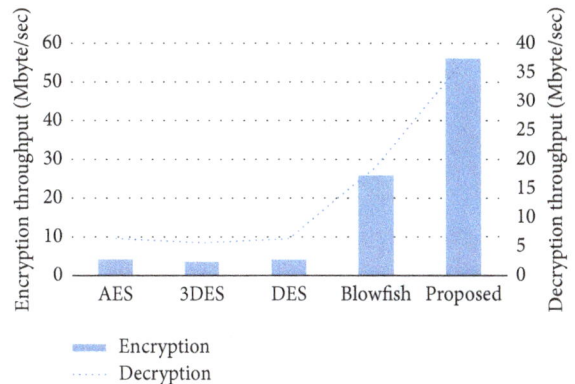

FIGURE 8: Encryption/decryption throughputs in Mbytes/sec.

encryption that would decode information into totally different messages. This resulted in the fact that permutation has good ability so by searching in the neighborhood of the permutation we do not expect to get permutations with better ability. As dynamic random encryption key is distributed at initial stage, permutation operation (encryption at source and decryption at destination) will give less computational overhead. So permutation encryption is lightweight in terms of computation.

The performance analysis of the proposed scheme is analyzed in built-in classes of Java. It uses managed packaging for AES, DES, and Blowfish which is available in http://java.cypto.com. The existing schemes used traditional cryptographic methods of cipher class that provide functionality of different cryptographic techniques used for encrypting and decrypting of data. It forms the core of the JCE framework. Simulation parameters are given in Table 2.

5.1. Storage Analysis. The proposed scheme is based on secret credentials to encrypt perturbing methods and acts as secure shared session key between end nodes. The pseudo random number generator (PRNG), a confusion function (f), permuting of selective data ($f(d/h)$) portion, and time stamp (T) are used as secret credentials. The memory overhead of the confusion key depends on the selective bits of data and other parameters. To overcome memory and computation overhead, the proposed scheme uses 32-bit key size which has lesser overhead from traditional cryptographic techniques like DES, AES, 3DES, and Blowfish. The application payload is 32 bits, which include 16 bits of session key and 16 bits of the random challenge. The performance analysis of Algorithms 1 and 2 shows the proposed scheme is efficient from traditional schemes and suitable for resource constrains devices. Performance comparison of different schemes is done on the basis of parameters like encryption time, encryption/decryption throughput, and energy consumption as shown in Table 3. We considered well known symmetric key encryption techniques for comparison with the proposed scheme. Data encryption standard (DES) has 128-bit key size and 64-bit block size. 3DES, which is an enhancement of DES, uses three 64-bit keys which makes 192-bit key size and has a 64-bit block size as well. AES has variable key lengths of 128, 192, or 256 bits and has a data block size of 128 bits. We consider 128-bit key.

Finally Blowfish uses variable key lengths of 32 bits to 448 bits and 64-bit data block size. We used 64-bit key size.

The appraisal is intended to assess the results by using block ciphers. Consequently, the load data (plaintext) is divided into smaller data generation size as per algorithm settings given in Table 3. De Meulenaer et al. [32] evaluated energy of wireless nodes in terms of communication cost, whereas Abdul Elminaam et al. [33] evaluated various symmetric cryptography algorithms used in MANETs in terms of energy cost. The energy is calculated by using the following equation: $E = P \times t$, where E is the energy in joules, P is the nominal power in watts, and t is the time duration in seconds. Operation cost and transmission of 1 byte is 5.76 μJ, reception of 1 byte is 6.48 μJ, and AES-128 encryption of 16 bytes is 42.88 μJ. On the basis of these assumptions we have computed energy consumption at processing and transmission level of proposed scheme. In addition, we analyzed the key size as in Figure 7.

5.2. Throughput. As the encryption time decreases, more data can be processed which results in larger throughput. Encryption time has to be considered while calculating throughput of an encryption algorithm. The throughput of an algorithm is obtained by dividing plain text in kilobytes by encryption time in milliseconds. From the graphical illustration of encryption and decryption throughput in Figure 8, the proposed scheme outperforms other schemes in terms of throughput. The gap between the proposed scheme and AES, DES, and 3DES is significantly large because of their higher

TABLE 3: Performance comparison of schemes.

Algorithms	Key size (Bit)	Data block (bit)	Encryption time (milliseconds)	Encryption throughput Mbyte/sec	Decryption throughput (Mbyte/sec)
AES	128	128	374	4.17	6.452
3DES	192	64	452	3.45	5.665
DES	128	64	389	4.01	6.347
Blowfish	64	64	60.3	25.8	18.7
Proposed scheme	32	32	31	56	37

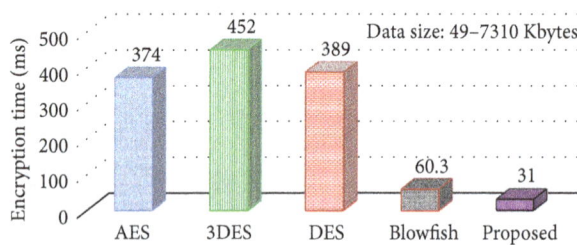

FIGURE 9: Data encryption time in milliseconds.

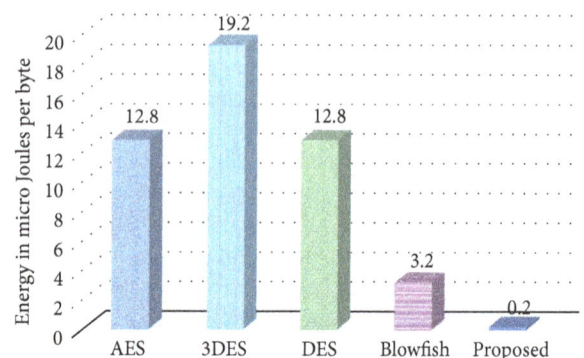

FIGURE 10: Energy consumption in μJ.

encryption time. Data throughput on encryption and decryption is depicted in graphical presentation. The throughput of the proposed scheme is calculated and then compared with other schemes according to the formula throughput = plain text size/encryption time. We have calculated the encryption and decryption time of various algorithms as per the throughput formula. We considered plaintext size to be the average of plain text size and the average time taken. So 1598.7 kbytes/374 ms = 4.174 Mbytes/s for AES while for our proposed scheme it is calculated as 1598.7 kbytes/31 ms = 56 Mbytes/s. In the same way, the decryption throughput is calculated via dividing the plain text size by decryption time.

5.3. Encryption Time. Encryption time is calculated as time taken by any device in executing the encryption algorithm. Encryption time is calculated by taking into account the size of plain text, key size, and block size. We consider the algorithm key sizes as given in Table 3. From Figure 9 we can clearly see that the proposed scheme has very less encryption time as compared with Blowfish algorithm which is a light weight encryption scheme. From Table 3 we can clearly see that when the key size increases, time for encryption increases as well. As for the proposed scheme the key size as well as the block size is smaller than the other algorithms, that is, 32 bits; the encryption time comes out to be 31 ms, which is significantly lower than other schemes.

5.4. Energy Consumption. The energy consumption is calculated as the average energy consumed during the process by an algorithm. The cost of the proposed scheme is based on energy consumption during encoding, transmitting, and receiving cost of data at sinks. On the basis of [32] we elaborate energy consumption of the proposed scheme and

compare it with the existing techniques. Less time taken for encryption means there are few cycles which gives lower energy consumption. As described before, the proposed scheme has less encryption time as compared with other encryption schemes. As shown in Figure 10, the energy consumption of the proposed scheme comes out to be 0.2 μJ which is significantly lower as compared with other schemes. The energy consumption is calculated by considering block size in bytes, and multiplying it with energy consumption for symmetric key comes out to be 0.8 μJ/bit. As described in the scheme, confusion key plays a vital role in the proposed scheme in which secret credentials are used for the session key. A node has to compute 16-bit PRNG, 8-bit time stamp, and 8-bit permuting of selective data to secure perturbing function. On the assumption of [32] the computation cost is 0.25 μJ/bit and our proposed scheme needs to compute 32-bit data for confusion key which means total energy is 32 × 0.25 = 8.0 μJ. In this scheme the confusion key computation cost depends on the random amount of selective data from n-bit data length. We assume that the proposed scheme computes on 32-bit block generation of data.

6. Conclusion

In this paper, we proposed a partial permutation encryption algorithm for network coded MANETs. Instead of permuting the whole packet as in the previous P-Coding scheme, the proposed scheme permutes only GEVs which decreases the computational complexity, making it an efficient encryption scheme in terms of energy, computation, and cost. To guarantee that the proposed scheme is secure against various attacks, we proposed dynamic key generation mechanism for our random key generation. We analyzed the proposed scheme by taking into account different parameters and concluded that our partial permutation scheme is efficient and lightweight. The proposed scheme outperforms other analyzed schemes in terms of efficiency and cost. The proposed scheme has lower encryption time and greater throughput which resulted in 117% improvement from the Blowfish algorithm for MANETs. Blowfish algorithm has 5 times greater throughput than DES. The proposed scheme has 16 times lesser energy consumption than the Blowfish algorithm that makes it an efficient encryption technique for energy constraint devices.

Conflicts of Interest

The authors declare that there are no conflicts of interest regarding the publication of this article.

Acknowledgments

This work was supported in part by the National Natural Science Foundation of China under Grant no. 61471034 and by the Ministry of Education of China under Grant no. 6141A02033307.

References

[1] L. Gruenwald, M. Javed, and M. Gu, "Energy—efficient data broadcasting in mobile ad hoc networks," in *Proceedings of the International Database Engineering and Applications Symposium (IDEAS '02)*, July 2002.

[2] L. Junhai, X. Liu, and Y. Danxia, "Research on multicast routing protocols for mobile ad-hoc networks," *Computer Networks*, vol. 52, no. 5, pp. 988–997, 2008.

[3] S. Giordano and W. W. Lu, "Challenges in mobile ad hoc networking," *IEEE Communications Magazine*, vol. 39, no. 6, p. 129, 2001.

[4] I. Chlamtac, M. Conti, and J. J.-N. Liu, "Mobile ad hoc networking: imperatives and challenges," *Ad Hoc Networks*, vol. 1, no. 1, pp. 13–64, 2003.

[5] J. Loo, J. Lloret, and J. H. Ortiz, "Mobile Ad Hoc Networks: Current Status and Future Trends," 2011.

[6] R. Ahlswede, N. Cai, S. R. Li, and R. W. Yeung, "Network information flow," *IEEE Transactions on Information Theory*, vol. 46, no. 4, pp. 1204–1216, 2000.

[7] S. R. Li, R. W. Yeung, and N. Cai, "Linear network coding," *IEEE Transactions on Information Theory*, vol. 49, no. 2, pp. 371–381, 2003.

[8] R. Koetter and F. R. Kschischang, "Coding for errors and erasures in random network coding," *IEEE Transactions on Information Theory*, vol. 54, no. 8, pp. 3579–3591, 2008.

[9] L. Chen, T. Ho, M. Chiang, S. H. Low, and J. C. Doyle, "Congestion control for multicast flows with network coding," *IEEE Transactions on Information Theory*, vol. 58, no. 9, pp. 5908–5921, 2012.

[10] D. Annapurna, N. Tejas, K. B. Raja, K. R. Venugopal, and L. M. Patnaik, "An energy efficient multicast algorithm for an

[11] E. Altman, L. Sassatelli, and F. De Pellegrini, "Dynamic control of coding for progressive packet arrivals in DTNs," *IEEE Transactions on Wireless Communications*, vol. 12, no. 2, pp. 725–735, 2013.

[12] A. M. Sheikh, A. Fiandrotti, and E. Magli, "Distributed scheduling for scalable P2P video streaming with network coding," in *Proceedings of the 32nd IEEE Conference on Computer Communications (IEEE INFOCOM '13)*, pp. 11–12, April 2013.

[13] R. R. Rout and S. K. Ghosh, "Enhancement of lifetime using duty cycle and network coding in wireless sensor networks," *IEEE Transactions on Wireless Communications*, vol. 12, no. 2, pp. 656–667, 2013.

[14] Q. Yan, M. Li, Z. Yang, W. Lou, and H. Zhai, "Throughput analysis of cooperative mobile content distribution in vehicular network using symbol level network coding," *IEEE Journal on Selected Areas in Communications*, vol. 30, no. 2, pp. 484–492, 2012.

[15] P. Zhang, C. Lin, Y. Jiang, Y. Fan, and X. Shen, "A lightweight encryption scheme for network-coded mobile Ad Hoc networks," *IEEE Transactions on Parallel and Distributed Systems*, vol. 25, no. 9, pp. 2211–2221, 2014.

[16] Y. Wu, P. A. Chou, and S.-Y. Kung, "Minimum-energy multicast in mobile ad hoc networks using network coding," *IEEE Transactions on Communications*, vol. 53, no. 11, pp. 1906–1918, 2005.

[17] M. Čagalj, J.-P. Hubaux, and C. Enz, "Minimum-energy broadcast in all-wireless networks: NP-completeness and distribution issues," in *Proceedings of the 8th Annual International Conference on Mobile Computing and Networking*, pp. 172–182, September 2002.

[18] C. Fragouli, J. Widmer, and J.-Y. Le Boudec, "A network coding approach to energy efficient broadcasting: from theory to practice," in *Proceedings of the 25th IEEE International Conference on Computer Communications (INFOCOM '06)*, pp. 1–11, April 2006.

[19] L. Li, R. Ramjee, M. Buddhikot, and S. Miller, "Network coding-based broadcast in mobile ad hoc networks," in *Proceedings of the 26th IEEE International Conference on Computer Communications (IEEE INFOCOM '07)*, pp. 1739–1747, May 2007.

[20] K. Bhattad and K. R. Narayanan, "Weakly secure network coding," in *Proceedings of the Workshop on Network Coding, Theory, and Applications (NetCod '05)*, Riva del Garda, Italy, 2005.

[21] L. Lima, M. Medard, and J. Barros, "Random linear network coding: a free cipher?" in *Proceedings of the IEEE International Symposium on Information Theory (ISIT '07)*, pp. 546–550, Nice, France, June 2007.

[22] J. Wang, J. Wang, K. Lu, B. Xiao, and N. Gu, "Optimal linear network coding design for secure unicast with multiple streams," in *Proceedings of the IEEE INFOCOM*, pp. 1–9, IEEE, San Diego, Calif, USA, March 2010.

[23] Z. Yu, Y. Wei, B. Ramkumar, and Y. Guan, "An efficient signature-based scheme for securing network coding against pollution attacks," in *Proceedings of the 27th IEEE Communications Society Conference on Computer Communications (INFOCOM '08)*, pp. 2083–2091, April 2008.

[24] J. P. Vilela, L. Lima, and J. Barros, "Lightweight security for network coding," in *Proceedings of the IEEE International*

Conference on Communications (ICC '08), pp. 1750–1754, May 2008.

[25] Y. Fan, Y. Jiang, H. Zhu, and X. Shen, "An efficient privacy-preserving scheme against traffic analysis attacks in network coding," in *Proceedings of the 28th Conference on Computer Communications (IEEE INFOCOM '09)*, pp. 2213–2221, Rio de Janeiro, Brazil, April 2009.

[26] J. Benaloh, "Dense probabilistic encryption," in *Proceedings of the Workshop on Selected Areas in Cryptography*, pp. 120–128, August 1994.

[27] Y. Wei, Z. Yu, and Y. Guan, "Efficient weakly-secure network coding schemes against wiretapping attacks," in *Proceedings of the IEEE International Symposium on Network Coding (NetCod '10)*, pp. 1–6, IEEE, Ontario, Canada, June 2010.

[28] P. Dutta and D. Culler, "Practical asynchronous neighbor discovery and rendezvous for mobile sensing applications," in *Proceedings of the 6th ACM Conference on Embedded Networked Sensor Systems (SenSys '08)*, pp. 71–83, November 2008.

[29] S. Tan, X. Li, and Q. Dong, "Trust based routing mechanism for securing OSLR-based MANET," *Ad Hoc Networks*, vol. 30, pp. 84–98, 2015.

[30] U. Venkanna, J. K. Agarwal, and R. L. Velusamy, "A cooperative routing for MANET based on distributed trust and energy management," *Wireless Personal Communications*, vol. 81, no. 3, pp. 961–979, 2015.

[31] M. Takeuchi, E. Kohno, T. Ohta, and Y. Kakuda, "Improving assurance of a sustainable route-split MANET routing by adapting node battery exhaustion," *Telecommunication Systems*, vol. 54, no. 1, pp. 35–45, 2013.

[32] G. De Meulenaer, F. Gosset, F.-X. Standaert, and O. Pereira, "On the energy cost of communication and cryptography in wireless sensor networks," in *Proceedings of the 4th IEEE International Conference on Wireless and Mobile Computing, Networking and Communication (WiMob '08)*, pp. 580–585, October 2008.

[33] D. S. Abdul Elminaam, H. M. Abdul Kader, and M. M. Hadhoud, "Performance evaluation of symmetric encryption algorithms," *Communications of the IBIMA*, vol. 8, pp. 54–64, 2009.

20

Secure-Network-Coding-Based File Sharing via Device-to-Device Communication

Lei Wang[1] and Qing Wang[2]

[1]*School of Computer, Nanjing University of Posts and Telecommunications, Nanjing 210023, China*
[2]*School of Tongda, Nanjing University of Posts and Telecommunications, Yangzhou 225127, China*

Correspondence should be addressed to Lei Wang; leiwang@njupt.edu.cn

Academic Editor: Arun K. Sangaiah

In order to increase the efficiency and security of file sharing in the next-generation networks, this paper proposes a large scale file sharing scheme based on secure network coding via device-to-device (D2D) communication. In our scheme, when a user needs to share data with others in the same area, the source node and all the intermediate nodes need to perform secure network coding operation before forwarding the received data. This process continues until all the mobile devices in the networks successfully recover the original file. The experimental results show that secure network coding is very feasible and suitable for such file sharing. Moreover, the sharing efficiency and security outperform traditional replication-based sharing scheme.

1. Introduction

Sharing large scale files such as high-resolution videos with many friends through mobile devices at the same time is becoming a popular application. Smart phones are always used to upload and download the shared files through WiFi, 3G, or LTE, but these ways will naturally incur high expense and security threat when large scale data needs to be shared. Actually, it is unnecessary to share data via commercial networks in some scenarios. If the devices are located in the same area, the users could share the files through direct link between devices so that the traffic fee could be saved. The mobile devices can be strategically switched to soft AP (Access Point) mode so that the other users could connect to it and receive the files. However, there are two constraints in this method. First, from the technical perspective, the number of users connecting a soft AP is often limited from four to eight. Second, some users cannot connect to the soft AP within one hop. Therefore, after the users close to the source receive the files, they are supposed to share the data with their neighbors by switching to a new soft AP. Through the sharing of multiple hops, all the users in the network could obtain the files.

When sharing files with many users, sharing efficiency and security should be focused. When a large scale data needs to be shared, it would be better to split the original file into multiple slices before sharing because the direct link between devices may be disconnected during the transmission. After splitting the files into multiple slices, as long as the other users receive all the slices, the original file could be recovered. However, this method has a drawback which could be optimized. When a node requires a specific block of the original file, its neighbors may not have it either. Therefore, they have to wait until the slice is received. In order to overcome this drawback, network coding [1] could be introduced in such applications. Network coding has been considered as a promising technology in big data transmission. Network coding has been proposed for more than ten years, and it has attracted much attention of researchers. Li et al. [2] proposed linear network coding, and then Ho et al. [3] and Jaggi et al. [4] proposed RLNC (Random Linear Network Coding) and DLNC (Deterministic Linear Network Coding), respectively. Network coding has been studied in many areas, such as information security [5], distributed storage [6], video communication [7], and content sharing [8].

The main feature of network coding is that it requires reencoding operation at the intermediate devices of the network. Benefiting from the reencoding operations, the network performance could be increased such as bandwidth and energy efficiency. Moreover, the data is highly mixed at the source node and intermediate nodes, which means that the data transmitted in the channel is no longer the original data. Therefore, the security is significantly increased. However, it is very difficult to change the traditional network architecture, which impedes the development of network coding because traditional intermediate devices such as routers and switches cannot perform additional computational operation. Currently, the development of mobile devices and 4G/5G networks makes the computational operation at the mobile devices feasible. Therefore, network-coding-based applications are becoming more and more popular in mobile devices [9, 10].

D2D communication is a key supporting technology for the fifth-generation communication network. In D2D communication, the mobile devices could communicate with others directly via physical links without the relay of the base station, and it is feasible to perform network coding operation at the devices. Therefore, 5G network is a perfect place to apply network coding. The aim of this paper is to model and analyze the sharing efficiency of large scale data in D2D communication when network coding is introduced.

The remainder of this paper is organized as follows. In Section 2, the authors introduce some closely related studies. In Section 3, the authors model the secure network-coding-based file sharing scheme. In Section 4, the authors evaluate the proposed scheme. Finally, the conclusion is made in Section 5.

2. Related Works

There are some existing papers closely related to this study. M. Yang and Y. Yang [11] proposed a network-coding-based file sharing scheme for peer-to-peer networks. They encode the original files and then deploy the encoded subfiles on a web server. All the clients not only download the encoded subfiles but also forward the encoded subfiles for each other. Their scheme achieves 15%–20% higher throughput than previous schemes, and it achieves good reliability and robustness to link failure. Their scheme shows that network coding is promising in the file sharing application on the Internet. Our research is for future wireless networks. Moreover, the network model in their study is abstracted as a combination network. Based on the network model, they proposed a deterministic algorithm to encode the files, while the network model in our system is based on RLNC.

Lin et al. [12] presented a stochastic analytical framework to study the performance of epidemic routing using network coding in opportunistic networks. They showed that network coding is superior when bandwidth and node buffers are limited. The application scenario they described is similar to ours. This paper made some modification based on the traditional epidemic model. Moreover, our scheme is designed for the mobile devices. In order to establish the network, the devices in our scheme have to switch between ordinary mode

and AP mode. Therefore, even if some devices are very close to each other, they may be unable to communicate.

There are also some studies [13, 14] on the ad hoc networks in which the nodes are mobile devices. In these studies, the mobile devices can connect to each other through working in ad hoc mode. BATMAN [14] is a representative protocol in such application. However, a precondition for this protocol is that those devices in the network have to be rooted because there are very rare commercial released operation systems which could work in ad hoc networking mode. Therefore, this paper studies the file sharing scheme for mobile devices without the support of ad hoc mode.

The contribution of this paper can be summarized as follows. First, the authors analyze and model the secure network-coding-based file sharing scheme for the network with a number of mobile devices. In the scheme, the mobile devices are not required to be rooted before sharing files, which is more realistic. Second, the authors evaluate the scheme and show that file sharing among mobile devices is an ideal place to apply network coding.

3. Proposed Scheme

In order to accelerate the sharing rate, this paper proposes a principle data sharing scheme which is based on network coding. The source device needs to encode the original data with network coding. When an intermediate device receives some or all of the data slices, it could reencode the received data with RLNC and then spread the data to its neighbors via D2D communication. Because the data is highly mixed during the reencoding operation, each device could receive and decode the data as long as sufficient slices are received with high probability.

It is feasible to implement direct communication between devices for mobile devices via IEEE 802.11n. When a device X receives part of the encoded slice, it could configure itself as a soft AP and then allow other devices to connect for data transmission. When a device Y joins the network of X, device X reencodes the slices it received and then forwards the slices after reencoding to device Y. Through strategically switching between AP mode and ordinary mode, all the devices in the network could receive and decode the original data.

3.1. Network Model. Instead of the traditional store-and-forward working mode of network devices, network coding technology uses the storage-coding-forwarding working mode at intermediate devices. Through the operation at the intermediate network devices, it can effectively improve the file transmission rate.

Network coding could work in unicast or multicast networks [15]. In most cases, network coding works in multicast networks. However, after a mobile device configures itself to AP mode, the device could not forward multicast message. Some authors consider that multiple devices could overhear the data transmitting to someone at the same time via unencrypted wireless channel [16], but that is another story. Therefore, this paper assumes that the device sends data to its neighbors via unicast connections.

(a) Traditional file transmission

(b) Linear-network-coding-based data sharing

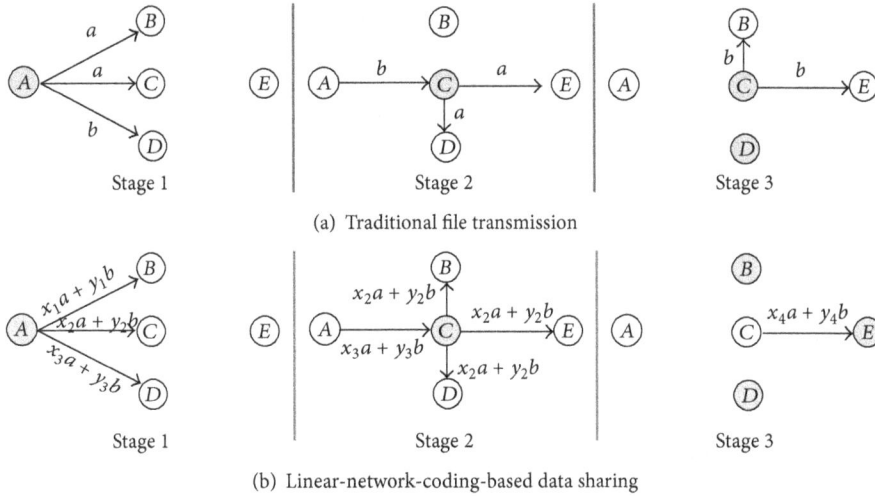

FIGURE 1: Data sharing in different schemes.

Figure 1 shows the advantages of network coding in sharing files.

The example in Figure 1 shows the principle why network coding helps accelerate the rate of data sharing. Nodes A–E in Figure 1 are a subset of a network. When an intermediate device needs to forward data to its neighbors, it switches to a soft AP. In the second stage of Figure 1(a), after node C switches to AP mode, node B can no longer receive data from C because they have the same data a. Node B can only receive data b at stage 3. After using network coding, this performance could be significantly improved. Node B could decode data a and b in stage 2. From an overall perspective, nodes B, D, and E could become soft APs and spread data to their own neighbors in stage 2 of Figure 1(b), while only node D could become the AP in stage 2 of Figure 1(a). In the third stage of Figure 1(b), nodes B, D, and E could work as soft APs and spread data to its own neighbors, while only node D could become a soft AP in Figure 1(a). Therefore, the sharing rate in the network-coding-based scheme is faster than that in traditional way.

3.2. Network Coding Strategy. Network coding scheme could be divided into linear network coding and nonlinear network coding. RLNC is a practical scheme, and RLNC is suitable for the network with dynamical topology. Deterministic algorithm has higher computation efficiency compared with randomized algorithm, but it is heavily dependent on the network topology. In our scheme, the mobile devices may change their modes from ordinary mode to AP mode, which will change the network logical topology. Therefore, we select RLNC in our scheme. First, the device who starts the sharing process needs to equally split the original file p into k slices $p_1, p_2, \ldots, p_k$. In each transmission session, this device randomly selects k elements $a_{i1}, a_{i2}, \ldots, a_{ik}$ from the finite field GF(256) and then obtains the encoded slice X_I with

$$X_i = a_{i1} \times p_1 + a_{i2} \times p_2 + \cdots + a_{ik} \times p_k. \tag{1}$$

The reencoding operation at the intermediate nodes could increase the performance of network transmission, including the throughput and security. For each intermediate device, when it needs to transmit a slice to its neighbor, it has to follow the same strategy. It randomly selects m ($m \leq k$) elements to be coefficients from the field GF(256) and then linearly reencodes the m slices it received with the m coefficients. After the reencoding operation, the linear dependency of the data is reduced. Therefore, the receiver could obtain a linearly independent slice with high probability.

As long as a device successfully accumulates k linearly independent slices, it could recover the original files with Gauss-Jordan elimination method.

3.3. Analysis Model. Through the analysis for Figures 1(a) and 1(b), we observe that both the schemes transmit data in a complex network environment. The second scheme is more complex because the data are linearly mixed at the source device and intermediate devices. In order to clarify the difference of the two schemes, we consider this kind of problem as complex-network-based epidemic model and then model the two schemes.

3.3.1. Classical Propagation Model. There are many disease propagation models proposed by previous researchers. In the researches about complex network, the most widely used models are SIS (Susceptible-Infected-Susceptible) model and SIR (Susceptible-Infected-Removed) model. This paper assumes that each device is a node in the network and then makes analysis for both the two models.

When SIS model is used, the nodes could be divided into two categories. One is the mobile devices that have become soft APs, and the other is the devices which are working in AP mode but switched to ordinary mode soon afterwards. However, after using network coding, there exists the third category, namely, the devices that received part of encoded slice but have not switched to AP mode. The devices of this kind cannot be expressed in SIS mode.

Compared with SIS model, there is one more category in SIR model, namely, removal individual. Removal individual is equivalent to the devices which become AP nodes, and then its neighbors received all the data. Finally, these devices permanently close the AP mode. In other words, the devices leave the network permanently.

In accordance with the above analysis, both the two schemes lack the expression for the devices that receive part of encoded data but have not become soft AP. Therefore, traditional SIS and SIR models cannot be directly used in our network environment. We have to make some improvement based on the SIR mode for our scheme.

3.3.2. Analysis Model for the Proposed Scheme. In our model, the concept of hidden nodes is introduced to indicate the devices which could switch to AP mode even if only part of encoded slices is received. Moreover, for any device in the network, it is not allowed to stay in suspended mode, which means that the devices neither switch to AP mode nor receive data from others at that state. Therefore, the switch time of AP mode is very important during the transmission. The transform is described in (2) in which m_j refers to the occupied cache of node j, n_j refers to the cache size of node j, and α_j refers to the proportion of received data

$$\alpha_j = \frac{m_j}{n_j} \quad (\alpha_j \leq 0.5). \tag{2}$$

Theoretically speaking, any intermediate node could switch to AP mode at any time in the ad hoc network. In order to guarantee the efficiency, when an intermediate node is receiving data, it cannot switch to AP mode. Only when the condition $\alpha_j \leq 0.5$ is satisfied can the device be allowed to start sharing. So the number of mobile devices working in AP mode in the network shows a kind of dynamic distribution. A node in the network will experience the following states:

(1) The data is transmitted from the source node to its neighbor.

(2) The neighbors receive part of encoded data.

(3) Some nodes receive part of data and switch to AP mode.

(4) The nodes decode and recover the original data.

Then this paper makes the following analysis.

(a) All the N mobile devices are divided into three categories, in which N is a dynamic value, and each device is randomly distributed.

(i) For the devices that have recovered all the data and switched to AP, we called them infected group.

(ii) For the devices that have received part of encoded data but been switched to AP mode, we called them hidden group.

(iii) For the devices that have not received any data, we called them healthy group.

(b) Due to the random distribution of mobile devices, the number of adjacent nodes of each device is different. We assume that all the mobile devices are subject to uniform distribution, and each device has λ neighbors. Moreover, this paper assumes that the number of devices working in AP mode at time t is $I(t)$, and the number of devices working in ordinary mode is $S(t)$.

(c) We assume that all the hidden devices could become infected group with a probability P. P is a variable related to the generation depth K, total resource number L, and time t.

Traditional file transmission mode is very different from the mode of RLNC in generation depth. When we set k to be 1, the scheme based on linear network coding will be degenerated to traditional scheme. Therefore, we make the analysis in two kinds of conditions.

(1) When k equals 1, the network-coding-based scheme is equivalent to traditional file sharing scheme. The probability that the hidden AP devices could recover the original file will be influenced by the total resource number L and the transmission time.

This paper assumes that the received data at hidden devices j cannot exceed local cache capacity n_j. When the generation number K is great, the hidden node has to receive all the data so that it could recover the original data. Therefore, the probability that the hidden AP node could recover the original file decreases as K increases. The relation can be expressed by the following equation:

$$P_1 \propto \frac{1}{L}. \tag{3}$$

As time passed, hidden nodes receive more and more slices it requires, and then the probability of successfully decoding will accordingly increase:

$$P_1 \propto t. \tag{4}$$

Through the analysis above, we observe that the transform probability P_1 and time t in traditional scheme have the following relation:

$$P_1 = \left(\frac{C_1}{L} \times t \right) \times \alpha_i. \tag{5}$$

(2) When K does not equal 1, all the data transmitted on the network is encoded with RLNC, and the intermediate devices have to send the linear combinations to its neighbors. Then the transfer probability P_2 will be influenced by the generation K and the transmission time t.

When K becomes greater, hidden nodes have to receive more slices to decode and recover the original file. Therefore, the probability of recovering the original file in a specific time will be reduced.

As time goes on, the probability that the slices required by a node exist in its neighbor will increase.

$$P_2 = \frac{C_2}{k} \times \alpha_j \times t \times L. \tag{6}$$

We assume each soft AP could make $\lambda S(t)$ ordinary devices become soft AP and then set up differential equations

$$N\frac{dI(t)}{d(t)} = \lambda S \times I \times N, \tag{7}$$

$$Y = N \times I \times P,$$

$$S(t) + I(t) = 1$$

$$P_2 = \frac{C_2}{k} \times \alpha_j \times L \quad (K \neq 1)$$

$$P_1 = \left(\frac{C_1}{L} \times t\right) \times \alpha_j \quad (K = 1) \tag{8}$$

$$Y_0 = 1$$

$$\alpha_j \geq \frac{1}{2}.$$

The meaning of the parameters in (8) is listed in the Abbreviations.

4. Evaluation Result

According to the differential equations and the constraints in (7) and (8), the relation between the time t and the probability $I(t)$ could be expressed in

$$I(t) = \frac{1}{1 + (1/i_0 - 1)e^{-\lambda t}}. \tag{9}$$

No matter what the transmission mode we used is, only the data is different, and the transmission frameworks are the same. We then make the simulation based on this model. When the number of adjacent nodes $\lambda = 2$, the initial ratio $I_0 = 1$:

$$\frac{d_i}{d_t} = \lambda i(1 - i), \tag{10}$$

$$I_0 = 0.1.$$

MATLAB is adopted to calculate the function in (10) and (11), and then we obtain the relation between the ratio i (the number of soft APs/the number of all devices), time t, and d_i/d_t, which are displayed in Figure 2 and Figure 3.

According to Figures 2 and 3, when sharing data in a network with N nodes, the number of nodes in the infected group reaches half of the whole nodes, and the sharing rate would reach the highest level which makes the number of successful devices increase at the highest rate.

When $K = 1$, the network-coding-based scheme degenerated to traditional replicate-based scheme. The relation between time t and the number of successful devices that could recover all the original data is calculated for both network-coding-based scheme $K > 1$ and traditional replicate-based scheme $K = 1$, respectively, which is shown in

$$Y(t)$$

$$= \begin{cases} \dfrac{N}{1 + (1/i_0 - 1)e^{-\lambda t}} \times \left[\left(\dfrac{C_1}{L} \times t\right) \times \alpha_j\right] & (K = 1) \\[4mm] \dfrac{N}{1 + (1/i_0 - 1)e^{-\lambda t}} \times \left[\dfrac{C_2}{K} \times \alpha_j \times t \times L\right] & (K \neq 1). \end{cases} \tag{11}$$

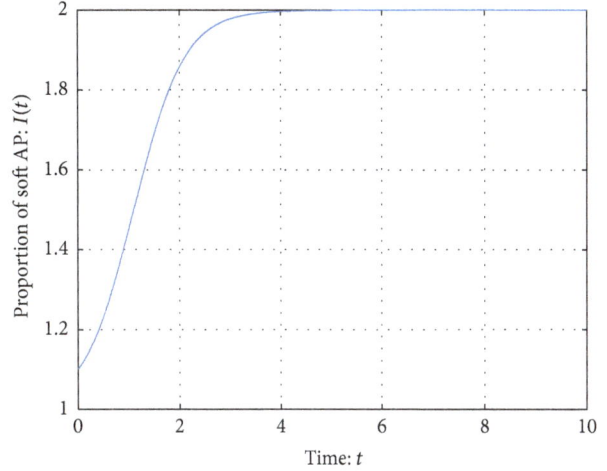

FIGURE 2: Relation between $I(t)$ and t.

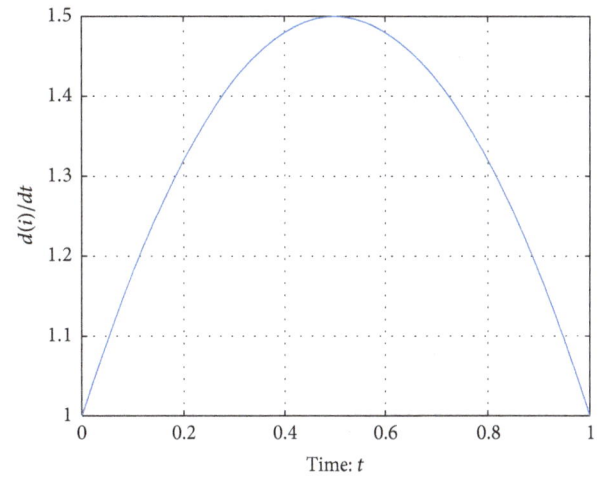

FIGURE 3: Relation between d_i/d_t and t.

Figure 4 is calculated with MATLAB. In the calculation, N is set to be 10, and file size L is set to be 4 M.

As shown in Figure 4, the network-coding-based scheme outperforms traditional replicate-based sharing scheme.

When network coding is used, the parameter K has influence on the data sharing efficiency.

It is clearly evident from Figure 5 that the sharing rate increases as K increases. However, a drawback is that the computational overhead would increase as K increases.

5. Conclusion

In order to realize the large scale date sharing in future networks, this paper studies a scheme based on secure network coding via D2D communication. Part of the mobile devices in the system may be switched to soft AP mode, and linear network coding operation will be performed on the AP before forwarding the file slices. Through the evaluation of analysis model, the authors observe that the time required for file sharing among multiple devices is less than that in traditional networks. In the future, the authors will implement the scheme in mobile devices such as smartphone networks.

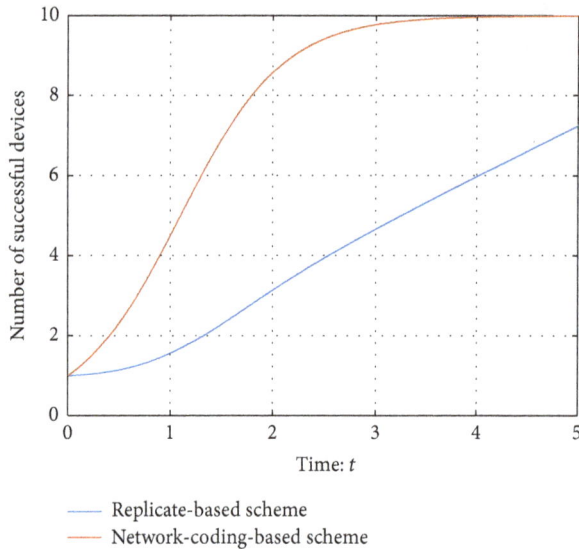

FIGURE 4: Performance of network-coding-based scheme and replicate-based scheme.

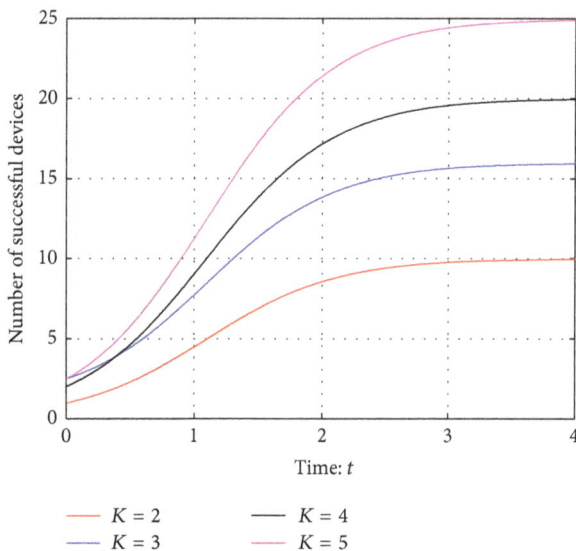

FIGURE 5: Influence of the generation size K.

Abbreviations

N: The number of nodes in the network
S: The proportion of devices that have no data but work in soft AP mode
I: The proportion of mobile devices working in soft AP mode
Y: The number of devices that can recover data
λ: The average number of neighbors for each device
ε: The probability of recovering data
K: The generation of transmission
L: The file size
α: Proportion of cache occupation
P: Probability of successful recovery
C_1, C_2:   Constant.

Conflicts of Interest

The authors declare that they have no conflicts of interest.

Acknowledgments

This work was supported by the Project for Science and Technology of Jiangsu Province (BK20150846), the Scientific Research Foundation of NUPTSF (214167), the Huawei Innovation Research Program (HIRP20140116), the Postdoctoral Fund of Jiangsu (1501041C), and the Postdoctoral Fund of China (2016M590484).

References

[1] R. Ahlswede, N. Cai, S. Y. R. Li, and R. W. Yeung, "Network information flow," *IEEE Transactions on Information Theory*, vol. 46, no. 4, pp. 1204–1216, 2000.

[2] S. Y. Li, R. W. Yeung, and N. Cai, "Linear network coding," *IEEE Transactions on Information Theory*, vol. 49, pp. 371–381, 2003.

[3] T. Ho, M. Médard, R. Koetter, M. Effros, J. Shi, and B. Leong, "A random linear network coding approach to multicast," *IEEE Transactions on Information Theory*, vol. 52, no. 10, pp. 4413–4430, 2006.

[4] S. Jaggi, P. Sanders, P. A. Chou et al., "Polynomial time algorithms for multicast network code construction," *IEEE Transactions on Information Theory*, vol. 51, no. 6, pp. 1973–1982, 2005.

[5] M. Effros, S. El Rouayheb, and M. Langberg, "An equivalence between network coding and index coding," *IEEE Transactions on Information Theory*, vol. 61, no. 5, pp. 2478–2487, 2015.

[6] K. V. Rashmi, N. B. Shah, K. Ramchandran, and P. V. Kumar, "Regenerating codes for errors and erasures in distributed storage," in *Proceedings of the 2012 IEEE International Symposium on Information Theory (ISIT '12)*, pp. 1202–1206, July 2012.

[7] B. Saleh and D. Qiu, "Performance analysis of network–coding–based P2P live streaming systems," *IEEE/ACM Transactions on Networking*, vol. 24, no. 4, pp. 2140–2153, 2016.

[8] D. Li, H. Zhao, F. Tian, H. Bo, Y. Xu, and G. Zhang, "Multipath network coding and multicasting for content sharing in wireless P2P networks: a potential game approach," *Computer Communications*, vol. 96, pp. 17–28, 2016.

[9] D. Ferreira, R. A. Costa, and J. Barros, "Real–Time network coding for live streaming in hyper–dense," *IEEE Journal on Selected Areas in Communications*, vol. 32, no. 4, pp. 773–781, 2014.

[10] L. F. Xie, P. H. J. Chong, I. W.-H. Ho, and H. C. B. Chan, "Virtual overhearing: an effective way to increase network coding opportunities in wireless ad-hoc networks," *Computer Networks*, vol. 105, pp. 111–123, 2016.

[11] M. Yang and Y. Yang, "Peer–to–peer file sharing based on network coding," in *Proceedings of the 28th International Conference on Distributed Computing Systems (ICDCS '08)*, pp. 168–175, IEEE, 2008.

[12] Y. Lin, B. Li, and B. Liang, "Stochastic analysis of network coding in epidemic routing," *IEEE Journal on Selected Areas in Communications*, vol. 26, no. 5, pp. 794–808, 2008.

[13] J. Thomas, J. Robble, and N. Modly, "Off grid communications with android meshing the mobile world," in *Proceedings of the IEEE Conference on Technologies for Homeland Security (HST '12)*, pp. 401–405, July 2012.

[14] R. Sanchez-Iborra, M.-D. Cano, and J. Garcia-Haro, "Performance evaluation of BATMAN routing protocol for VoIP services:
a QoE perspective," *IEEE Transactions on Wireless Communications*, vol. 13, no. 9, pp. 4947–4958, 2014.

[15] M. Médard, F. H. P. Fitzek, M.-J. Montpetit, and C. Rosenberg,
"Network coding mythbusting: why it is not about butterflies
anymore," *IEEE Communications Magazine*, vol. 52, no. 7, pp.
177–183, 2014.

[16] L. Keller, A. Le, B. Cici, H. Seferoglu, C. Fragouli, and A. Marko-
poulou, "MicroCast: cooperative video streaming on smartphones," in *Proceedings of the 10th International Conference on
Mobile Systems, Applications, and Services (MobiSys '12)*, pp. 57–
70, June 2012.

21

Data Selective Rake Reception for Underwater Acoustic Communication in Strong Multipath Interference

Shingo Yoshizawa,[1] Hiroshi Tanimoto,[1] and Takashi Saito[2]

[1]Department of Electrical and Electronic Engineering, Kitami Institute of Technology, Kitami, Japan
[2]Mitsubishi Electric TOKKI Systems Corporation, Kanagawa, Japan

Correspondence should be addressed to Shingo Yoshizawa; yosizawa@mail.kitami-it.ac.jp

Academic Editor: George S. Tombras

In underwater acoustic communication (UAC), very long delay waves are caused by reflection from water surfaces and bottoms and obstacles. Their waves interfere with desired waves and induce strong multipath interference. Use of a guard interval (GI) is effective for channel compensation in OFDM. However, a GI tends to be long in shallow-water environment because a guard time is determined by a delay time of multipath. A long GI produces a very long OFDM frame in several seconds, which is disadvantageous to a response speed of communication. This paper presents a method of keeping good communication performance even for a short GI. We discuss influence of intercarrier interference (ICI) in OFDM demodulation and propose a method of data selective rake reception (DSRake). The effectiveness of the proposed method is discussed by received signal distribution and confirmed by simulation results.

1. Introduction

Remotely operated underwater vehicle (ROV) and autonomous underwater vehicle (AUV) are widely used in current marine surveys [1, 2]. Wireless communication is an important underlying technology in remote control and information gathering for ROV and AUV. Since light and electromagnetic waves have large attenuation in seawater, use of sound waves is suitable for long range communication. Underwater acoustic communication (UAC) has been studied for a long time as well as radio communication. For instance, a communication unit of single-sideband amplitude modulation (SSB-AM) was developed in the 1950s. Digital modulation schemes of spread spectrum [3, 4], OFDM [5–7], and MIMO [8, 9] have been studied in recent studies.

Demodulation is affected by multipath interference and Doppler in UAC, which degrade communication performance. Doppler compensation has been discussed in [10–13]. We focus on the problem of multipath interference in this paper. Very long delay waves are caused by reflection from water surfaces and bottoms and obstacles. Their waves interfere with desired waves and induce strong multipath

interference. For mitigation of multipath interference, OFDM with a guard interval (GI) (also named as a cyclic prefix (CP)) is adopted. As far as a delay time of multipath is less than a guard time, influence of delay waves can be expressed by channel coefficients for every frequency bin. These channel coefficients can be estimated and equalized by frequency domain equalization (FDE). Effectiveness of OFDM using a GI has been verified by sea trials in [5–7].

The drawback of using a GI is decrease of communication efficiency because a GI itself is redundant. In shallow-water environment, a long GI is required when a guard time is determined by a delay time of multipath. The delay time ranges from several milliseconds to 100 milliseconds in underwater acoustic propagation, being dependent on surrounding environments. In the sea trial presented by Berger et al. [7], the GI and FFT length were set to 48 ms and 491 ms. OFDM frame duration runs up to 5.4 seconds, which would be undesirable in terms of a response speed of communication.

This paper presents a method of keeping good communication performance even for a short GI. Strong multipath interference is assumed in our study, where arrival

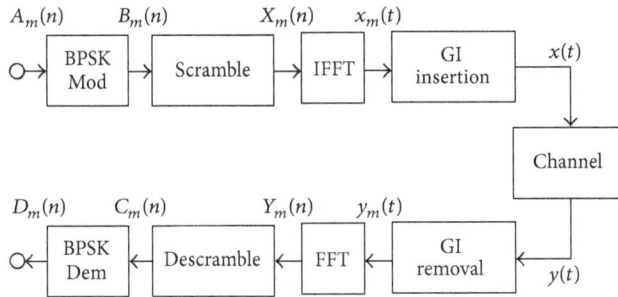

FIGURE 1: Basic OFDM model.

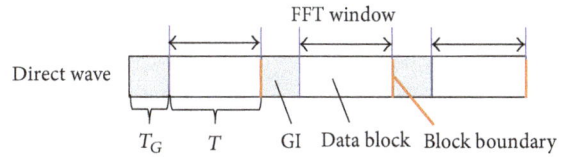

FIGURE 2: OFDM frame structure and timing positions for FFT windowing.

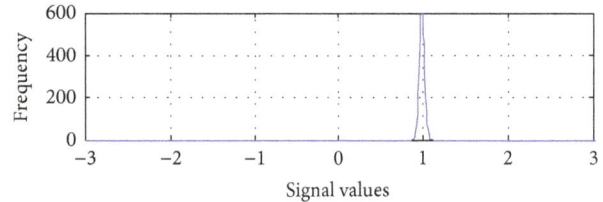

FIGURE 3: Received signal distribution for only direct wave.

times of large delay waves exceed a guard time. First, we discuss the influence of interblock interference (IBI) and intercarrier interference (ICI) in received signal distribution. Although IBI always interferes with demodulation, ICI can be suppressed by taking an appropriate FFT window timing. Next, we propose a new idea of data selective rake reception (DSRake) according to the above discussion. DSRake takes multiple fingers by changing FFT window timing for every OFDM block. The best finger with the least ICI is selected by checking data errors for all fingers. With regard to QPSK modulation, the mitigation of ICI has an impact on avoiding error floor in BER performance. This paper discusses OFDM as communication scheme. As for single carrier frequency domain equalization (SC-FDE), we briefly report it in [14].

This paper is organized as follows. Section 2 discusses the influences of IBI and ICI by received signal distribution. Section 3 proposes DSRake for the mitigation of ICI. Section 4 reports simulation results evaluating DSRake in strong multipath interference. Section 5 summarizes our work.

2. Received Signal Distribution

2.1. OFDM Model. We discuss the influences of IBI and ICI by received signal distribution. Theoretical symbol error rates (SERs) of PSK and QAM can be obtained by probability density function (PDF) when we observe received signal amplitudes in noisy propagation channels. We use a basic OFDM model illustrated in Figure 1. In the transmitter side, all transmitted data are set to zero, given by $A_m(n) = 0$ ($0 \leq m \leq M - 1$, $0 \leq n \leq N - 1$). m denotes a block number for M OFDM blocks. n is an subcarrier index for N OFDM subcarriers. A transmitted symbol becomes $B_m(n) = 1$ after BPSK modulation. The transmitted symbol is converted into 1 or -1 by multiplying random patterns of $S_m(n)$ in the scramble block, which becomes $X_m(n)$. A time-domain signal block is given by $x_m(t)$ after IFFT operation, where t is a discrete sample time. A transmitted signal is expressed by $x(t)$ after GI insertion and parallel to serial conversion. We presuppose that this GI is given by a cyclic prefix.

In the receiver side, a received signal block of $y_m(t)$ is obtained by cutting out a received signal of $y(t)$ by a FFT window having a rectangular shape. A frequency domain signal block is given by $Y_m(n)$ after FFT operation. $C_m(n)$ is obtained by multiplying the random patterns of $S_m(n)$ used in transmitter side. Received data of $D_m(n)$ are obtained after

BPSK demodulation. We set lengths of a data block, GI, and OFDM block to $T(= N)$, T_G, and $T + T_G$.

We use a two-path channel model consisting of direct and delay waves. A relation between transmitted and received signals is expressed as

$$y(t) = x(t) + \alpha x(t - \tau) + n(t), \tag{1}$$

where α is a propagation channel coefficient ($|\alpha| < 1$) for the delay wave and τ is an arrival time difference between direct and delay waves. $n(t)$ denotes noise signal component determined by a metric of the carrier to noise ratio (CNR).

Figure 2 shows an OFDM frame structure and timing positions for FFT windowing. This figure shows the case of receiving only a direct wave. When timing synchronization is perfect, their positions are the same of those of data blocks, not overlapping with GIs. The block boundary is emphasized between OFDM blocks.

The received signal distribution for a 30-dB CNR is shown in Figure 3. We set a data block length and a guard time to $T = 256$ and $T_G = 64$, respectively. The signal distribution for the received BPSK symbols of $C_m(n)$ is plotted. The total number of received BPSK symbols is $256 \times 20 = 5,120$. In BPSK demodulation, a symbol error occurs when $C_m(n)$ has a negative value. All the signals in Figure 3 locate around 1, which indicates the error-free demodulation of $C_m(n) \approx B_m(n)$.

2.2. Influence of Interblock Interference (IBI). Let us consider the influence of IBI as a long delay wave overlaps with a direct wave. Figure 4 shows the relations between direct and delay waves where their arrival time differences of $\tau = 320$ and $\tau = 330$. The propagation channel coefficient is set to $\alpha = 0.7$ for a delay wave. IBI happens due to the collision of different data blocks for direct and delay waves.

The received signal distributions for Figure 4(a) are shown in Figure 5. In Figure 4(a), $(m-1)$th block of the delay wave exactly overlaps with mth block of the direct wave in the FFT window period. The received symbol of $C_m(n)$ can be

(a) $\alpha = 0.7$, $\tau = 320$, 30-dB CNR

(b) $\alpha = 0.7$, $\tau = 330$, 30-dB CNR

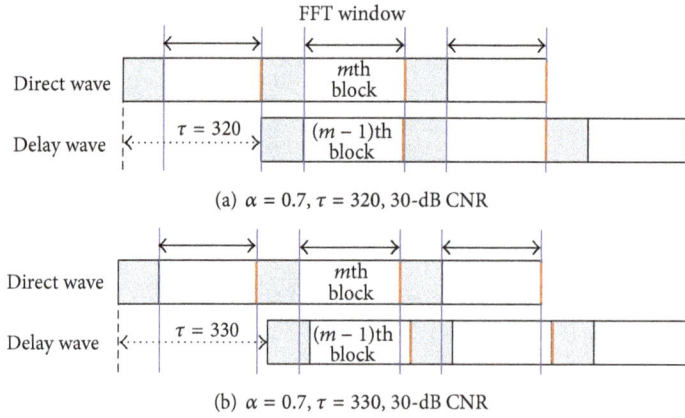

FIGURE 4: Relations between direct and delay waves ($\tau = 320$ and $\tau = 330$).

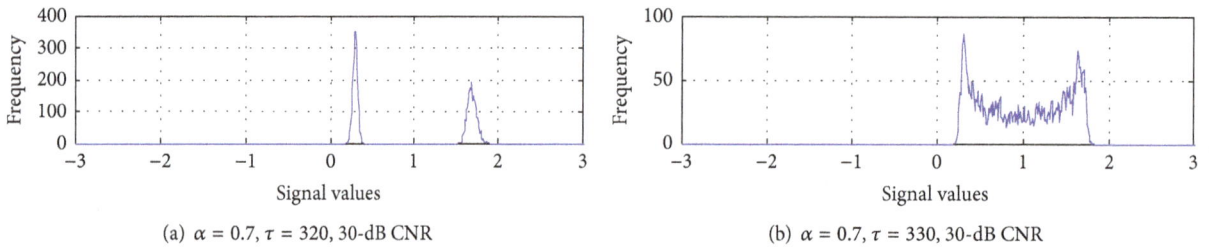

(a) $\alpha = 0.7$, $\tau = 320$, 30-dB CNR

(b) $\alpha = 0.7$, $\tau = 330$, 30-dB CNR

FIGURE 5: Received signal distributions for long delay waves.

introduced from the following equations, omitting the noise component of $n(t)$.

$$y_m(t) = x_m(t) + \alpha x_{m-1}(t) \tag{2}$$

$$Y_m(n) = X_m(n) + \alpha X_{m-1}(n) \tag{3}$$

$$Y_m(n) = S_m(n) B_m(n) + \alpha S_{m-1}(n) B_{m-1}(n) \tag{4}$$

$$C_m(n) = B_m(n) + \alpha S_m(n) S_{m-1}(n) B_{m-1}(n). \tag{5}$$

$$C_m(n) = 1 + \alpha S_m(n) S_{m-1}(n). \tag{6}$$

Since $S_m(n)$ and $S_{m-1}(n)$ are random patterns consisting of 1 or -1, (6) gives $C_m(n) \in \{0.3, 1.7\}$. This signal distribution can be observed in Figure 5(a). Although the signal values of $C_m(n)$ do not concentrate on 1, all of them are positive. A symbol error does not occur in Figure 4(a).

In Figure 4(b), $(m-1)$th block of the delay wave is slightly deviated from mth block of the direct wave. $C_m(n)$ can be introduced by

$$y_m(t) = x_m(t) + \alpha x_{m-1}(t - \tau_d) \tag{7}$$

$$Y_m(n) = X_m(n) + \alpha X_{m-1}(n) e^{j2\pi n(\tau_d/N)} \tag{8}$$

$$\mathrm{Re}\left[C_m(n)\right]$$
$$= \mathrm{Re}\left[B_m(n) + \alpha S_m(n) S_{m-1}(n) B_{m-1}(n) e^{j2\pi n(\tau_d/N)}\right] \tag{9}$$

$$\mathrm{Re}\left[C_m(n)\right] = 1 + \mathrm{Re}\left[\alpha S_m(n) S_{m-1}(n) e^{j2\pi n(\tau_d/N)}\right], \tag{10}$$

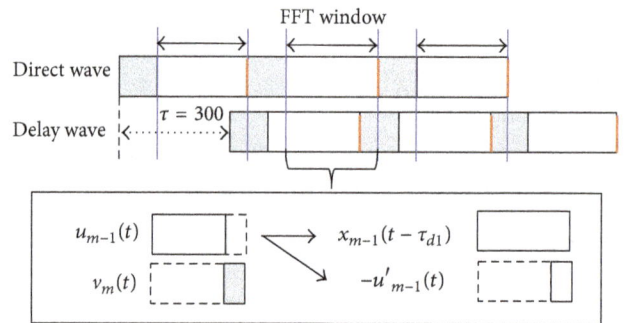

FIGURE 6: Relation between direct and delay waves ($\tau = 300$).

where we apply $\tau_d = \tau - T_G + T$ from circular shift property. The signal values of $\mathrm{Re}[C_m(n)]$ range from 0.3 to 1.7 as shown in Figure 5(b). This case also does not induce a symbol error.

The IBI does not take a symbol error as long as a high CNR condition is kept as for this observation. The same phenomenon would be observed even in QPSK transmission. Improvement of SNR using antenna arrays is practical rather than keeping a high CNR, where Zheng presented MRC diversity in SIMO-OFDM as a measure against insufficient guard interval in [15].

2.3. Influence of Intercarrier Interference (ICI).
Let us consider the influence of ICI by giving another arrival time difference of $\tau = 300$. The relation between direct and delay waves is shown in Figure 6. Different from Figure 4, $(m-1)$th

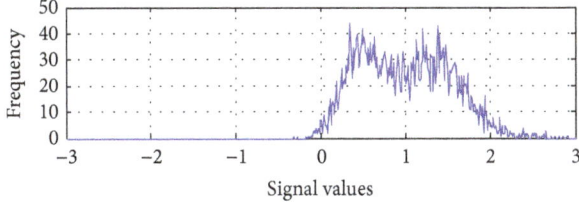

FIGURE 7: Received signal distribution affected by ICI.

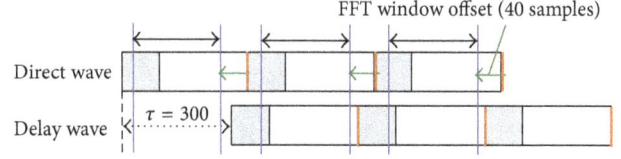

FIGURE 8: Adjustment of FFT windowing.

data block and mth GI of the delay wave overlap with mth data block of the direct wave. This signal distribution is shown in Figure 7. Since some of $C_m(n)$ have a negative value, a symbol error occurs.

We introduce $C_m(n)$ as well as Section 2.2. First, the received signal of $y_m(t)$ is given by

$$y_m(t) = x_m(t) + \alpha \left(u_{m-1}(t) + v_m(t) \right). \tag{11}$$

We decompose a received signal of the delay wave into $u_{m-1}(t)$ and $v_m(t)$ as shown in Figure 6. Their functions are given by

$$u_{m-1}(t) = \begin{cases} x_{m-1}(t - \tau_{d1}) & \text{if } 0 \le t \le \tau_{d1} - 1 \\ 0 & \text{if } \tau_{d1} \le t \le N - 1 \end{cases}$$

$$v_m(t) = \begin{cases} 0 & \text{if } 0 \le t \le \tau_{d1} - 1 \\ x_m(t - \tau_{d2}) & \text{if } \tau_{d1} \le t \le N - 1, \end{cases} \tag{12}$$

where we apply $\tau_{d1} = \tau - T_G$ and $\tau_{d2} = \tau_{d1} - T_G$ from circular shift property. $u_{m-1}(t)$ can be replaced with $x_{m-1}(t - \tau_{d1}) - u'_{m-1}(t)$. $u'_{m-1}(t)$ is given by

$$u'_{m-1}(t) = \begin{cases} 0 & \text{if } 0 \le t \le \tau_{d1} - 1 \\ x_{m-1}(t - \tau_{d1}) & \text{if } \tau_{d1} \le t \le N - 1. \end{cases} \tag{13}$$

$C_m(n)$ can be expressed as

$$\begin{aligned} y_m(t) = & \, x_m(t) + \alpha x_{m-1}(t - \tau_{d1}) \\ & + \alpha \left(-u_{m-1}(t) + v_m(t) \right) \end{aligned} \tag{14}$$

$$\begin{aligned} \mathrm{Re}\left[C_m(n)\right] = & \, 1 + \mathrm{Re}\left[\alpha S_m(n) S_{m-1}(n) e^{j2\pi n(\tau_{d1}/N)}\right] \\ & + \mathrm{Re}\left[\alpha S_m(n) \left(-U'_{m-1}(n) + V_m(n)\right)\right]. \end{aligned} \tag{15}$$

The received signal distribution of (15) would be almost the same as that of (10) if $U'_{m-1}(n)$ and $V_m(n)$ are excluded.

$U'_{m-1}(n)$ and $V_m(n)$ can be expressed by using inverse discrete Fourier transform (IDFT) and DFT as

$$U'_{m-1}(n)$$
$$= \sum_{t=\tau_{d1}}^{N-1} \left[\frac{1}{N} \sum_{n=0}^{N-1} S_{m-1}(n) e^{j(2\pi tn/N)} e^{-j(2\pi \tau_{d2} n/N)} \right] \tag{16}$$
$$\cdot e^{-j(2\pi nt/N)}$$

$$V_m(n) = \sum_{t=\tau_{d1}}^{N-1} \left[\frac{1}{N} \sum_{n=0}^{N-1} S_m(n) e^{j(2\pi tn/N)} e^{-j(2\pi \tau_{d2} n/N)} \right] \tag{17}$$
$$\cdot e^{-j(2\pi nt/N)}.$$

The interferences of (16) and (17) are added for every subcarrier, which corresponds to ICI. Assuming that the average amplitude for the OFDM transmit signals after IDFT is $1/N$ (i.e., calculation within the square bracket in (16)), the average of deviations caused by $U'_{m-1}(n)$ and $V_m(n)$ is roughly calculated as

$$\pm 2\alpha \frac{N - \tau_{d1}}{N} \simeq \pm 0.11. \tag{18}$$

These deviations would be observed by comparing the received signal distributions in Figures 5 and 7. The difference between Figures 4 and 6 is whether a block boundary is included within a FFT window.

2.4. Adjustment of FFT Window. The ICI can be avoided by changing FFT window timings, whose adjustment is illustrated in Figure 8. The time positions of FFT windows have been shifted by 40 samples ahead. The block boundaries for the delay wave are not included for their FFT windows. Although this adjustment induces a phase rotation after FFT operation in frequency domain, the phase rotation can be detected and compensated by FDE. The received signal distribution after the FFT window adjustment is shown in Figure 9, where the phase rotation can be compensated before descramble. This distribution looks like Figure 5(b) owing to the ICI avoidance.

3. Data Selective Rake Reception (DSRake)

The ICI avoidance is achieved when the arrival time of delay wave is perfectly known. Note that arrival times of individual delay waves are almost unknown in the actual environment. We introduce an OFDM rake reception as an alternative method, whose scheme is shown in Figure 10. Since the arrival

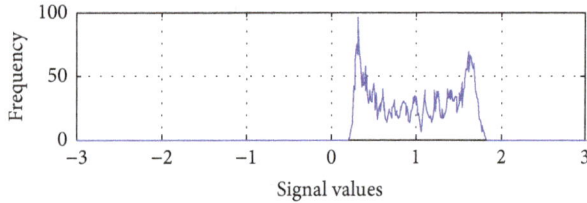

FIGURE 9: Received signal distribution after FFT window adjustment.

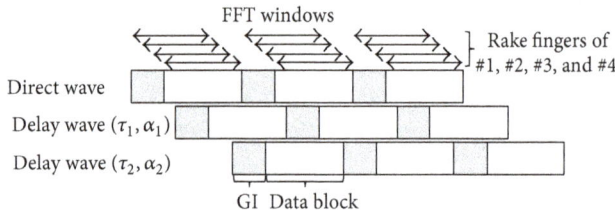

FIGURE 10: OFDM rake reception.

TABLE 1: Results of delay profiles.

(a) 8-m distance

	ch1	ch2	ch3	ch4
Ave. delay time [ms]	4.1	3.3	3.6	2.8
RMS delay spread [ms]	9.4	8.1	8.7	7.4

(b) 20-m distance

	ch1	ch2	ch3	ch4
Ave. delay time [ms]	6.6	5.8	5.2	5.9
RMS delay spread [ms]	12.1	11.0	10.5	11.4

times (τ_1 and τ_2) and magnitude (α_1 and α_2) of delay waves are unknown, we take multiple FFT window timings for OFDM demodulation, that is, rake fingers.

Original rake reception itself is used as path diversity in spread spectrum [16]. In general, OFDM and rake reception for path diversity are not compatible. Received symbols in rake fingers have high correlation with each other as far as multipath delay time is less than a guard time. The improvement of received SNR is little considering increase of computational complexity in demodulation. We use the rake reception to find the best rake finger that is not affected by ICI so much. It does not aim at path diversity. The selection of rake fingers is achieved by checking data errors after demodulation, where the proposed scheme of data selective rake reception (DSRake) is shown in Figure 11. In the transmitter side, cyclic redundancy check (CRC) codes are inserted in binary data before forward error correcting (FEC) coding. In the receiver side, multiple OFDM demodulators accept received signals in rake fingers and output decoded data blocks. The best data block having no error is selected as final data by observing the CRC results in the data selection unit. If all fingers have data errors, the final data are generated by merging all decoded data in bit level.

DSRake would not be adopted in general OFDM systems such as IEEE802 WLANs and LTE in RF communication due to considerable increase in computational complexity. Note that the bandwidth of UAC is much narrower than that of RF. The increase of computational complexity for UAC does not become a problem from the viewpoint of implementation in RF. The overhead of CRC is trivial because its length is enough for 16 bits (CRC-16) in typical usage.

DSRake belongs to selection combining (SC) in diversity combining. Maximal ratio combining (MRC) should be discussed as another method. The alternative scheme of MRC rake reception (MRCRake) is shown in Figure 12. The received symbols in rake fingers are synthesized after OFDM demodulation. Generally, a diversity gain of MRC is higher than that of SC. However, MRCRake is inferior to DSRake in

terms of the mitigation of ICI. The synthesis of rake fingers takes in undesirable received symbols affected by ICI and the effect is limited. The superiority of DSRake will be confirmed by our simulation in the next section.

4. Simulation

4.1. Channel Model. As an example of underwater acoustic propagation, we use two channel models measured in a swimming pool. The delay profiles were measured on the condition of horizontal link where one transmitter and four receiver hydrophones horizontally face each other. The location of hydrophones is drawn in Figure 13. The pool length and width are 25 m and 13 m and the water depth is 1.2 m. The distances between transmitter and receiver hydrophones are 8 m and 20 m. The space of four hydrophones is 5 cm.

The delay profiles for 8 m and 20 m distances are shown in Figures 14 and 15. A direct wave is located at 0 on the time axis and has normalized magnitude of 0 dB. Delay waves are expressed by individual values of relative magnitude and delay time. Several clusters of delay waves are periodically observed around 30 to 35 ms, 65 to 70 ms, and 97 to 102 ms in Figure 15. These clusters come from several round trip reflections at the side walls. The delay waves of more than −10 dB (i.e., less than 10 dB in desired to undesired signal ratio (DUR)) range from 0 seconds to 35 ms. Since we set a guard time to 12.8 ms in our simulation, the delay waves beyond the GI induce IBI and ICI. If a guard time is more than 110 ms (i.e., more than 20 dB DUR), the influences of IBI and ICI would be small. However, we must keep in mind that a long GI is undesirable in terms of a response speed of communication.

Summary of the delay profiles is reported in Table 1. The 20 m distance shows larger values in average delay time and RMS delay spread than the 8 m distance. The results of average delay time and RMS delay spread are different among receiver channels to some extent. The signal correlation among received antennas would not be very high as having different propagations. Space diversity using antenna arrays is effective to improve a received SNR in this case. RMS delay spread is helpful in the determination of a GI length as long as the magnitude of delay waves is exponentially decaying. However, the magnitude of delay waves does not always fade as time goes on as shown in Figures 14 and 15. Even though the RMS delay spread is less than the GI length, the strong interference of delay waves should be considered.

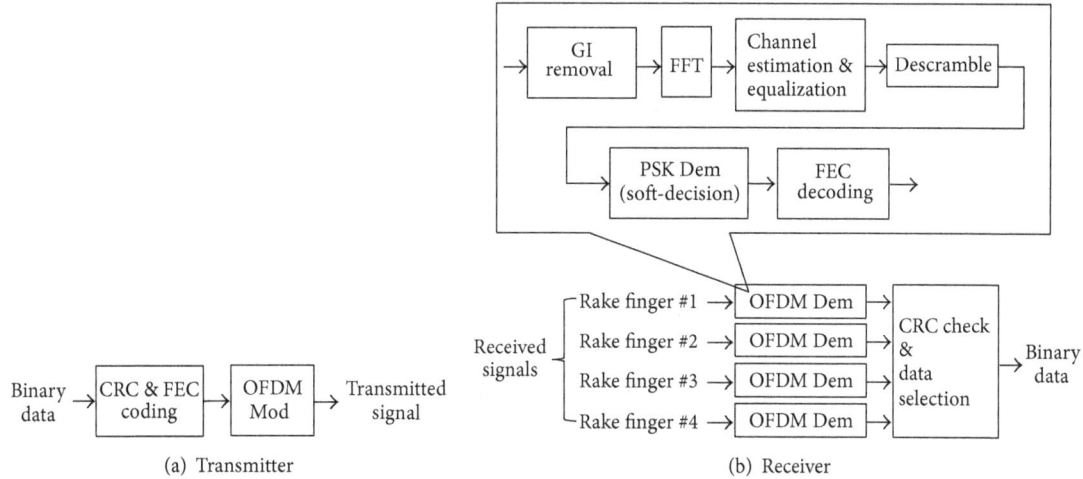

FIGURE 11: Data selective rake reception (DSRake).

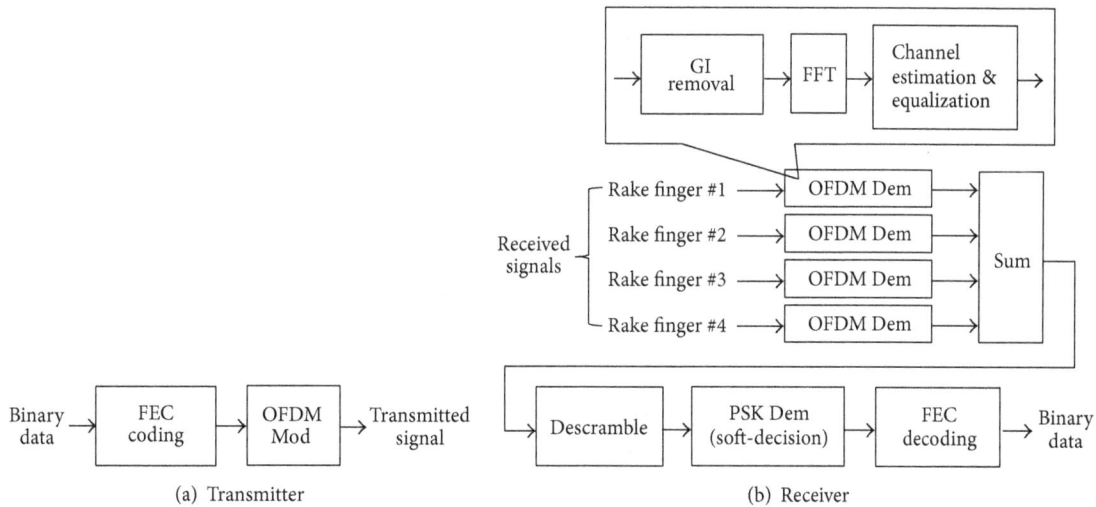

FIGURE 12: MRC rake reception (MRCRake).

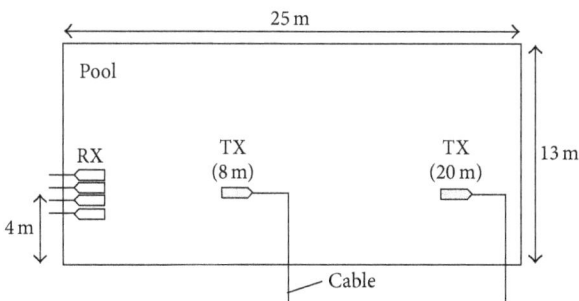

FIGURE 13: Location of transmitter and receiver hydrophones.

4.2. Simulation Parameters. The simulation parameters are enumerated in Table 2. The baseband OFDM signals with a frequency band of −10 kHz to 10 kHz are modulated by a carrier wave of 50 kHz. One-tap frequency domain linear equalization based on MMSE criterion is used in channel equalization. The GI length is set to 12.8 ms, corresponding to 256 samples in baseband domain. Two training data blocks are added to the beginning of an OFDM frame, where the frame format is shown in Figure 16. The two long training fields (LTFs) are used for channel estimation. Since the LTFs are located at the head of frame, they do not have the influence of IBI and ICI. The number of rake fingers is set to 64 for DSRake and MRCRake. We have used convolutional coding with a coding rate of 1/2. The transmit data rate is about 13.3 kbps considering the overhead of LTFs and GIs. Although the overhead of CRC codes (CRC-16) might be counted for DSRake, this overhead is very small (less than 2%).

We apply space diversity using array antennas for the mitigation of IBI. The scheme of OFDM space diversity is shown in Figure 17. Space diversity combining based on MRC is performed after channel equalization. Space diversity combining and OFDM rake reception of DSRake or MRCRake are compatible. The diversity block is inserted into the OFDM demodulation units in Figures 11 and 12.

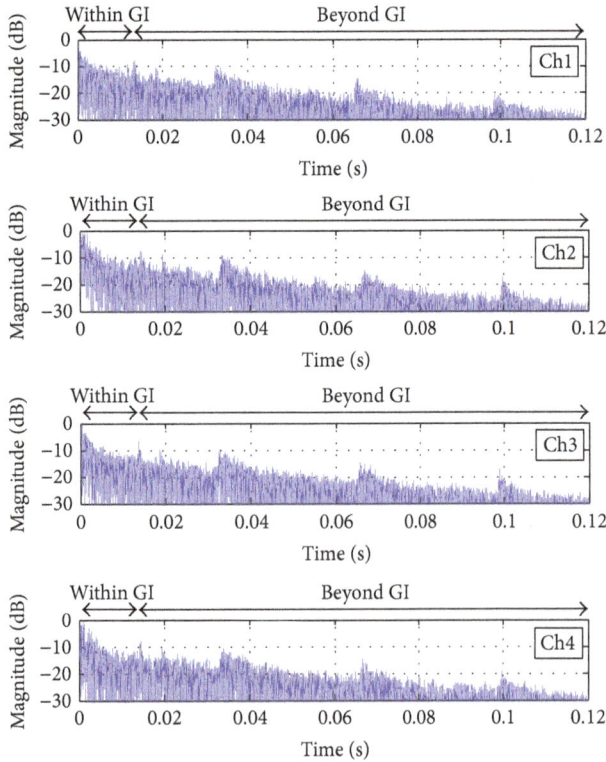

FIGURE 14: Delay profile for 8 m distance.

TABLE 2: Simulation parameters.

Modulation	QPSK-OFDM
Sampling frequency [kHz]	200
Center frequency [kHz]	50
Frequency band [kHz]	40 to 60
FFT size	1024
Number of data subcarriers	1024
OFDM symbol length [ms]	51.2
GI [ms]	12.8
Number of OFDM symbols	10
Number of training OFDM symbols	2
OFDM frame length [ms]	768
OFDM frame data size [bytes]	1280
FEC	Convolutional coding & Viterbi decoding
Coding rate	0.5
Number of antennas	1 (TX)/4 (RX)
Number of OFDM rake fingers	64
Timing synchronization	Perfect
Number of evaluated OFDM frames	100

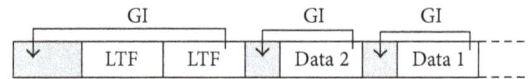

FIGURE 16: OFDM frame format.

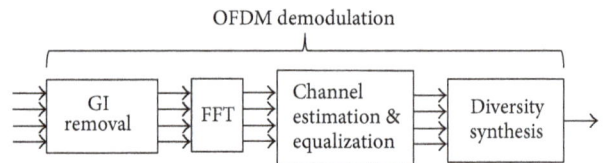

FIGURE 17: OFDM space diversity.

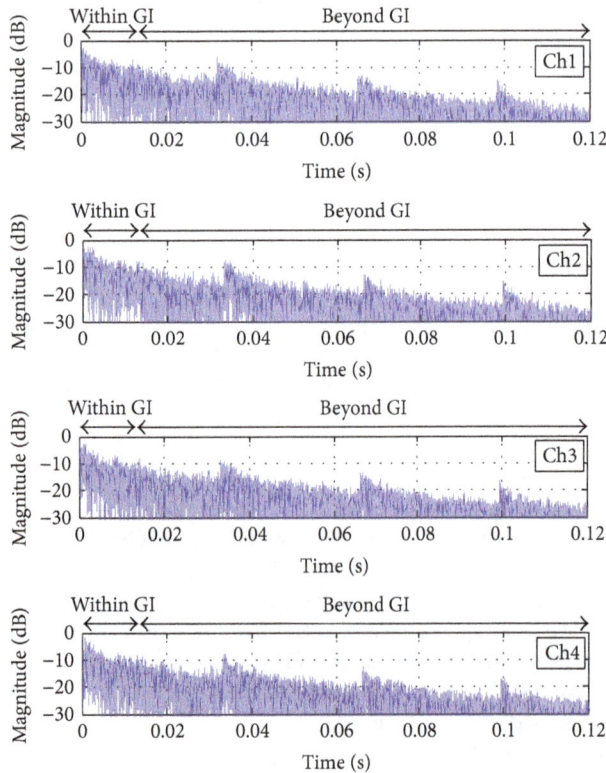

FIGURE 15: Delay profile for 20 m distance.

4.3. Simulation Results. Bit error rates (BERs) for the 8 m and 20 m distances are plotted in Figures 18 and 19. We have evaluated the schemes of single channel reception (average of four channels), space diversity, DSRake, and MRCRake. Both DSRake and MRCRake are given by the combination of space diversity and rake reception. The single channel reception has the BER floor of 10^{-2} due to strong multipath interference. The space diversity decreases the BER floor from 10^{-2} to 10^{-3} as shown in both figures. The influence of IBI would be decreased by space diversity combining to some extent. DSRake and MRCRake show further improvement of decreasing BER floor. DSRake is clearly superior to MRCRake from the BER results. The ICI mitigation contributes to the improvement of communication quality rather than taking path diversity. DSRake can eliminate a BER floor for the 8 m distance and decrease by up to 2×10^{-4} for the 20 m distance.

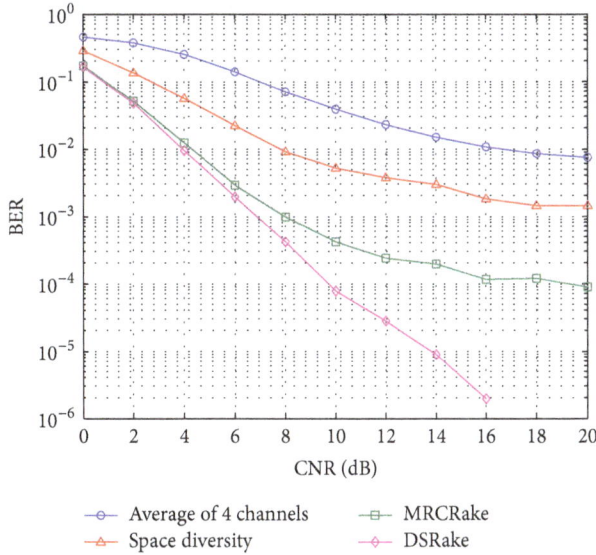

FIGURE 18: BER results for 8 m distance.

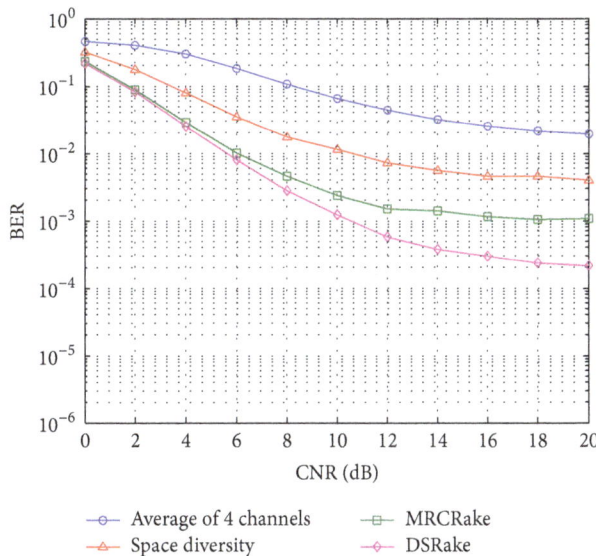

FIGURE 19: BER results for 20 m distance.

The effectiveness of DSRake in strong multipath interference has been observed from this simulation.

5. Conclusion

This paper presents a new method of OFDM rake reception in strong multipath interference. Very long delay waves beyond GI induce IBI and ICI. The influence of IBI and ICI is discussed by received signal distribution. Regarding ICI, we reported that the ICI avoidance can be achieved by changing FFT window timing. According to the idea of ICI avoidance, we have proposed DSRake as one of rake reception techniques. Original rake reception is used for obtaining path diversity. However, our rake reception aims at the mitigation of ICI. We have explained that

selection combining by DSRake is superior to maximal ratio combining by MRCRake. The effectiveness of DSRkae has been confirmed by the simulation results based on actual underwater propagation models. In our future work, we will investigate communication performance of DSRake when Doppler effect is added.

Conflicts of Interest

The authors declare that there are no conflicts of interest regarding the publication of this paper.

Acknowledgments

The authors would like to thank the staff of Kitami City Board of Education. This work was supported by JSPS KAKENHI Grants nos. 16K18099 and 15K06048.

References

[1] L. E. Freitag and J. A. Catipovic, "A signal processing system for underwater acoustic ROV communication," in *Proceedings of the 6th International Symposium on Unmanned Untethered Submersible Technology Technology*, pp. 34–41, June 1989.

[2] J. Borden and J. Dearruda, "Long range acoustic underwater communication with a compact AUV," in *IEEE OCEANS*, October 2012.

[3] G. Loubet, V. Capellano, and R. Filipiak, "Underwater spread-spectrum communications," in *Proceedings of the MTS/IEEE Conference OCEANS*, pp. 574–579, October 1997.

[4] K. G. Kebkal and R. Bannasch, "Implementation of a sweep-spread function for communication over underwater acoustic channels," in *Proceedings of the MTS/IEEE Conference and Exhibition OCEANS*, vol. 3, pp. 1829–1837, October 2000.

[5] C. M. Anil, *Underwater acoustic communications in warm shallow water channels, Thesis*, Doctor of Philosophy, National University of Singapore, 2006.

[6] F. Frassati, C. Lafon, P. Laurent, and J. Passerieux, "Experimental assessment of OFDM and DSSS modulations for use in littoral waters underwater acoustic communications," in *Proceeding of the Europe Oceans*, vol. 2, pp. 826–831, June 2005.

[7] C. R. Berger, J. Gomes, and J. M. F. Moura, "Sea-trial results for cyclic-prefix OFDM with long symbol duration," in *IEEE OCEANS*, June 2011.

[8] S. Roy, T. M. Duman, V. McDonald, and J. G. Proakis, "High-rate communication for underwater acoustic channels using multiple transmitters and space—time coding: receiver structures and experimental results," *IEEE Journal of Oceanic Engineering*, vol. 32, no. 3, pp. 663–688, 2007.

[9] P. Bouvet and A. Loussert, "An analysis of MIMO–OFDM for shallow water acoustic communications," in *Proceedings of the IEEE OCEANS*, pp. 1–5, September 2011.

[10] B. S. Sharif, J. Neasham, O. R. Hinton, and A. E. Adams, "A computationally efficient doppler compensation system for underwater acoustic communications," *IEEE Journal of Oceanic Engineering*, vol. 25, no. 1, pp. 52–61, 2000.

[11] B. Li, S. Zhou, M. Stojanovic, L. L. Freitag, and P. Willett, "Multicarrier communication over underwater acoustic channels with nonuniform Doppler shifts," *IEEE Journal of Oceanic Engineering*, vol. 33, no. 2, pp. 198–209, 2008.

[12] C.-H. Hwang, K.-M. Kim, S.-Y. Chun, and S.-K. Lee, "Doppler estimation based on frequency average and remodulation for underwater acoustic communication," *International Journal of Distributed Sensor Networks*, vol. 2015, Article ID 746919, 8 pages, 2015.

[13] A. E. Abdelkareem, B. S. Sharif, C. C. Tsimenidis, and J. A. Neasham, "Compensation of linear multiscale doppler for OFDM-based underwater acoustic communication systems," *Journal of Electrical and Computer Engineering*, vol. 2012, 16 pages, 2012.

[14] S. Yoshizawa, H. Tanimoto, and T. Saito, "SC-FDE vs OFDM: performance comparison in shallow-sea underwater acoustic communication," in *Proceedings of the IEEE International Symposium on Intelligent Signal Processing and Communication Systems (ISPACS '16)*, 5 pages, October 2016.

[15] L. T. Phuc, Y. Zheng, and Y. Karasawa, "A simplified propagation channel model for evaluating MRC diversity characteristics in SIMO OFDM with insufficient guard interval," in *Proceedings of the United States National Committee of URSI National Radio Science Meeting (USNC-URSI NRSM '16)*, usa, January 2016.

[16] J.-H. Son, E.-H. Jeon, K.-M. Kim, D.-W. Lee, and T.-D. Park, "Alternative approach for combination of fingers in underwater acoustic communication," *International Journal of Distributed Sensor Networks*, vol. 2016, 16 pages, 2016.

22

An Acquisition Algorithm with NCCFR for BOC Modulated Signals

Yongxin Feng,[1] Fang Liu,[2] Xudong Yao,[2] and Xiaoyu Zhang[2]

[1]*Communication and Network Institute, Shenyang Ligong University, Shenyang, Liaoning, China*
[2]*School of Information Science and Engineering, Shenyang Ligong University, Shenyang, Liaoning, China*

Correspondence should be addressed to Yongxin Feng; fengyongxin@263.net

Academic Editor: Iickho Song

With the development of satellite navigation technology, BOC (Binary Offset Carrier) signals are proposed and applied in navigation system. However, in the advantages of enhancing the utilized rating of the band resource, some new problems are also emerging in the acquisition processing. On the basis of analyzing the limitations of the existing methods in suppressing side peaks, a NCCFR (New Cross-Correlation Function Reconstruction) algorithm is proposed, in which different modulation coefficients are used to construct correlation function with a shifter phase. The simulation results show that the new algorithm can suppress first side peaks and restrain other side peaks.

1. Introduction

With the continuous development of satellite communication technology, the limited frequency band resources become increasingly scarcity, so BOC modulation signal is proposed and used in satellite navigation to solve these problems. On the basis of traditional PSK signal, BOC signal owns superior spectrum splitting characteristics, which can solve the frequency resource shortage problem; meanwhile it has better acquisition precision. However, the BOC signal also brings some problems, such as the acquisition ambiguity problem caused by multipeak characteristics [1].

Many effective acquisition methods [2, 3] have been proposed to solve some acquisition problems, and some tracking technology [4] is proposed for BOC signals. In addition, some methods are proposed to remove side peaks, such as the BPSK-like method [5], ASPeCT method, and RQCC (Remove Quadratic BOC Cross-Correlation) reconstruction algorithm. In BPSK-like method, the frequency domain of BOC signals is regarded as the two BPSK signals. On the basis of filtering and frequency transform to the side-lobe of BOC signal, the acquisition processing is accomplished by using conventional BPSK acquisition method [6, 7]. In ASPeCT method, the cross-correlation function of the signal and PRN code is applied, which is constructed by pseudorandom code to restrain the side peaks [8]. Using appropriate phase shift of the cross-correlation function and nonrelated accumulation about the shifted correlation function, the filtered method can enhance the main peak and restrain the side peaks [9]. In RQCC method, the new QBOC (Quadratic Binary Offset Carrier) auxiliary signal is reconstructed by using the characteristics of same autocorrelation function peak width between the auxiliary signal and BOC signals to solve the problem that the side peaks restrained weakly, which is caused by different correlation function peak width between the PRN auxiliary signal and BOC signals [10]. On the basis of the shifted correlation function phase, by introducing QBOC auxiliary signal and adding the shifted results to eliminate the side peaks, these methods can solve some multipeak problems, but they cannot restrain the multipeak problem in different modulation coefficient and meanwhile ensure main peak sharply.

Therefore, considering the acquisition ambiguity problems and the shortage of the existing methods, a NCCFR (New Cross-Correlation Function Reconstruction) acquisition algorithm is proposed. Using different modulation coefficient to make a different fixed phase shifter correlation function, the algorithm can efficiently restrain the first side peaks and improve the main peak.

2. NCCFR Acquisition Algorithm

Firstly, an QBOC auxiliary signal [11, 12] can be got by $\pi/2$ shifting the phase of local BOC signals, and then the BOC signal and the QBOC signal are defined in formulas (1) and (2). Then the autocorrelation function of receiving signal $R_X(\tau)$ will be got through correlation processing, and the cross-correlation function $R_{X/Q}(\tau)$ will be got by correlation processing between the auxiliary signal and received signal, which is shown in formulas (3) and (4), respectively, where $C(n)$ is the baseband code, $SN(n)$ is the square wave, and $\text{tri}_x(\tau)$ is the correlation peak whose position is x.

$$BOC\,(m,n) = C\,(n)\,SN\,(2\pi\omega n + \theta) \tag{1}$$

$$QBOC\,(m,n) = C\,(n)\,SN\left(2\pi\omega n + \frac{\pi}{2} + \theta\right) \tag{2}$$

$$R_X\,(\tau) = \sum_{i=1}^{n-1}\left(\left[(-1)^i\,\text{tri}_{-i/n}\,(\tau) + (-1)^i\,\text{tri}_{i/n}\,(\tau)\right]\right.$$

$$\left. \times \frac{n-i}{n}\right) + \text{tri}_0\,(\tau) = \sum_{i=-(n-1)}^{n-1}(-1)^{|i|}\,\text{tri}_{-i/n}\,(\tau)\,\frac{n-|i|}{n} \tag{3}$$

$$R_{X/Q}\,(\tau) = \sum_{i=1}^{n}\left(\left[(-1)^i\,\text{tri}_{-(2i-1)/2n}\,(\tau)\right.\right.$$

$$\left.\left. + (-1)^{i+1}\,\text{tri}_{(2i-1)/2n}\,(\tau)\right] \times \frac{2n-(2i-1)}{2n}\right). \tag{4}$$

Due to the difference of correlation peak phase between $R_{X/Q}(\tau)$ and $R_X(\tau)$, the main peak and some side peaks, which is similar to $R_X(\tau)$, can be reconstructed. Therefore, the nonconstant phase shift of $R_{X/Q}(\tau)$ is taken both early and late, $((n-1)T_c)/2n$, where n is the modulation coefficient, T_c is the period of one chip, and τ is the chip delay. As formulas (5) and (6) show, the superfluous side peaks will be eliminated as much as possible by taking addition and subtraction, which is shown in formula (7) as follows:

$$R_{X/Q}\left(\tau - \frac{(n-1)\,T_c}{2n}\right) = \sum_{i=1}^{n}\left(\left[(-1)^i\,\text{tri}_{-(2i+n-2)/2n}\,(\tau)\right.\right.$$

$$\left.\left. + (-1)^{i+1}\,\text{tri}_{(2i-n)/2n}\,(\tau)\right] \times \frac{2n-(2i-1)}{2n}\right) \tag{5}$$

$$R_{X/Q}\left(\tau + \frac{(n-1)\,T_c}{2n}\right) = \sum_{i=1}^{n}\left(\left[(-1)^i\,\text{tri}_{-(2i-n)/2n}\,(\tau)\right.\right.$$

$$\left.\left. + (-1)^{i+1}\,\text{tri}_{(2i+n-2)/2n}\,(\tau)\right] \times \frac{2n-(2i-1)}{2n}\right) \tag{6}$$

$$M = \left|R_{X/Q}\left(\tau - \frac{(n-1)\,T_c}{2n}\right)\right| + \left|R_{X/Q}\left(\tau\right.\right.$$

$$\left.\left. + \frac{(n-1)\,T_c}{2n}\right)\right| - \left|R_{X/Q}\left(\tau - \frac{(n-1)\,T_c}{2n}\right)\right|$$

$$+ R_{X/Q}\left(\tau + \frac{(n-1)\,T_c}{2n}\right)\right| = \frac{1}{n}\left(\text{tri}_{-1/2}\,(\tau)\right.$$

$$\left. + \text{tri}_{1/2}\,(\tau)\right) + \sum_{i=-n/2+1}^{n/2-1}\text{tri}_{-i/n}\,(\tau)\left(\frac{n+1}{n} - \frac{|2i|}{n}\right), \tag{7}$$

where $\text{tri}_{\pm i/n}(\tau)$ represents the correlation peak in location $\pm i/n$.

There are many correlation peaks similar to $R_X(\tau)$ in formula (7), especially the main peak and the correlation peak before $n/2$. Therefore, the reconstructed signal can improve main peak and restrain the side peaks of the correlation function and then can obtain formula (8) as follows:

$$R_X\,(\tau) + M$$

$$= \sum_{i=-(n-1)}^{n-1}(-1)^{|i|}\,\text{tri}_{-i/n}\,(\tau)\,\frac{n-|i|}{n}$$

$$+ \frac{1}{n}\left(\text{tri}_{-1/2}\,(\tau) + \text{tri}_{1/2}\,(\tau)\right)$$

$$+ \sum_{i=-n/2+1}^{n/2-1}\text{tri}_{-i/n}\,(\tau)\left(\frac{n+1}{n} - \frac{|2i|}{n}\right) \tag{8}$$

$$= \sum_{i=-n/2+1}^{n/2-1}\text{tri}_{-i/n}\,(\tau)\left[(-1)^{|i|}\,\frac{n-|i|}{n} + \frac{n+1}{n} - \frac{|2i|}{n}\right]$$

$$+ \left(\frac{1}{n} + \frac{1}{2}\right)\left(\text{tri}_{-1/2}\,(\tau) + \text{tri}_{1/2}\,(\tau)\right)$$

$$+ \sum_{i=n/2+1}^{n-1}(-1)^{|i|}\,\frac{n-|i|}{n}\left(\text{tri}_{-i/n}\,(\tau) + \text{tri}_{i/n}\,(\tau)\right).$$

According to formula (8), when $i = 0$, $\text{tri}_0(\tau)$ represents the main correlation peak of the signal, and the main peak will promote $(n+1)/n$. Not only will the signal synchronization acquisition rate of receiver be greatly improved, but also the acquisition ambiguity of the signal will be reduced effectively, when $i = 1$, and the results of both $\text{tri}_{1/n}(\tau)$ and $\text{tri}_{-1/n}(\tau)$ are 0, indicating that the first side peak is eliminated. In the odd condition, the absolute value of $(-1)^{|i|}((n-|i|)/n) + (n+1)/n - |2i|/n$ is less than $(-1)^{|i|}((n-|i|)/n)(-1)^{|i|}((n-|i|)/n)$. Therefore, the odd correlation peak before $n/2$ after the reconstruction can be restrained, and the even correlation peak can be enhanced.

Because there are no high rate subcarrier signals mixed, making the peak width $R_{X/P}(\tau)$ of both auxiliary signal PRN and the received BOC signals wider than $R_X(\tau)$, which can restrain the side peaks in some extent, auxiliary signal PRN can be introduced to restrain the even peak and the correlation peak after $n/2$. The cross-correlation function is shown in formula (9).

$$R_{X/P}\,(\tau) = \sum_{i=1}^{n/2}\left(\left[-\text{tri}_{-(2i-1)/n}\,(\tau) + \text{tri}_{(2i-1)/n}\,(\tau)\right] \times \frac{1}{n}\right). \tag{9}$$

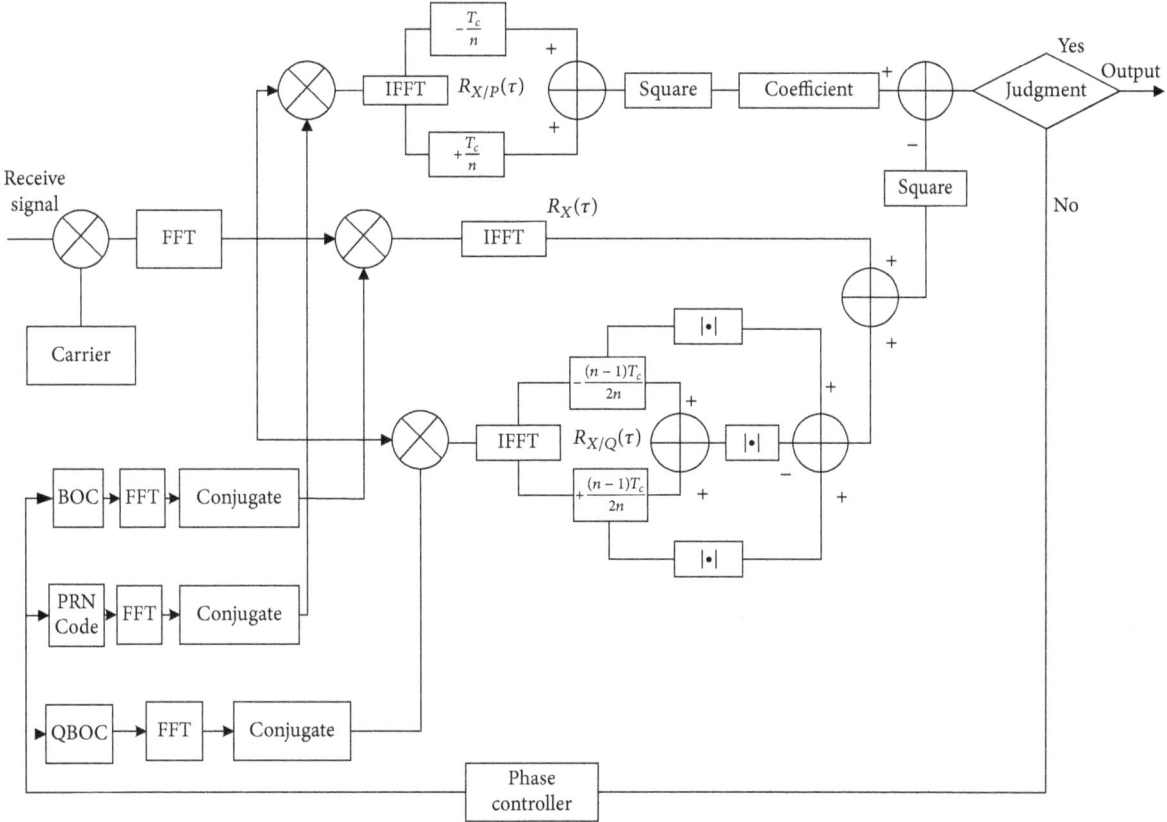

FIGURE 1: The principle of NCCFR.

In order to get the reconstructed signal which only consists of the even peak, by phase shift of the cross-correlation function (CCF), we can get formula (10).

$$\left| R_{X/P}\left(\tau + \frac{T_c}{n}\right) + R_{X/P}\left(\tau - \frac{T_c}{n}\right) \right|$$

$$= \sum_{i=1}^{n/2-1} \left(\left[\mathrm{tri}_{-2i/n}(\tau) + \mathrm{tri}_{2i/n}(\tau) \right] \times \frac{2}{n} \right) + \frac{1}{n} \tag{10}$$

$$\times \left(\mathrm{tri}_1(\tau) + \mathrm{tri}_{-1}(\tau) \right).$$

According to the characteristics that there is only even peak appearance in formulas (8) and (10), the constant coefficient, coef, can adjust the correlation peak value in formula (10); moreover the acquisition results of the correlation side peaks will be restrained. Therefore the final formula of NCCFR can be expressed as follows by the similar processing as formula (11) between formulas (8) and (10):

$$\left[R_X(\tau) + \left(\left| R_{X/Q}\left(\tau - \frac{(n-1)T_c}{2n}\right) \right| \right. \right.$$

$$+ \left| R_{X/Q}\left(\tau + \frac{(n-1)T_c}{2n}\right) \right|$$

$$\left. - \left| R_{X/Q}\left(\tau - \frac{(n-1)T_c}{2n}\right) \right| \right)$$

$$+ R_{X/Q}\left(\tau + \frac{(n-1)T_c}{2n}\right) \right| \Big)\Big]^2 - \left[\mathrm{coef} \times \left| R_{X/P}\left(\tau\right. \right. \right.$$

$$\left. \left. - \frac{T_c}{n}\right) + R_{X/P}\left(\tau - \frac{T_c}{n}\right) \right| \Big]^2. \tag{11}$$

Combined with the principle of NCCFR, the principle based on FFT (Fast Fourier Transform) is shown in Figure 1.

3. Algorithm Simulation and Test

3.1. Comparison Analysis of the Reconstruction Algorithm. In order to verify the performance of the new algorithm in restraining side peaks, according to the design model of Figure 1, reconstruction comparison analysis of different algorithm is fulfilled based on several typical BOC signals. The simulation parameters are as follows:

① BOC $(2, 2)$ signal, pseudorandom code rate is $2 \times 1.023e^6$, and subcarrier rate is $2 \times 1.023e^6$;

② BOC $(8, 4)$ signal, pseudorandom code rate is $4 \times 1.023e^6$, and subcarrier rate is $8 \times 1.023e^6$;

③ BOC $(6, 1)$ signal, pseudorandom code rate is $1 \times 1.023e^6$, and subcarrier rate is $6 \times 1.023e^6$.

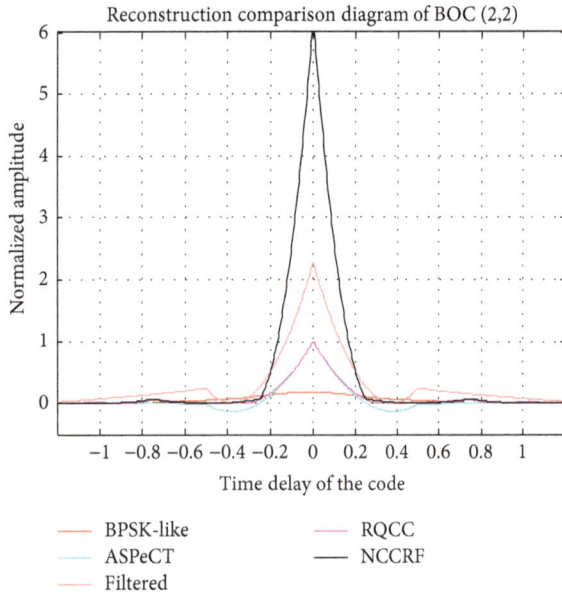

FIGURE 2: The reconstruction comparison of BOC $(2, 2)$.

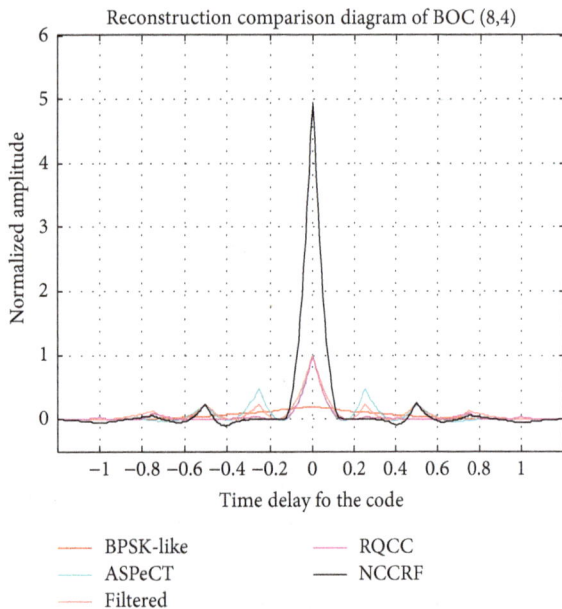

FIGURE 4: The reconstruction comparison of BOC $(6, 1)$.

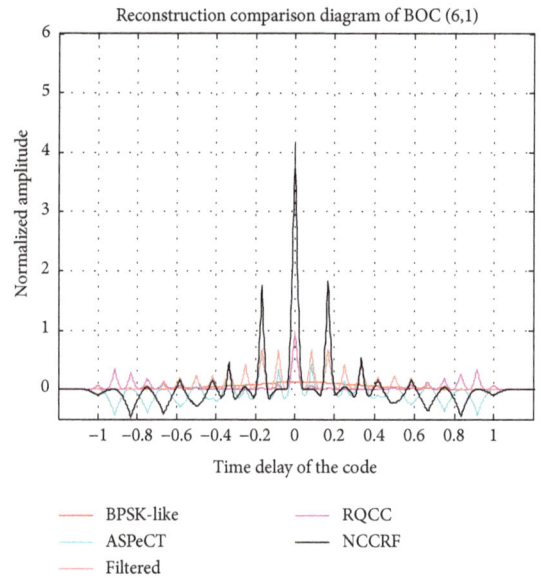

FIGURE 3: The reconstruction comparison of BOC $(8, 4)$.

The carrier rate and sampling rate are $30 \times 1.023e^6$ and $480 \times 1.023e^6$, respectively; the simulation results are shown in Figures 2–4.

They show the side peaks restrained performance promoted in NCCFR correlation reconstruction for BOC $(2, 2)$ signal from Figure 2, in which the side peaks have almost been eliminated. It can be seen from Figure 3 that the first side peak has been eliminated in NCCFR for BOC $(8, 4)$ signal, although new side peaks are produced, whose ratio is smaller than the promotion of the main peak. From Figure 4 it can be seen that the first side peak has been eliminated in NCCFR for BOC $(6, 1)$ signal, but it also produces new larger side peaks. In conclusion, NCCFR has very good performance in side peaks restrained for low order modulation coefficient of BOC signals and weaker peaks restrained performance for high order modulation coefficient.

3.2. Comparison Analysis of the Correlation Value. Moreover, in order to validate the adaptability of the algorithm in noisy environment, comparison analysis should be done. And MPMR is defined, which is the ratio between the main peak and the mean peak, MSPR is defined, which is the ratio between the main peak and the side peak, and MFSPR is defined, which is the ration between the main peak and the first side peak, where the first side peaks represent the second high degree of the peak [13].

First of all, the main peak enhanced performance is analyzed from MPMR [14–16]. The simulation results are shown in Figures 5–7.

It shows that the main peak enhanced performance promoted in NCCFR correlation reconstruction for both BOC $(2, 2)$ and the BOC $(8, 4)$ from Figures 5 and 6, combined with the formula analysis, and NCCFR algorithm enhance $(n+1)/n$ of the main peak, so main peak enhanced performance will be shown for low order BOC signal. But it is weaker than RQCC in high order modulation coefficient that is because the algorithm will produce new correlation side peaks by using ASPeCT acquisition algorithm which will restrain the correlation peak after $n/2$.

Secondly, the correlation peak enhanced performance also contains the restrain performance of the side peaks. Therefore, acquisition performance can be analyzed from MSPR, which is shown in Figures 8–10.

The results show that all MSPR for different BOC signals in NCCFR is the biggest, indicating that the algorithm has good performance in restraining the side peaks and enhancing the main peak for different order modulation coefficient BOC signals; combined with the formula analysis,

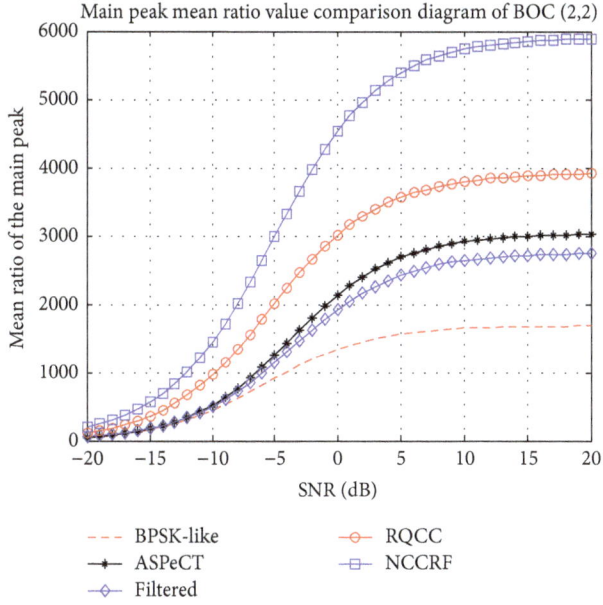

FIGURE 5: The MPMR comparison of BOC (2, 2).

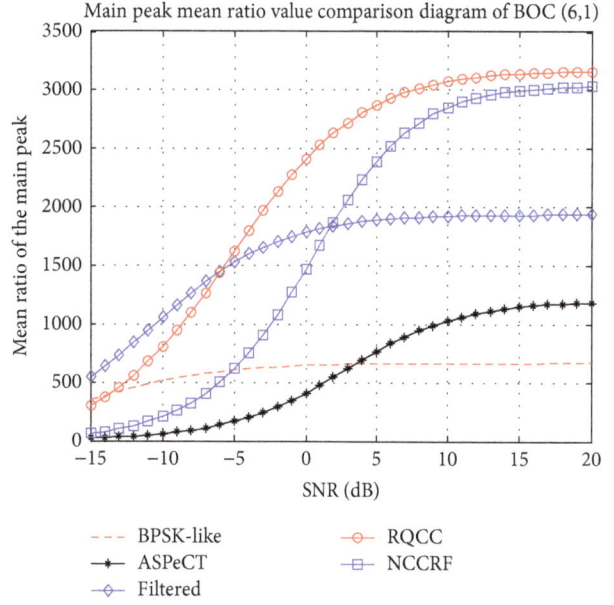

FIGURE 7: The MPMR comparison of BOC (6, 1).

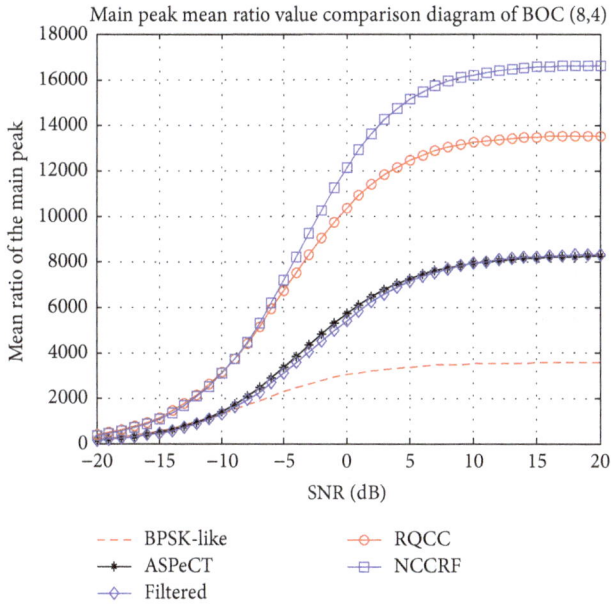

FIGURE 6: The MPMR comparison of BOC (8, 4).

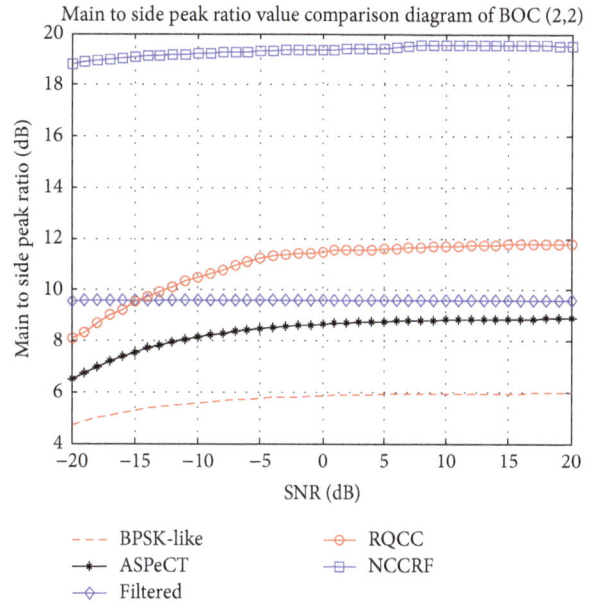

FIGURE 8: The MSPR comparison of BOC (2, 2).

the new algorithm has better side peaks restrained ability; that is, NCCFR has the best enhanced performance on MSPR.

Finally, the first side peaks restrain performance is analyzed for MFSPR in different conditions of SNR. Considering that there are no other side peaks in NCCFR reconstruction for low order BOC (2, 2), the MFSPR is analyzed for both BOC (8, 4) and BOC (6, 1), which is shown in Figures 11 and 12.

The results show that MFSPR for BOC (8, 4) in NCCFR is the biggest, indicating that the algorithm has good performance in restraining the first side peaks. At the same time, performance in restraining the first side peaks of NCCFR is weaker than RQCC for BOC (6, 1), but in general the new algorithm has a better first side peaks restrained performance.

3.3. Comparison Analysis of the Detection Probability. Signal acquisition probability is another important capability for signal acquisition, including the detection probability and false alarm probability [17, 18]. Noise will exist to the received signal during the transmission, which probability usually obeys the noncentral chi square distribution; however pure

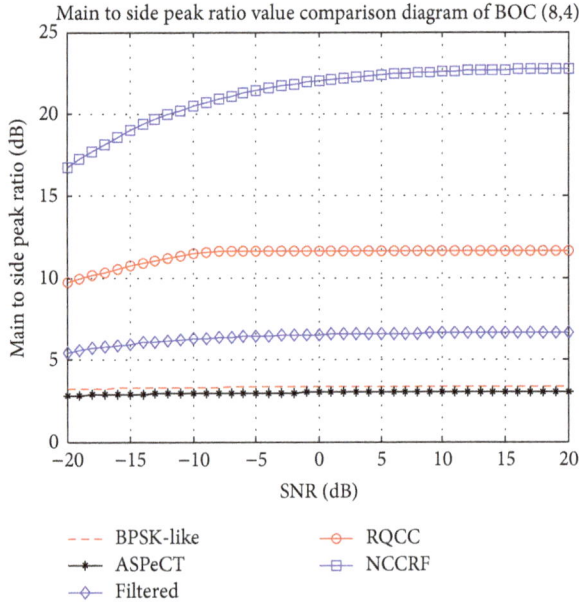

FIGURE 9: The MSPR comparison of BOC $(8, 4)$.

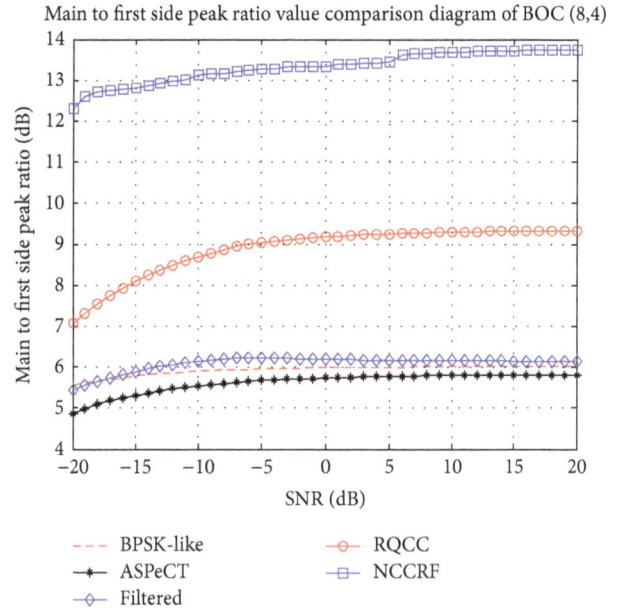

FIGURE 11: The MFSPR comparison of BOC $(8, 4)$.

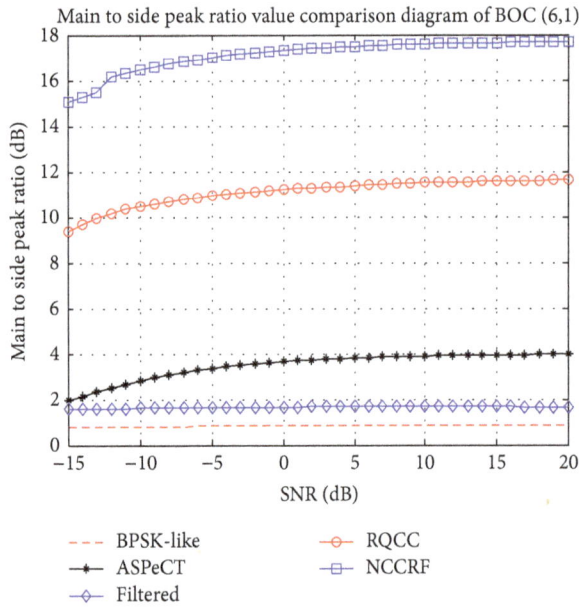

FIGURE 10: The MSPR comparison of BOC $(6, 1)$.

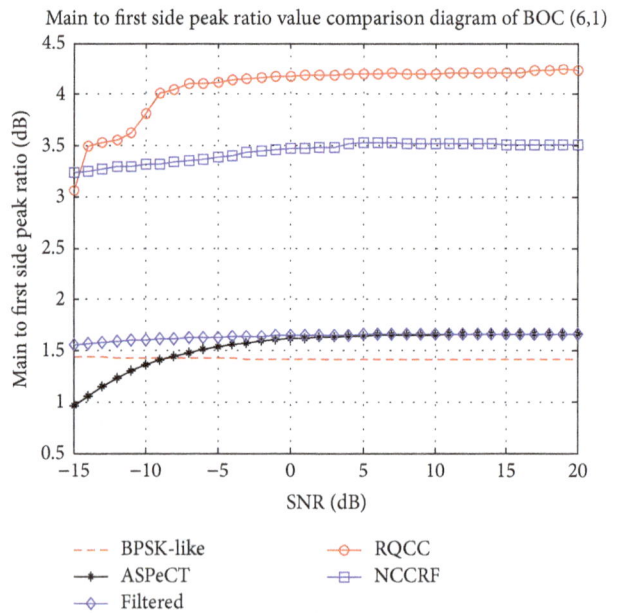

FIGURE 12: The MFSPR comparison of BOC $(6, 1)$.

noise or nonsynchronized signal usually obeys the noncentral chi square distribution [19]. Furthermore, in order to validate the acquisition performance of NCCFR, setting the parameter of false alarm probability, pfa = 0.01, which is a constant value, comparison analysis has been done from the correlation peak detection probability according to several typical signals in different SNR conditions, which mainly focuses on the main peak detection probability (MPDR), the side peaks detection probability (SPDR), and the first side peaks detection probability (FSPDR).

Firstly, the MPDR has been analyzed in different SNR conditions, which are shown in Figures 13–15.

The results show that MPDR performance for both $(2, 2)$ and BOC $(8, 4)$ in NCCFR is better than others and MPDR performance of NCCFR is weaker than RQCC for BOC $(6, 1)$, but in general the new algorithm has a better MPDR performance.

Secondly, according to the ambiguity problem of the side peaks, the SPDR has been analyzed in different SNR conditions, which are shown in Figures 16–18.

The results show that SPDR for NCCFR in different SNR conditions is the smallest one, indicating that the algorithm

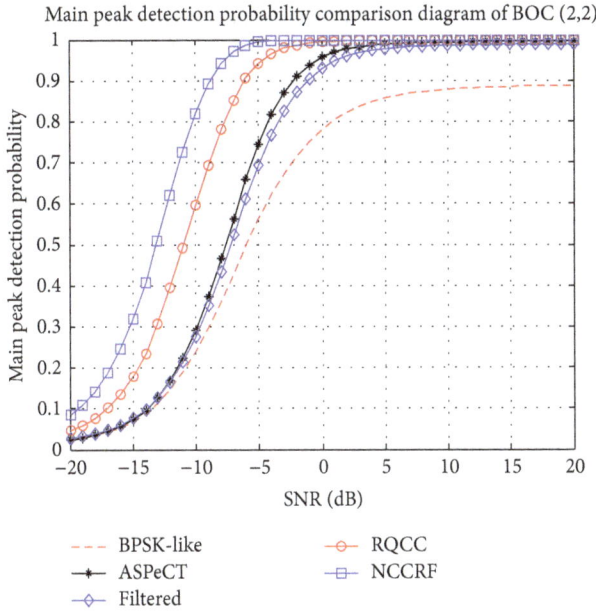

FIGURE 13: The MPDR comparison of BOC (2, 2).

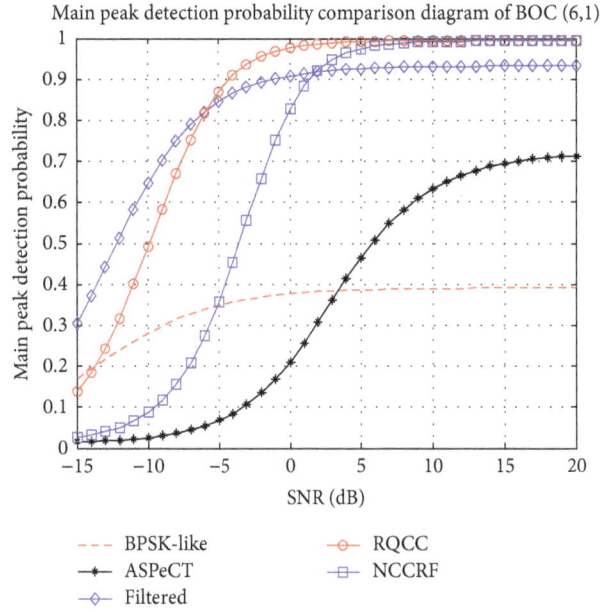

FIGURE 15: The MPDR comparison of BOC (6, 1).

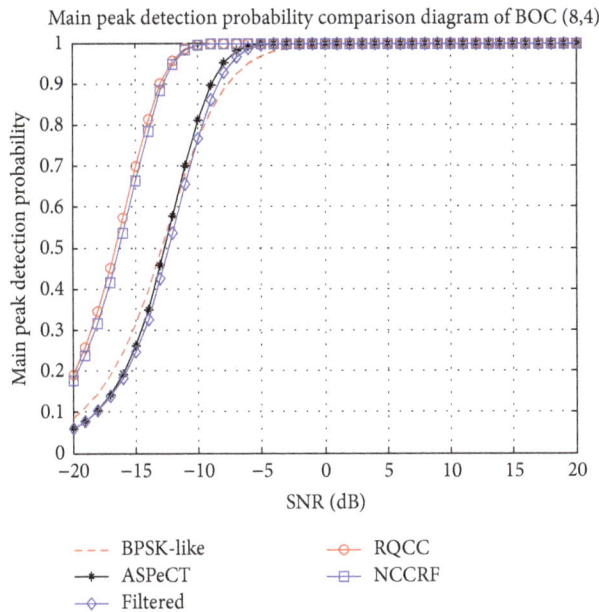

FIGURE 14: The MPDR comparison of BOC (8, 4).

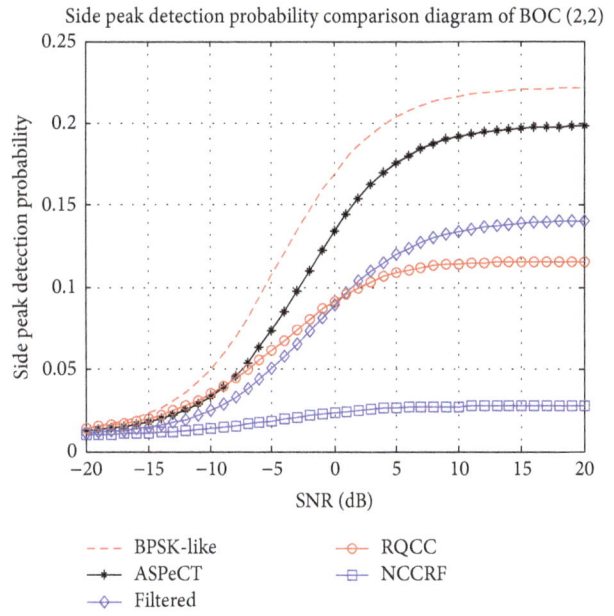

FIGURE 16: The SPDR comparison of BOC (2, 2).

has good performance in restraining the side peaks for BOC signals.

Finally, according to the problem of producing new side peaks in reconstruction algorithm, the acquisition performance has been analyzed in different SNR conditions focusing on FSPDR, which are shown in Figures 19-20.

The results show that algorithm has good performance in restraining the first side peaks for BOC (8, 4) signal. From Figure 20 it can be seen that the performance in restraining the first side peaks of NCCFR is better than others when the SNR is low, but weaker than RQCC when the SNR is

high. In general, the new algorithm has better first side peaks restrained performance in different modulation coefficient for all BOC signals.

4. Conclusions

On the basis of analyzing the limitations for the existing correlation reconstruction methods, a method of NCCFR is been proposed. In order to verify the acquisition performance for the new algorithm, the simulation comparison has been analyzed by using different acquisition methods in different

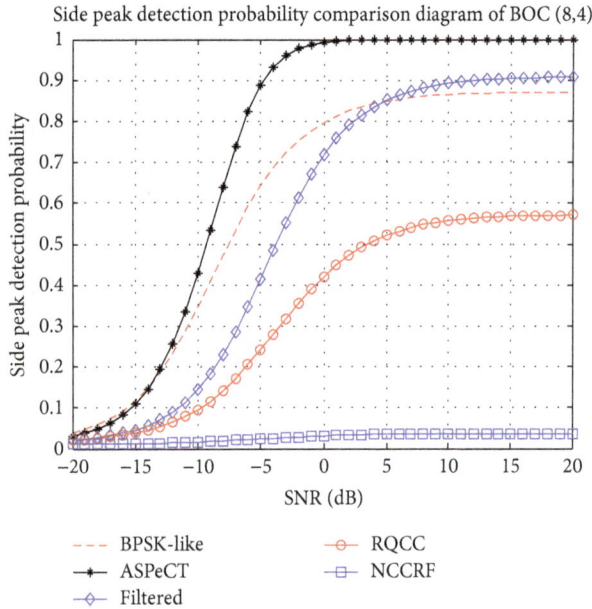

FIGURE 17: The SPDR comparison of BOC $(8, 4)$.

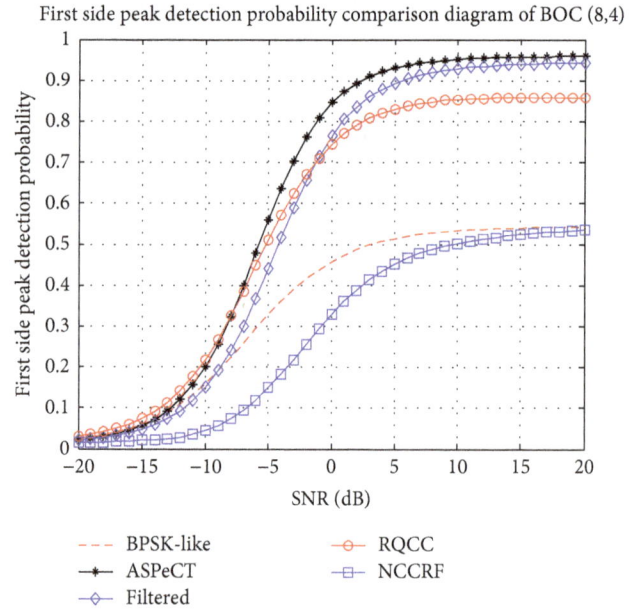

FIGURE 19: The FSPDR comparison of BOC $(8, 4)$.

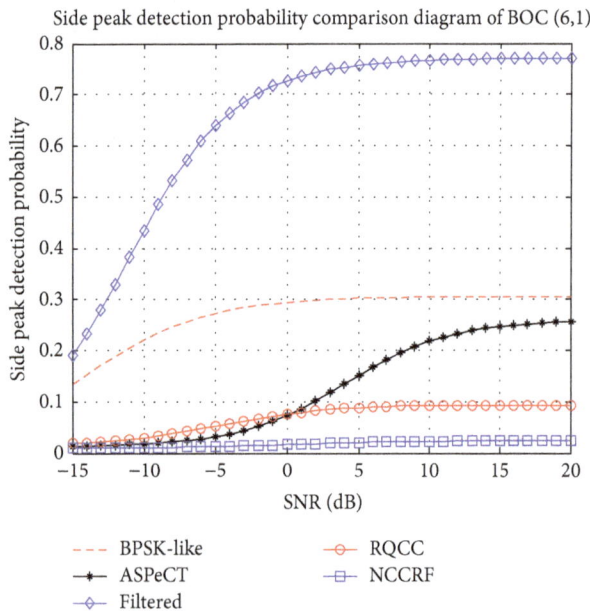

FIGURE 18: The SPDR comparison of BOC $(6, 1)$.

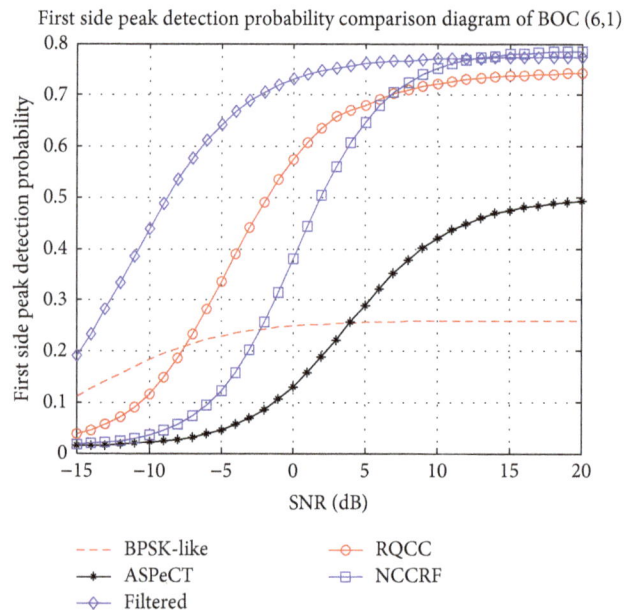

FIGURE 20: The FblackSPDR comparison of BOC $(6, 1)$.

SNR conditions based on BOC $(2, 2)$, BOC $(8, 4)$, and BOC $(6, 1)$. The simulation results show that the proposed algorithm is better than all of the other algorithms in enhancing the main peak and restraining the side peaks, when the modulation coefficients are of low order. And the results show that the new algorithm has better performance than other algorithms when the modulation coefficients are of high order.

Conflicts of Interest

The authors declare that they have no conflicts of interest.

Acknowledgments

This work was supported by the National Natural Science Foundation of China (no. 61501309), the Program for Liaoning Innovative Research Team in University (no. LT2011005), New Century Program for Excellent Talents of Ministry of Education of China (no. NCET-11-1013), and the China Postdoctoral Science Foundation (no. 2015M580231; no. 2017T100185).

References

[1] P. Li, F. Gao, and Q. Li, "An improved unambiguous acquisition scheme for BOC (n,n) signals," in *Proceedings of the International Conference on Wireless Communications Signal Processing WCSP '15*, pp. 1–6, 2015.

[2] S. Yoon, S. C. Kim, J. Heo, I. Song, and S. Y. Kim, "Twin-cell detection (TCD): a code acquisition scheme in the presence of fractional doppler frequency offset," *IEEE Transactions on Vehicular Technology*, vol. 58, no. 4, pp. 1797–1803, 2009.

[3] D. Chong, Y. Lee, I. Song, and S. Yoon, "A two-stage acquisition cheme based on multiple correlator outputs for UWB signals," *IEICE Electronics Express*, vol. 8, no. 7, pp. 436–442, 2011.

[4] Y. Lee, D. Chong, I. Song, S. Y. Kim, G.-I. Jee, and S. Yoon, "Cancellation of correlation side-peaks for unambiguous BOC signal tracking," *IEEE Communications Letters*, vol. 16, no. 5, pp. 569–572, 2012.

[5] H. Chen, W. Jia, and M. Yao, "Cross-correlation function based multipath mitigation technique for cosine-BOC signals," *Journal of Systems Engineering and Electronics*, vol. 24, no. 5, Article ID 00087, pp. 742–748, 2013.

[6] W.-L. Mao, C.-S. Hwang, C.-W. Hung, J. Sheen, and P.-H. Chen, "Unambiguous BPSK-like CSC method for Galileo acquisition," in *Proceedings of the 18th International Conference on Methods and Models in Automation and Robotics (MMAR '13)*, pp. 627–632, IEEE, Międzyzdroje, Poland, August 2013.

[7] A. Burian, E. S. Lohan, and M. Renfors, "BPSK-like methods for hybrid-search acquisition of galileo signals," in *Proceedings of the IEEE International Conference on Communications (ICC '06)*, pp. 5211–5216, IEEE, Istanbul, Turkey, July 2006.

[8] O. Julien, C. Macabiau, M. E. Cannon, and G. Lachapelle, "ASPeCT: unambiguous sine-BOC(n,n) acquisition/tracking technique for navigation applications," *IEEE Transactions on Aerospace and Electronic Systems*, vol. 43, no. 1, pp. 150–162, 2007.

[9] E. F. Brickell, D. M. Gordon, K. S. Mccurley et al., "Fast exponentiation with precomputation," in *Proceedings of the Eurocrypt on Advances in Cryptology*, vol. 658, pp. 200–207, Springer-Verlag, New York, NY, USA, 1993.

[10] L. Yanzan, *The Fast Acquisition Technology of Multi-Mode Navigation Signals Based on The BOC Signals*, Dalian University, Dalian, China, 2013.

[11] W. Cui, D. Zhao, J. Liu, S. Wu, and J. Ding, "A novel unambiguous acquisition algorithm for BOC(m,n) signals," in *Proceedings of the 5th IEEE International Conference on Signal Processing, Communications and Computing, ICSPCC '15*, IEEE, Ningbo, China, September 2015.

[12] Y. Zhang, W. Luy, and D. Yu, "A fast acquisition algorithm based on FFT for BOC modulated signals," in *Proceedings of the 35th IEEE Region 10 Conference, TENCON '15*, IEEE, Macao, China, November 2015.

[13] Z. Yang, Z. Huang, and S. Geng, "Unambiguous acquisition performance analysis of BOC(m,n) signal," in *Proceedings of the International Conference on Information Engineering and Computer Science, ICIECS '09*, IEEE, Wuhan, China, December 2009.

[14] F. Liu, Y.-X. Feng, and M.-H. Tian, "The main peak estimate algorithm based on BOC(2n,n) signal," in *Proceedings of the 3rd IEEE International Conference on Advanced Computer Control, ICACC '11*, pp. 165–168, IEEE, Harbin, China, January 2011.

[15] O. M. Mubarak, "Performance comparison of multipath detection using early late phase in BPSK and BOC modulated signals," in *Proceedings of the 7th International Conference on Signal Processing and Communication Systems, ICSPCS '13*, IEEE, Carrara, VIC, Australia, December 2013.

[16] L. Yang, C.-S. Pan, Y.-X. Feng, and Y.-M. Bo, "A new algorithm for synchronous main lobe detection for BOC modulated navigation signals," *Journal of Astronautics*, vol. 31, no. 8, pp. 2008–2014, 2010.

[17] A. Wang, J. Wang, and B. Xue, "Acquisition of BOC(n, n) with large Doppler," in *Proceedings of the 7th International Conference on Wireless Communications, Networking and Mobile Computing, WiCOM '11*, IEEE, Wuhan, China, September 2011.

[18] F. Shen, G. Xu, and D. Xu, "Unambiguous acquisition technique for cosine-phased binary offset carrier signal," *IEEE Communications Letters*, vol. 18, no. 10, pp. 1751–1754, 2014.

[19] Y. Chuanxi, *Performance Analysis of Ranging Codes And Study on Acquisition Algorithm of BOC Signal*, Chinese Academy of Science (National Time Service Center), Beijing, China, 2013.

23

Improved Design of Bit Synchronization Clock Extraction in Digital Communication System

Huimin Duan (ID)**, Hui Huang, and Cuihua Li**

Department of Electronic and Electrical Engineering, Hefei University, Hefei, China

Correspondence should be addressed to Huimin Duan; duanhuimin@foxmail.com

Academic Editor: Jose R. C. Piqueira

An improved method is proposed in this design to reduce the phase jitter after the synchronization or the random noise induced phase jitter in a bit synchronization clock extraction circuit. By using a newly added digital filter between the phase detector and the controller, the phase difference pulses from the phase detector are counted and processed, before being transmitted to the controller for adjusting the phase of the output clock. The design is completed by using FPGA chip and VHDL hardware description language and performs the simulation verification on Quartus II. The results show that the improved system performs the accurate extraction of bit synchronized clock, reduces the phase jitter problem, improves the system running efficiency and the ability of anti-interference, and guarantees the synchronization performance of the digital communication system.

1. Introduction

In digital communication systems, information is transmitted in a series of code sequences; the receiver must know the starting and ending time of each code [1, 2], so it needs to have a bit timing pulse sequence for the sampling decision, which has the same repetition frequency as the code rate of transmit end and the same phase as the optimal decision time. The process of extracting the timing pulse sequence is called bit synchronization [3], which is implemented by an external or a self-synchronization method. The self-synchronization method extracts the synchronization information from the code in the receiving end without inserting the pilot signal at the transmitter end, making it the most commonly used method in modern digital communication [4, 5].

In the self-synchronizing digital communication system, the methods to extract the bit synchronized signal mainly includes filtering method, enveloping "collapse" method, and DPLL (Digital Phase-Locked Loop) method. The DPLL method is mature and widely used [6–8]. At present, the system that uses phase-locked loop to extract bit synchronous clock has complex extraction process, slow running speed, or lack of ability to resist random noise. This work is to present an improved DPLL based design for bit synchronized

clock extraction in digital communication system. The main idea is to implement a digital filter between the PD (Phase Detector) and the control module, which is used to process the leading or lagging control pulses corresponding to the phase difference of the PD output before sending them to the control module for a phase adjustment to its clock. This implementation can be used to avoid the alternation of leading and lagging pulses after the synchronous locking and to improve the phase jitter problem caused by the random noise effect [9–13]. In this paper we omit rigorous mathematical proof of local and global stability of proposed circuit. Local and global analysis can be done with reference to classic PLL-based circuit [14, 15].

2. The Principle of Bit Synchronization Clock Extraction Based on DPLL

The idea of DPLL method to extract bit synchronized clock is shown in Figure 1, which includes four main steps of crystal oscillator signal shaping, frequency division, phase detection, and control. The output of crystal oscillator is transmitted into two clock signals with a duty ratio of about 1/4 and a phase difference of π, which are used together with the

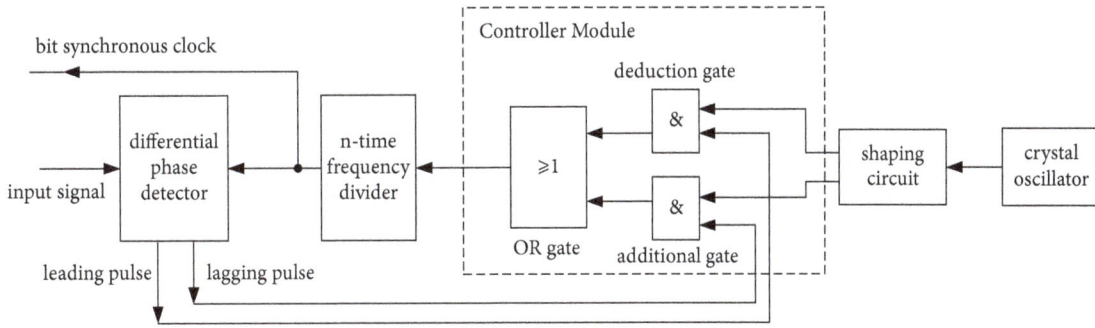

FIGURE 1: Structure diagram of DPLL.

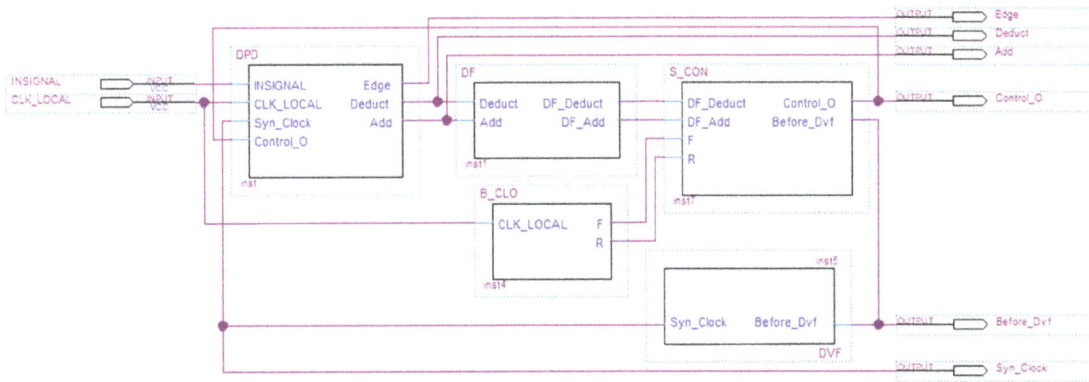

FIGURE 2: Schematic of bit synchronization clock extraction system based on DPLL.

leading and lagging control pulses as input signals to the control module. The controller in the figure consists of a deduction gate, an additional gate, and an OR gate circuit. The output of the control module is sent to the n-time frequency divider for frequency division. The final generated bit synchronized clock signal is used both for the sampling decision of the receiver of the digital communication system and as the feedback signal input to the differential PD [6, 16].

If the Baud rate of the receiving code is F Baud, the frequency of the bit synchronized clock sequence must also be F Hz. As shown in Figure 1, if the oscillation frequency of crystal oscillator is set as nF Hz, the frequency of the narrow pulses after shaping is $nF/2$ Hz, and the frequency of the adjusted signal after the controller is nF Hz; after n-time frequency division, the bit synchronous clock signal with a frequency of F Hz is obtained. If the clock signal is not exactly in the same frequency and phase as the receiving signal, we have to use the controller to adjust the input of the frequency divider according to the phase difference signal generated from the PD. In this way, the phase of the bit synchronized clock can be changed and adjusted constantly in the phase-locked loop until the accurate synchronization signal is obtained [3].

3. The Implementation of Bit Synchronous Clock Extraction System Based on FPGA

According to the principle mentioned above, the typical extraction circuit requires four parts for bit synchronized

signal extraction. In order to reduce the jitter problem after synchronization, a digital filter module is implemented is this design, and VHDL (Very-high-speed Integrated Circuit Hardware Description Language) with FPGA (Field-Programmable Gate Array) is used for the system design and performs the simulation verification on Quartus II [17–20]. The top layer schematic of the system, as shown in Figure 2, consists of five modules: differential phase detector (DPD), digital filter (DF), biphasic clock source (B_CLO), controller (S_CON), and frequency divider (DVF).

3.1. Differential Phase Detector Module. The differential phase detector circuit consists of a differential circuit and a PD circuit, as shown in Figure 3. The input digital sequence (INSIGNAL) is transformed into the corresponding rising edge detection narrow pulse signal (Edge) after the differential circuit., which is then added to the deduction gate (AND2) and the additional gate (AND3), respectively.

If the phase of the bit synchronized clock (Syn_Clock) is ahead of the edge detection signal, or the input sequence, the additional gate closes, and the deduction gate outputs at the same time a control signal (Deduct) to the follow-up circuit to deduct a clock pulse before frequency division and therefore to weaken the leading state. The simulation result is shown in Figure 4.

On the contrary, if the Syn_Clock is lagging behind, then the deduction gate closes, and the additional gate

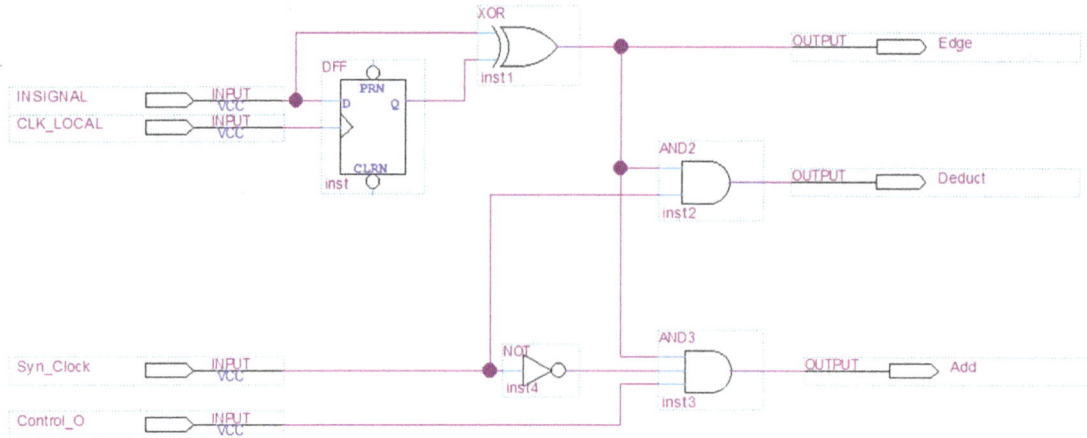

FIGURE 3: Schematic of differential phase detector.

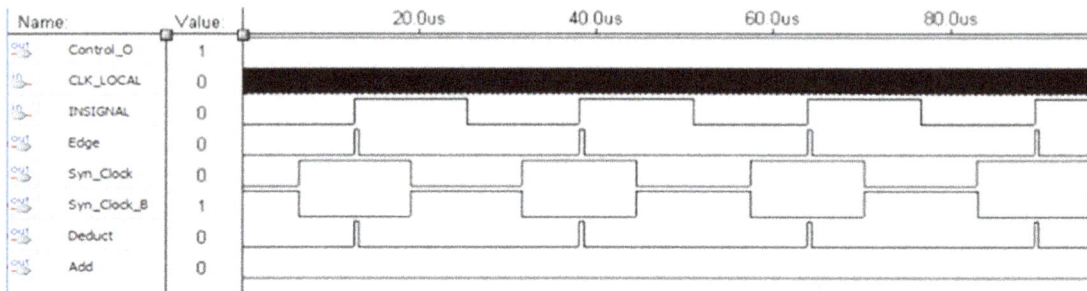

FIGURE 4: Simulation result of differential phase detector in phase lead.

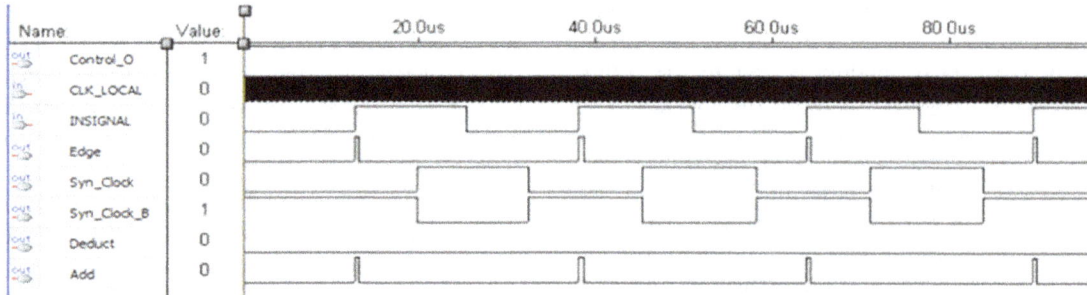

FIGURE 5: Simulation result of differential phase detector in phase lag.

generates a control signal (Add) to the follow-up circuit to add a clock pulse before frequency division to weaken the lagging state. The simulation result is shown in Figure 5.

3.2. Biphasic Clock Source Module. The circuit schematic of the biphasic clock source module is shown in Figure 6. On the one hand, this module is used to shape the crystal oscillator outputs to be with an output duty ratio of about 1/4. On the other hand, this module is used to generate two clock signals (F and R) which have a half frequency of the crystal oscillator output signal and with a phase difference of π. The simulation result is shown in Figure 7.

3.3. Control Module. The circuit schematic of the control module implemented by FPGA is shown in Figure 8, which mainly includes D flip-flop, constant-opened deduction gate (inst5), and constant-closed add gate (inst4).

When the phase of the synchronized clock is ahead, the leading control signal (DF_ Deduct) is a positive pulse, which is converted into low level after passing the trigger and NOT gate, making the constant-opened gate closed for one trigger period and deducting a pulse from the F signals, as shown in Figure 9. To the contrary, when the phase of the synchronized clock is lagged, the lagging control signal (DF_ Add) is a positive pulse, and a high level is generated by

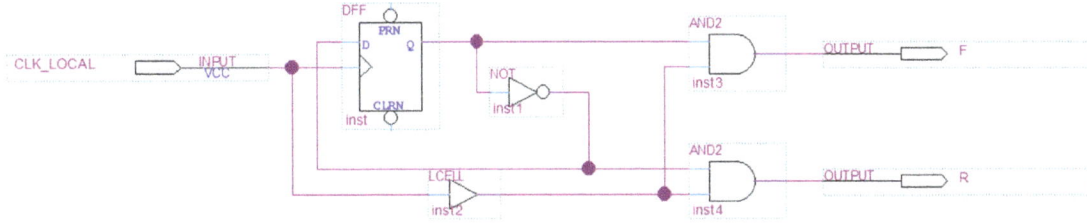

FIGURE 6: Schematic of biphasic clock source circuit.

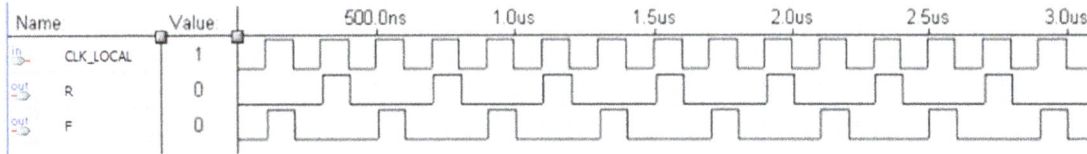

FIGURE 7: Simulation result of biphasic clock source circuit.

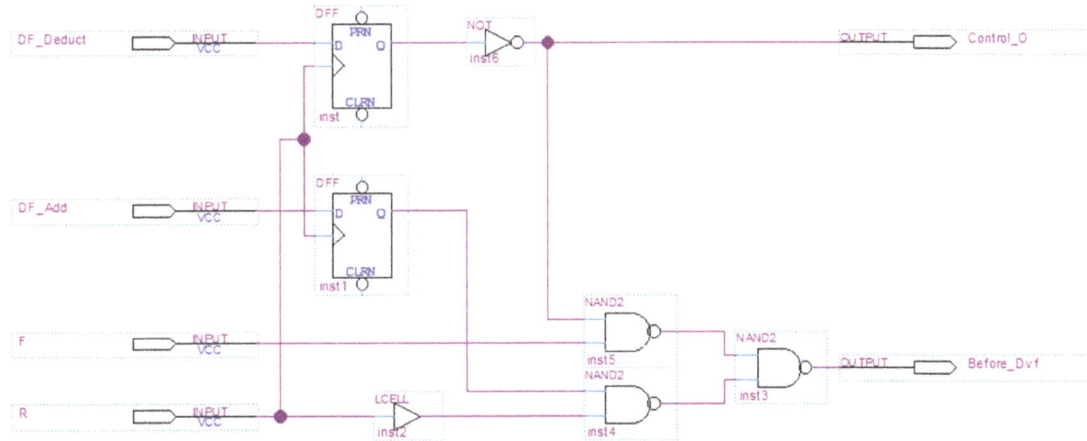

FIGURE 8: Schematic of the control module.

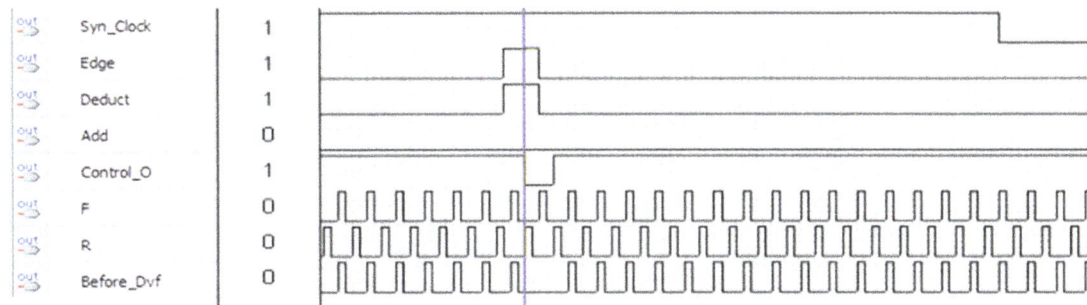

FIGURE 9: Simulation result of control module in phase lead.

the trigger, making the constant-closed gate opened for one trigger period and adding a pulse to the R signals, as shown in Figure 10.The adjusted biphasic clock (Before_ Dvf) is then sent to the divider after the OR gate, to achieve the goal of phase adjustment.

In this part of the design, an antiphase control signal (Control_O) output is added in the circuit to avoid the fake synchronization. When the descent edge of the bit synchronization signal is aligned with the edge detection signal, the input signal and the local synchronized clock signal are of the same frequency but reversed phase, making a synchronous illusion that the leading and lagging control pulses appear alternately in turns as shown in a real synchronization. In this case, the control module first deducts a pulse and then adds another, causing the phase of the Before_Dvf clock signal to be unchanged, therefore making it impossible to adjust the phase of the bit synchronized clock. This situation can be avoided by the Control_O signal output from the controller, with a

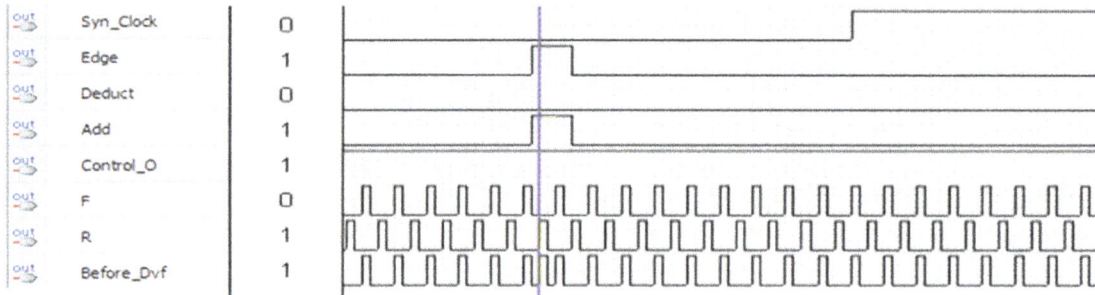

FIGURE 10: Simulation results of control module in phase lag.

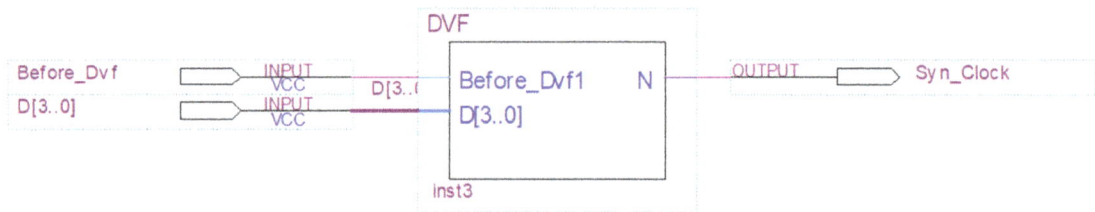

FIGURE 11: Schematic of numerically controlled frequency divider circuit.

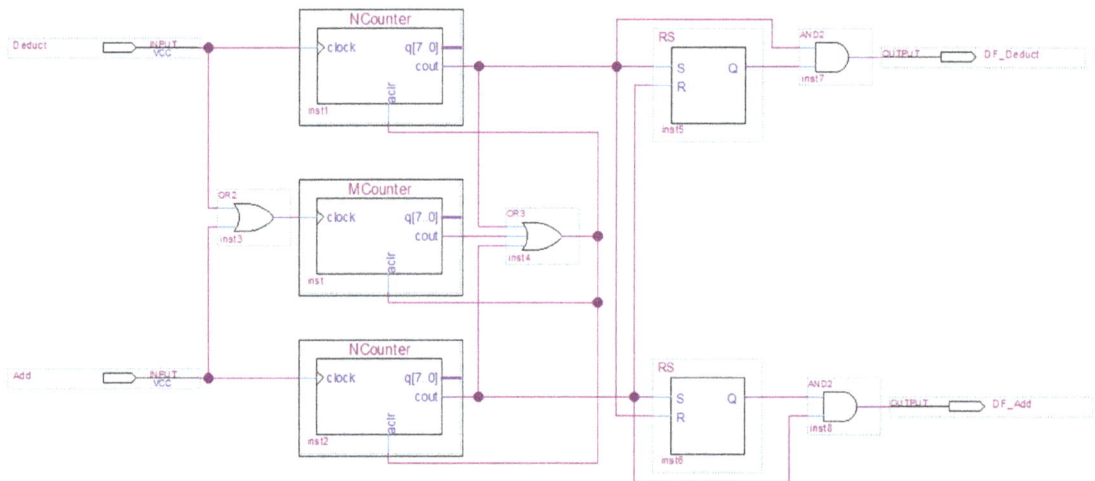

FIGURE 12: Schematic of digital filter circuit.

working process of the following: once the leading control signal (DF_Deduct) is sent into the controller, Control_O is turned to low level and the signal is fed back to stop adding gate (AND3) in the differential PD, so that control pulses with added pulses (ADD) are not generated, making the entire phase-locked loop continue to work properly.

3.4. Frequency Division Module.

The numerically controlled frequency divider module is shown as in Figure 11, Before_Dvf is the input clock, d[3..0] is the port for setting the frequency divider's coefficient, and n is the clock signal after frequency division, or the required bit synchronized clock.

3.5. Digital Filter Module.

Without this implementation of a digital filter module, the output leading control signal (Deduct) and lagging control signal (Add) from the differential PD will be sent directly to the corresponding port of the control module, to perform a deducting or an adding to the biphasic pulse. When a digital filter module is added, the leading and lagging control signals are first processed by the digital filter before being transmitted to the controller [21]. The details are described below.

As shown in Figure 12, the digital filter module sends the receiving leading and lagging controls signals to their respective N counters and sends their sum to the M counter, so that N and M satisfy the $N \leq M \leq 2N$. When any of the three counters is full, the carry pulse is output to the asynchronous zero end of the three counters. The three counters are cleared to zero at the same time and the counting restarts.

If the phase of the bit synchronized clock is ahead of the input signal, the continuous output of differential PD will

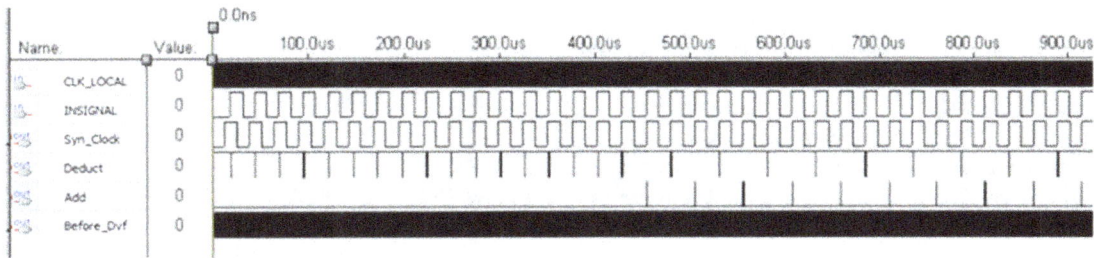

FIGURE 13: System simulated results without digital filter. CLK_LOCAL: the local clock, INSIGNAL: the input signal, Syn_Clock: the synchronization clock, Deduct: the leading control pulse of PD, Add: the lagging control pulse of PD, and Before_Dvf: control module output/frequency divider input.

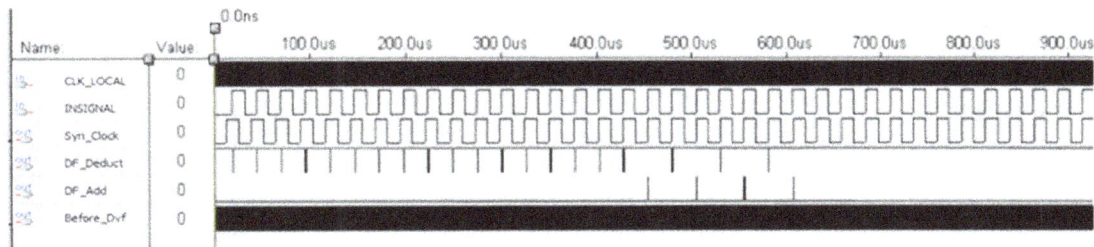

FIGURE 14: System simulated results with adding digital filter. CLK_LOCAL: the local clock, INSIGNAL: the input signal, Syn_Clock: the synchronization clock, DF_Deduct: the leading control pulse of digital filter module, DF_Add: the lagging control pulse of digital filter module, Before_Dvf: control module output/frequency divider input.

fill the N counter with the number of leading control pulse first. Then the trigger inst5 outputs high level and opens the AND gate inst7, through which the leading control pulse is output to the control module. If the filter continues to receive the leading control pulse, these leading pulses (DF_Deduct) will be continuously output as the inst7 gate is in the open state. Similarly, if the phase of the bit synchronized clock dose lags behind the input signal, the continuous output of differential PD will fill the counter with the number of lagging control pulse first. The trigger inst6 outputs high level and opens the AND gate inst8, and the lagging control pulse passes through inst8 gate to the control module. If the filter continues to receive the lagging control pulse, these lagging pulses (DF_Add) will be continuously output as the inst8 gate is in the open state.

When it is in the synchronous (locked) working state, as the phase difference between the input signal and the synchronized clock is small, the DPLL is toggling in both the leading and lagging states, as shown in Figure 13 for the Deduct and Add pulses. In addition, when a phase difference is observed between the input signal and the synchronized clock due to the noise, a leading state or a lagging state caused by the random errors will be observed with the same probability. It will also lead to the DPLL toggles in both the leading and lagging states, leading to the phase jitter of the bit synchronized clock. In both cases, either of the two N counters is not full, but three counters will be cleared as the M counter is already full, so the digital filter module dose not output any leading and lagging control pulses, and the subsequent controller does not make any adjustment to the phase of the local clock. As a result, the adjustment to the

phase of the synchronized clock in the dynamic stability state is relieved and the jitter problem is avoided. The simulation results are shown in Figure 14.

4. System Simulation Results and Analysis

The system simulation results without and with a digital filter are shown in Figures 13 and 14, respectively. By comparing the results shown in these two figures, we can see that the use of a digital filter eliminates the problem of the alternate occurrence of the leading and lagging control pulses after synchronous locking.

In Figure 13, a dynamic balance status is reached in the latter half of the simulation diagram. As the edge detection pulse has a fixed width, the jump edge of Syn_Clock is located in the middle of the edge detection pulse when compared to Syn_Clock. Therefore, the leading and lagging control pulses circulate alternately. The system deducts a pulse before the leading pulse is received and adds a pulse after lagging pulse is received, causing the entire system to be in a dynamic balance state, or a synchronous locking state, or the bit synchronized state that our design is pursuing.

However, when it is locked in the synchronous working state, the DPLL is just toggling in both leading and lagging states as the phase difference between the input signal and the bit synchronized clock is small. In addition, when a phase difference is observed between the input signal and the synchronized clock due to the noise, a leading state or a lagging state caused by the random errors will be observed with the same probability. It will also lead to the DPLL toggles in both the leading and lagging states, leading to the phase jitter of the bit synchronized clock. The above two conditions make the system inefficient and increase system consumption [22, 23].

In Figure 14, the required synchronization state is also reached in the latter half of the simulation diagram, but the toggling no longer occurs and the phase jittering of bit synchronized clock caused by the toggling is therefore avoided.

5. Conclusion

In this paper, the digital phase-locked technique is implemented in our design of extraction of the phase synchronization clock. The phase difference information of the phase detector is firstly processed with an improved method and then sent to the controller to adjust the phase of the output clock. As a result, accurate synchronization pulse extraction is obtained, while at the same time the jitter phenomenon is reduced, the system anti-interference ability is increased, and the operating efficiency is improved. Furthermore, antiphase control signals are also added in our design to prevent the fake sync state and ensure that the DPLL is able to work properly. FPGA chip, VHDL hardware description language, and Quartus II are used to perform the system design and simulation. The results show that this bit synchronous extraction circuit achieves the expected goal of improvement, which is expected to be better applied in the actual digital communication system.

Conflicts of Interest

The authors declare that they have no conflicts of interest.

Acknowledgments

This project was supported by Anhui Provincial Natural Science Research Key Project (KJ2017A532), Anhui Provincial Quality Engineering Project (2015mooc075), and Key Projects of Outstanding Young and Middle-Aged Backbone Talents in Anhui Universities and Colleges (gxfxzd2016223).

References

[1] H. Liu, R. F. Zhang, J. N. Liu, and M. Zhang, "Time synchronization in communication networks based on the Beidou foundation enhancement system," *Science China-Technological Sciences*, vol. 59, no. 1, pp. 9–15, 2016.

[2] J. M. Myers and F. H. Madjid, "Unpredictability and the transmission of numbers," *Quantum Information Processing*, vol. 15, no. 3, pp. 1057–1067, 2016.

[3] F. Zhang, B. Xu, and C. Wu o, *Principle of Communications*, Tsinghua University Press, Beijing, China, 2nd edition, 2016.

[4] Z. Zhang, "Design of bit synchronization circuit based on FPGA," *Modern Electronics Technique*, vol. 39, no. 4, pp. 132–134, 2016.

[5] Y. Zhou, R. Takahashi, N. Fujii, and T. Hikihara, "Power packet dispatching with second-order clock synchronization," *International Journal of Circuit Theory and Applications*, vol. 44, no. 3, pp. 729–743, 2016.

[6] B. Lin, L. Shi, and P. Lu, "Design of bit synchronization clock extraction circuit for wireless communication receiver," *Electronic Technology*, no. 4, pp. 36–39, 2017.

[7] Y. Tian, Y. Wang, H. Dai, H. Fang, and W. Liu, "Design of local clock source of satellite borne spectrometer based on digital phase locked loop," *Journal of Electronics and Information Technology*, vol. 39, no. 10, pp. 2397–2403, 2017.

[8] C. K. Ahn, P. Shi, and S. H. You, "A new approach on design of a digital phase-locked loop," *IEEE Signal Processing Letters*, vol. 23, no. 5, pp. 600–604, 2016.

[9] J. Zhu, "Analysis of period jitter induced by power supply noise of phase locked loop," *Electronic Science and Technology*, vol. 29, no. 7, pp. 102–105, 2016.

[10] Y. Zhao, X. Li, G. Zhao, and C. Sun, "Design of configurable digital filter and its ASIC implementation," *Journal of Central South University (Science and Technology)*, vol. 48, no. 4, pp. 990–995, 2017.

[11] A. Tonk and N. Afzal, "On advance towards sub-sampling technique in phase locked loops—a review," *Integration, the VLSI Journal*, vol. 59, pp. 90–97, 2017.

[12] S. Yang, Y. Jia, X. Zhang, and B. Yang, "Phase discriminator with dynamic frequency division for eliminating nonlinearity at high frequency based on FPGA," *Measurement Control Technology and Instruments*, vol. 43, no. 12, pp. 55–58, 2017.

[13] Y. Li, S. Bielby, A. Chowdhury, and G. W. Roberts, "A jitter injection signal generation and extraction system for embedded test of high-speed data I/O," *Journal of Electronic Testing-Theory and Applications*, vol. 32, no. 4, pp. 423–436, 2016.

[14] R. E. Best, N. V. Kuznetsov, G. A. Leonov, M. V. Yuldashev, and R. V. Yuldashev, "Tutorial on dynamic analysis of the Costas loop," *Annual Reviews in Control*, vol. 42, pp. 27–49, 2016.

[15] M. A. Kiseleva, E. V. Kudryashova, N. V. Kuznetsov et al., "Hidden and self-excited attractors in Chua circuit: synchronization and SPICE simulation," *International Journal of Parallel, Emergent and Distributed Systems*, vol. 32, pp. 1–12, 2017.

[16] G. Sun, Y. Liu, B. Li, Y. Zhu, and Y. Fan, "An all-digital phase-locked loop with compensating feedback unit delay," *Transactions of China Electrotechnical Society*, vol. 32, no. 20, pp. 171–177, 2017.

[17] Y. Liu and Y. Fan, "Design and analysis of all-digital full-hardware phase-locked loop," *Transactions of China Electrotechnical Society*, vol. 30, no. 2, pp. 172–179, 2015.

[18] S. Palnitkar, *Verilog HDL: A Guide to Digital and Syntheis*, translated by Y. Xia, Y. Hu, and L. Diao, Publishing House of Electronics Industry, Beijing, China, 2nd edition, 2013.

[19] M. Alcin, I. Pehlivan, and I. Koyuncu, "Hardware design and implementation of a novel ANN-based chaotic generator in FPGA," *Optik*, vol. 127, no. 13, pp. 5500–5505, 2016.

[20] Y. Jiang, H. Zhang, H. Zhang et al., "Design of mixed synchronous/asynchronous systems with multiple clocks," *IEEE Transactions on Parallel and Distributed Systems*, vol. 26, no. 8, pp. 2220–2232, 2015.

[21] B. Qu, H. Song, W. Zhou, S. Li, and Q. Meng, "Novel phase-locked loop with direct phase detection for two frequency different signals," *Journal of Xidian University (Science and Technology)*, vol. 41, no. 2, pp. 172–177, 2014.

24

Linear Processing Design of Amplify-and-Forward Relays for Maximizing the System Throughput

Qiang Wang, Tiejun Chen⬦, and Tingting Lan

Yulin Normal University, Yulin, China

Correspondence should be addressed to Tiejun Chen; chentiejun2000@163.com

Academic Editor: Jit S. Mandeep

In this paper, firstly, we study the linear processing of amplify-and-forward (AF) relays for the multiple relays multiple users scenario. We regard all relays as one special "relay", and then the subcarrier pairing, relay selection and channel assignment can be seen as a linear processing of the special "relay". Under fixed power allocation, the linear processing of AF relays can be regarded as a permutation matrix. Employing the partitioned matrix, we propose an optimal linear processing design for AF relays to find the optimal permutation matrix based on the sorting of the received SNR over the subcarriers from BS to relays and from relays to users, respectively. Then, we prove the optimality of the proposed linear processing scheme. Through the proposed linear processing scheme, we can obtain the optimal subcarrier paring, relay selection and channel assignment under given power allocation in polynomial time. Finally, we propose an iterative algorithm based on the proposed linear processing scheme and Lagrange dual domain method to jointly optimize the joint optimization problem involving the subcarrier paring, relay selection, channel assignment and power allocation. Simulation results illustrate that the proposed algorithm can achieve a perfect performance.

1. Introduction

Relay-assisted cooperative communication and orthogonal frequency-division multiplexing (OFDM) are core technologies for the next generation mobile communication system which attract tremendous attentions. Relay-assist cooperative communication can effectively extend communication coverage, and reduce the consumption of transmit power. It can also improve overall system throughput due to its exploiting spatial diversity and combating channel fading [1, 2]. OFDM divides a wideband channel into some orthogonal narrow band subcarriers. Then, the fast data flow can be transformed to a set of slow data flows. Furthermore, inner-symbol interference can be eliminated significantly. OFDM can also increase the flexibility for coding and modulation. Therefore, relay-assisted cooperative communication and OFDM are adopted by a lot of communication standards, such as 3GPP and IEEE 802.16e.

There are two common categories of relay strategies: amplify-and-forward (AF) and decode-and forward (DF). The DF relay can decode and re-encode the received signals, and then retransmit these signals to destination node.

Different from DF relay, the AF relay amplifies the received signals linearly and forwards these signals to destination node without decoding. The AF strategy is the most practical relay approach as the result of its low complexity transceiver design [3]. Moreover, AF relays are more transparent to adaptive modulation techniques than DF relays.

Due to the independent fading on each subcarrier, the incoming and outgoing subcarrier should be matched carefully to maximize the overall system throughput. This problem is also called subcarrier paring. Recently, subcarrier paring has gained a lot of attentions, such as [4–6]. Reference [4] proposes an ordered subcarrier pairing (OSP) method, in which it is proved that subcarrier paring according to the channel state is optimality for equal power allocation. The authors in [5] further prove that the OSP is still optimality for optimal power allocation. Reference [6] considers a linear processing design at the relay for amplified-and-forward (AF) relaying communication system without direct path and with direct path under fixed power gain. Different from [4, 5], [6] dose not only consider the cooperative mode without direct path, but also study the the cooperative mode with direct path. In [6], the authors separate the processing structure into two components which are the power amplification matrix and the unitary linear processing matrix, respectively. They show the optimal unitary processing matrix is of permutation structure.

The joint resource allocation for cooperative communication has been also attracted tremendous attentions. For the scenario with multiple relays multiple users, the joint optimization problem involves subcarrier pairing, relay selection, channel assignment and power allocation. Due to the combinational nature of the subcarrier pairing, relay selection and channel assignment, the joint optimization problem is difficult to solve when the number of subcarriers, relays and users are large. Therefore, most previous works only consider a subset of these issues.

In this paper, firstly, we exploit the linear processing scheme for multiple relays multiple users network to maximize the system throughput. We regard all relays as one special "relay", and the subcarrier pairing, channel assignment and relay selection can be seen as the linear processing of the special "relay" under fixed power allocation. It can be shown that the optimal linear processing matrix of the special "relay" is a promotion of the permutation matrix. To maximize the system throughput under fixed power allocation, the optimal linear processing matrix should be found out. To this end, we propose a linear processing scheme for the AF relays under fixed power allocation which can find out the optimal permutation matrix, and then we prove the optimality of this linear processing design. It is interesting that, through the proposed linear processing scheme, we can obtain the optimal subcarrier paring, relay selection and channel assignment with fixed power allocation in accordance with the ordered SNR over the incoming and outgoing channels in polynomial time. To the best of our knowledge, this paper is the first work to investigate the linear processing of the AF relays for the multiple relays multiple users network. Finally, based on the proposed linear process scheme and Lagrange dual domain method, we propose an iterative algorithm to solve the joint optimization problem involving subcarrier pairing, relay selection, channel assignment as well as power allocation. Simulation results illustrate that the proposed iterative algorithm is effective to find out an *asymptotically* optimal solution in polynomial time.

The rest of the paper is organized as follows. In Section 2, the previous related works are reviewed. We describe the system model in Section 3. The optimal linear processing under fixed power gain is introduced in Section 4. In Section 5, we extend the linear process scheme to the scenario with multiple relays multiple users. In Section 6, we describe the iterative algorithm based on the proposed linear process scheme and Lagrange dual domain method to solve the joint resource allocation problem. In Section 7, simulation results are provided to evaluate the performance of the proposed algorithm. Finally, we conclude this paper in Section 8.

2. Related Work

The resource allocation for the scenario with single BS, multiple relays and multiple users is a joint optimization problem which involves subcarrier paring, relay selection, channel assignment and power allocation. This joint optimization problem can be formulated as a mixed-integer programming due to the combinatorial nature of the subcarrier paring, relay selection and channel assignment. It becomes more and more intractable with the increase of the number of subcarriers, relays and users because the problem of subcarrier pairing, relay selection and channel assignment is NP hard [7]. To decrease the computational complexity, most previous works consider a subset of the joint optimization problems.

References [8–11] consider the scenario with one BS, one relay and one user. Reference [8] jointly optimizes the subcarrier pairing and power allocation, and a mixed-integer programming problem is formulated. Then, the authors in [8] transform the mixed integer programming problem into a convex optimization through continuous relaxation. [9] proposes a hybrid scheme in which the full-duplex and half-duplex relaying modes can be switched opportunistically to obtain a tradeoff between spectral efficiency and self-interference. For DF relay strategy, if the ratio of the time allocation in the first transmission phase over the whole period is not large enough, the DF relay can not decode the received signals accurately. It is necessarily to optimize the the ratio of the time allocation in the first transmission phase over the whole period. Therefore, reference [10] optimizes not only the power allocation but also time allocation for DF relay strategy to minimize the outage probability. The perfect channel state information (CSI) at the source is impractical, thus it is reasonable to study the resource allocation problem with limited feedback [11].

The resource allocation for the scenario with one BS, one relay and multiple users is studied by reference [7]. The authors in [7] propose an optimal algorithm for the joint resource allocation including subcarrier pairing, channel assignment and power allocation through transforming the original problem into some simple linear programming problems. In [12], a combination of transmit-receive weights and Tomlinson-Harashima precoding is used to cancel the interference, and then a low-complexity power allocation is proposed to achieve data rate fairness among different users.

The cooperative communication scenario with one BS, multiple relays and multiple users is studied by [13, 14], and the direct transmission mode is also considered at the same time. The authors in [13] transform the combinational optimization problem of subcarrier paring, channel assignment, relay selection and transmission mode selection into a minimum cost network flow problem, and apply the linear optimal distribution algorithm to solve the minimum cost network flow problem. Then, the Lagrange dual domain method is used to solve the power allocation. Different from [13], reference [14] only considers the relay selection and subcarrier assignment for multiuser cooperative network to decrease the computational complexity. In the first, [14] simplifies the original problem into a new problem through decreasing the number of variables which is easily to handle. Then, branch-and-cut is introduced to optimize the new problem.

There are also a lot of works on the resource allocation for the multiple-input and multiple-output (MIMO) cooperative communication networks, such as [3, 15–21].

Referencee [15] considers the problem of minimizing the total power consumption in a two-hop single-relay MIMO network with QoS requirements. In [15], a nonconvex power allocation problem is approximated with a convex problem, and then the convex problem is computed in closed-form through a multistep procedure. [16] studies the problem of resource allocation in relay-enhanced bidirectional MIMO-OFDM networks to minimize the total power consumption. The authors in [16] propose a green resource allocation scheme to jointly optimize the subchannel assignment, power allocation and phase duration assignment, in which both the separate-downlink (DL)-and-uplink (UL) and mixed-DL-and-UL relaying assignments and the linear block diagonalization (LBD) technique are adopted. Reference [17] study the optimal power allocation structure for the multiple relays network. The authors in [17] show that the power allocation at BS and each relay follow a matching structure. The cooperative communication with virtual MIMO has also attracted a lot of attentions, such as [18]. Reference [18] studies a wireless network where multiple users cooperate with each other. The authors in [18] propose a new auction-based power allocation framework with multiple auctioneers and multiple bidders to maximize the weighted sum-rates of the users.

Recently, energy efficient (EE) resource allocation for MIMO cooperative communication has become attractive, such as [22–27]. Reference [22] considers the energy efficient joint source-relay power allocation problem for MIMO AF relaying system. In [22], the objective of optimization is the number of the bits per second per hertz per Joule with the guarantee of the minimum spectral efficiency, and the objective function and the constraint are not convex. Firstly, [22] transforms the original problem into a pseudo-convex problem by employing the high signal-to-noise ratio (SNR) approximation. Then, the authors in [22] propose a relaxation method according to the Jensen inequality to solve the pseudo-convex problem. [23] studies the energy-spectral efficiency (EE-SE) trade-off of the uplink of a multi-user cellular virtual MIMO system. Finally, a heuristic resource allocation algorithm is proposed to optimize the EE-SE tradeoff in [23]. The energy-efficient resource allocation for OFDMA cellular networks with user cooperation is studied by [24]. In [24], a mixed-integer nonlinear programming problem is formulated, and then an optimal algorithm is proposed to solve the problem.

3. System Model

In this paper, for matrix M, we let $\text{Tr}\{M\}$, $M^\dagger$ and $|M|$ denote the trace, conjugate transpose and determinant of M, respectively. I denote the identity matrix. $\text{diag}\{M_1, \ldots, M_n\}$ represents the block diagonal matrix with $M_1, \ldots, M_n$ on its main diagonal. We assume that the perfect CSI is obtained by the technology of channel estimation, and then the accurate channel gain is known.

In this paper, we intend to exploit the optimal linear processing of relays in OFDMA AF-based relaying network consisting multiple relays multiple users, similar as Figure 1. For easy to exposition, we first study the linear processing

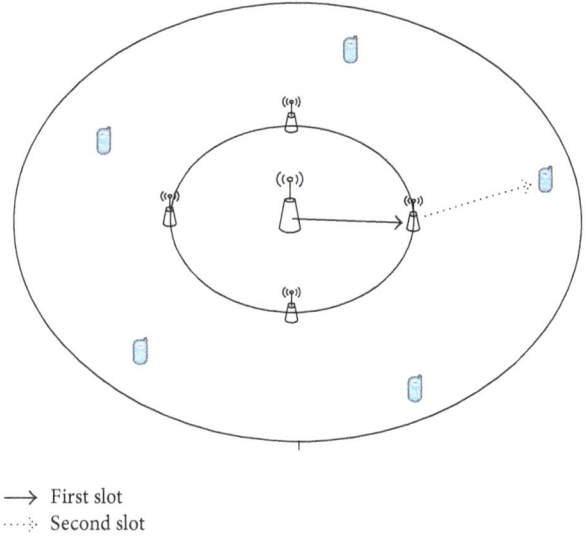

→ First slot
⋯⋯⋗ Second slot

FIGURE 1: Illustration of dual-hop multi-relay multi-user AF relaying.

scheme for the scenario with multiple relays and single user in this section. Then, we will show that our proposed linear processing scheme can be readily extended to the multiple relays multiple users scenario in the next section. In this work, we assume there are N subcarriers and M AF relays. The half-duplex relay is considered, that is, each relay can not transmit and receive the signals in the same subcarrier at the same time. The transmission duration is divided into two equal time slot. Let $H_{sr}^i \in \mathbb{C}^{N \times N}$ and $H_{rd}^i \in \mathbb{C}^{N \times N}$ denote the channel gain matrices between BS and relay i and between relay i and the user, respectively.

The received signals at the relay i in the first time slot are given by $y_r^i = H_{sr}^i D_{sr}^i s + n_r^i$, $i = 1, 2, \ldots, M$, where $s \in \mathbb{C}^{N \times 1}$ is the transmission symbol vector and $D_{sr}^i = \text{diag}\{\sqrt{p_{b,i,1}}, \ldots, \sqrt{p_{b,i,N}}\} \in \mathbb{C}^{N \times N}$ is the transmission power coefficient matrix from BS to relay i, and $n_r^i \in \mathbb{C}^{N \times 1}$ represents the AWGN at relay i, with $\sim \mathscr{CN}(0, \sigma_r^2 I)$. Next, we let $H_{sr} = \text{diag}\{H_{sr}^1, H_{sr}^2, \ldots, H_{sr}^M\} \in \mathbb{C}^{MN \times MN}$ denote the combined relay channel matrix between BS and all relays. Similarly, we let $D_{sr} = \text{diag}\{D_{sr}^1, D_{sr}^2, \ldots, D_{sr}^M\} \in \mathbb{C}^{MN \times MN}$, $s = \underbrace{[s^\dagger, \ldots, s^\dagger]^\dagger}_{M} \in \mathbb{C}^{MN \times 1}$ and $n_r = [n_r^{[1]\dagger}, \ldots, n_r^{[M]\dagger}]^\dagger \in \mathbb{C}^{MN \times 1}$. Thus, the combined received signal of all relays can be given as $y_r = H_{sr} D_{sr} s + n_r$.

In the second time slot, the AF relays process the received signals linearly and then retransmit these signals to the user. Since the relays cannot share their received signals with each other, we let $W = \text{diag}\{W_1, \ldots, W_M\} \in \mathbb{C}^{MN \times MN}$ denote the linear processing matrix of all relays. The processed signals at relay i can be represented as $x_r^{[i]} = W_i y_r^{[i]}$. Then, the overall relay signals can be given as $x_r = W y_r = W(H_{sr} D_{sr} s + n_r)$, where $x_r = [x_r^{[1]\dagger}, \ldots, x_r^{[M]\dagger}]^\dagger$.

Also, we let $H_{rd} = \text{diag}\{H_{rd}^1, \ldots, H_{rd}^M\} \in \mathbb{C}^{MN \times MN}$ represent the combined channel gain matrix between all relays and the user. Then, the received signals at user can be written as:

$$y_d = H_{rd}x_r + n_d = H_{rd}W\left(H_{sr}D_{sr}s + n_r\right) + n_d, \qquad (1)$$

where $n_d \in \mathbb{C}^{MN\times 1}$ is the AWGN, with $n_d \sim \mathscr{CN}(0, \sigma_d^2 I)$ at the user in the second time slot.

In the second time slot, the relays combine the received signals linearly and then retransmit the amplified version of the processed signal to the user. Therefore, the processing matrix W can be represented as $W = D_r U$, where $U \in \mathbb{C}^{MN\times MN}$ is a special permutation matrix in which there are only N rows and N columns with nonzero entry, which determines how to combine the incoming signals linearly and $D_r = \text{diag}\{D_r^1, \ldots, D_r^M\} \in \mathbb{C}^{MN\times MN}$ is the power amplification matrix of all relays, where $D_r^i = \text{diag}\{d_{i,1}, \ldots, d_{i,N}\}$ and $d_{i,n}$ means the power amplification factor for the nth subcarrier at relay i. In this section, we intend to investigate the linear processing matrix U with fixed D_s and D_r. That is, our goal is to investigate what form of the linear processing scheme leads to the optimal relay selection and subcarrier pairing. Since we are interesting in the permutation matrix U, the received signals at the user can be rewritten as

$$y_{dr} = \widetilde{H}(U)s + \tilde{n}(U), \qquad (2)$$

where $\widetilde{H}(U) \triangleq H_{rd}D_r UH_{sr}D_s$ and $\tilde{n}(U) \triangleq H_{rd}D_r Un_r + n_d$ are the functions of matrix U, and they can be seen as the equivalent channel matrix and noise term.

In fact, there are M relays and N subcarriers so that there are MN incoming subchannels and MN outgoing subchannels. To avoid interference, we can only select N incoming subchannels and N outgoing subchannels (each subchannel can only be assigned to one relay node at each time slot). Therefore, U is a non-full rank permutation matrix in which there only N rows and N columns with nonzero entry.

4. Overall System Throughput

As aforementioned, the end-to-end achievable rate can be given by

$$R(U) = \frac{1}{2}\log\left|I + R_n^{-1}\widetilde{H}(U)\widetilde{H}^\dagger(U)\right|, \qquad (3)$$

where $R_n \triangleq E[\tilde{n}(U)\tilde{n}^\dagger(U)]$ represents the covariance matrix of the equivalent noise term. The factor $1/2$ in (3) accounts for the two time slots required to complete a transmission. Due to the maximum transmission power limit at BS and each relay i, we have the following peak power constraints:

$$E\|D_s s\| \leq P_s, \qquad (4)$$

$$\text{Tr}\left\{W_i\left(H_{sr}^{[i]}D_s^{[i]2}H_{sr}^{[i]\dagger} + \sigma_r^2 I\right)W_i^\dagger\right\} \leq P_i, \qquad (5)$$
$$i = 1, \ldots, M,$$

where P_s and P_i are the maximum transmission power of BS and relay i, respectively.

In this section, our objective is to find the optimal U^* to maximize the system achievable $R(U)$, which can be formulated as the following optimization problem:

$$\max_U \quad R(U) = \frac{1}{2}\log\left|I + R_n^{-1}\widetilde{H}(U)\widetilde{H}^\dagger(U)\right| \qquad (6)$$
$$\text{s.t.} \quad (4), (5).$$

For the scenario with multiple relays and single user, relay selection and subcarrier pairing can be essentially represented as a kind of U. In this case, $U \in \mathbb{C}^{MN\times MN}$ is a non-full rank permutation matrix in which there are only N rows and N columns with nonzero entries.

Next, we give our method to find the optimal matrix U^*. In accordance with $|I + AB| = |I + BA|$, (6) can be rewritten as:

$$\max_U \quad R(U) = \frac{1}{2}\log\left|I + \widetilde{H}^\dagger(U)R_n^{-1}\widetilde{H}(U)\right|. \qquad (7)$$

Substituting $\widetilde{H}(U) \triangleq H_{rd}D_r UH_{sr}D_s$ and $\tilde{n}(U) \triangleq H_{rd}D_r Un_r + n_d$ into (7), we have

$$R(U) = \frac{1}{2}\log\left|I + \left(R_n^{-1/2}\widetilde{H}(U)\right)^\dagger R_n^{-1/2}\widetilde{H}(U)\right| = \frac{1}{2}$$
$$\cdot \log\left|I + \left(R_n^{-1/2}H_{rd}D_r UH_{sr}D_{sr}\right)^\dagger\right. \qquad (8)$$
$$\left.\cdot R_n^{-1/2}H_{rd}D_r UH_{sr}D_{sr}\right|.$$

Let $F = R_n^{-1/2}H_{rd}D_r$ and $T = H_{sr}D_{sr}$, and then (8) can be rewritten as

$$\max_U \quad \frac{1}{2}\log\left|I + (FUT)^\dagger FUT\right|. \qquad (9)$$

The jth diagonal entries of F and T are given by:

$$f_j = \frac{h_{rd}^j d_r^j}{\sqrt{\sigma_d^2 + \sigma_r^2\left|h_{rd}^j d_r^j\right|^2}}, \qquad (10)$$
$$t_j = h_{sr}^j d_s^j.$$

Let $\text{SNR}_{sr,j} = |h_{sr}^j|^2|d_s^j|^2/\sigma_r^2$ and $\text{SNR}_{rd,j} = |h_{rd}^j|^2|d_r^j|^2/\sigma_d^2$ be the received SNR from BS to relay and from relay to user respectively, and then there are

$$f_j^2 = \frac{\text{SNR}_{rd,j}}{1 + \sigma_r^2\text{SNR}_{rd,j}}, \qquad (11)$$
$$t_j^2 = \sigma_r^2\text{SNR}_{sr,j}.$$

It is clear that $\{f_j\}$ and $\{t_j\}$ have the same sorting order with $\{\text{SNR}_{rd,j}\}$ and $\{\text{SNR}_{sr,j}\}$. Therefore, f_j and t_j are equivalent to $\text{SNR}_{rd,j}$ and $\text{SNR}_{sr,j}$, respectively.

To maximize (9), in the first, we give the following Lemma.

Lemma 1. *Let* $P = \text{diag}\{p_1, \ldots, p_N\}$ *and* $Q = \text{diag}\{q_1, \ldots, q_N\}$ *be two diagonal matrices. We let* $\{|p_{(i)}|\}$ *and* $\{|q_{(i)}|\}$ *be the ordered sequences of* $\{|p_i|\}$ *and* $\{|q_i|\}$ *in descending order, respectively. The optimal* U *to maximize* (9) *is as follows:*

$$\left| I + (PU^*Q)^\dagger (PUQ) \right| = \prod_i^N (1 + |p_i| |q_i|). \qquad (12)$$

That is, U^* *is the permutation matrix that matches* p_i *and* q_i.

Proof. Please refer to Appendix   □

For the scenario with one relay and one user, U is a full rank permutation matrix since there is only subcarrier pairing that need to be optimized. According to Lemma 1, we can readily obtained the optimal subcarrier pairing with fixed power allocation, which is that the incoming subcarrier with the best SNR is paired with the outgoing subcarrier with the best SNR, and the incoming subcarrier with the worst SNR is paired with the outgoing subcarrier with the worst SNR, and so on. However, when the number of relays is more than one, the best incoming subcarrier and the best outgoing subcarrier may not belong to the same relay which makes these two subcarrier can not be matched. Therefore, the case with multiple relays differs significantly from the case with single relay. To exploit the optimal linear processing scheme for the scenario with multiple relays, in the following, we first give two constraints to avoid interference and make the linear processing scheme be in line with the reality, and then we will give a Theorem to exploit the optimal linear processing for multiple relays network.

Constraint 2. Only N orthogonal subchannels can be selected at each hop to avoid co-channel interference.

Constraint 3. Two subcarriers which are matched to one pair should belong to the same relay node.

Indeed, Constraint 2 means that each subcarrier can be only assigned to one relay at each transmission slot to avoid interference.

Next, we give a Theorem as follows:

Theorem 4. *Let* $F = \text{diag}\{f_1, f_2, \ldots, f_{MN}\}$ *and* $T = \text{diag}\{t_1, t_2, \ldots, t_{MN}\}$. *Let* $G = \{g_1, g_2, \ldots, g_{(MN)^2}\}$, *where* $g_i = f_k q_l, k, l \in \{1, 2, \ldots, MN\}$. *It is clear that the cardinality of* G *is* $(MN)^2$ *since there are* $(MN)^2$ *combinations of* f_k, $k = 1, 2, \ldots, MN$ *and* $t_l, l = 1, 2, \ldots, MN$. *Let* $\{|g_i|\}$ *be a ordered sequence in descending order. Under Constraints 2 and 3, the optimal* U^* *to maximize* (9) *is as follows:*

$$\left| I + (FU^*T)^\dagger FU^*T \right| = \prod_1^N (1 + |g_i|^2). \qquad (13)$$

Proof. We assume the optimal incoming and outgoing channels obtained from Theorem 4 are $\{f_{(1)}, f_{(2)}, \ldots, f_{(N)}\}$ and $\{t_{(1)}, t_{(2)}, \ldots, t_{(N)}\}$ respectively, which correspond to a set of $\{g_{(1)}, g_{(2)}, \ldots, g_{(N)}\}$, where $g_{(1)} = f_{(1)} t_{(1)}$, $g_{(2)} = f_{(2)} t_{(2)}$, and so on. Without loss of generality, we assume $\{g_{(i)}\}$ are ordered

sequence with descending order. Then, the corresponding permutation matrix U^* satisfies:

$$\left| I + (FU^*T)^\dagger FU^*T \right| = \prod_1^N (1 + |g_{(i)}|^2). \qquad (14)$$

For any other permutation matrix $\widetilde{U}$ which also satisfies Constraints 2 and 3, we assume the corresponding subcarrier pair set is $\{\tilde{g}_{(1)}, \tilde{g}_{(2)}, \ldots, \tilde{g}_{(N)}\}$. From Lemma 1, there is

$$\left| I + (F\widetilde{U}T)^\dagger F\widetilde{U}T \right| = \prod_1^N (1 + |\tilde{g}_{(i)}|^2). \qquad (15)$$

According to the selection criterion of Theorem 4, there is at least one $\tilde{g}_{(i)} \leq g_{(i)}$. So, there is

$$\prod_1^N (1 + |\tilde{g}_{(i)}|^2) \leq \prod_1^N (1 + |g_{(i)}|^2). \qquad (16)$$

Therefore, the permutation matrix U^* is optimal for (9).
                                                                        □

Theorem 4 gives an optimal linear processing scheme of relays corresponding to one optimal relay selection and subcarrier pairing scheme for the scenario with multiple relays. In fact, as aforementioned, when there is only one relay, the linear processing matrix "U" is a full rank permutation matrix which only corresponds to the subcarrier pairing. In accordance with Lemma 1, the optimal pairing strategy is based on the sorted received SNRs. However, for the scenario with multiple relays considered in this paper, the processing matrix "U" is a non-full rank matrix since each subcarrier can only be allocated to one relay node to avoid interference. Theorem 4 shows that the optimal relay selection and subcarrier pairing strategy is based on the sorted products of the incoming and outgoing subchannel's SNRs.

In fact, there are $M \times N$ "possible" incoming channels between BS and all relays and $M \times N$ "possible" outgoing channels between all relays and the user, and we can only select N incoming and N outgoing channels. There are $(MN)^2$ combinations of incoming and outgoing channels. Each combination corresponds to one product of one incoming and one outgoing subchannel's SNR. Theorem 4 means the optimal scheme is that, firstly, these combinations are ordered according to their corresponding products. Then, N best combinations satisfying Constraints 2 and 3 would be selected from the ordered $(MN)^2$ combinations.

For example, if the number of subcarriers, relays and users are 2, 2 and 1 respectively, there are 4 "possible" incoming channels: $\{f_{11}^1, f_{12}^1, f_{21}^1, f_{22}^1\}$, and 4 "possible" outgoing channels: $\{q_{11}^1, q_{12}^1, q_{21}^1, q_{22}^1\}$. There are 16 combinations of the incoming and outgoing channels. Assume $f_{11}^1 = 5$, $f_{12}^1 = 4$, $f_{21}^1 = 2$, $f_{22}^1 = 1$, $q_{11}^1 = 2$, $q_{12}^1 = 3$, $q_{21}^1 = 4$, $q_{22}^1 = 1$. Then, $\{|g_i|\} = \{f_{11}^1 q_{21}^1 = 20, f_{12}^1 q_{21}^1 = 16, f_{11}^1 q_{12}^1 = 15, f_{12}^1 q_{12}^1 = 12, f_{11}^1 q_{11}^1 = 10, f_{12}^1 q_{11}^1 = 8, f_{21}^1 q_{21}^1 = 8, f_{21}^1 q_{12}^1 = 6, f_{11}^1 q_{22}^1 = 5, f_{12}^1 q_{22}^1 = 4, f_{21}^1 q_{11}^1 = 4, f_{22}^1 q_{21}^1 = 4, f_{22}^1 q_{12}^1 = 3,$

$f_{21}^1 q_{22}^1 = 2$, $f_{22}^1 q_{11}^1 = 2$, $f_{22}^1 q_{22}^1 = 1$}. Therefore, {f_{11}^1, f_{22}^1} and {q_{12}^1, q_{21}^1} are selected because another channels do not satisfy Constraints 2 and 3. Thus, the optimal transmission paths are $(2, 1, 1, 1)$ and $(1, 2, 1, 1)$, which mean that the subcarrier pair $(2, 1)$ is assigned to user 1 with the help of relay 1 and the subcarrier pair $(1, 2)$ is assigned to user 1 with the help of relay 1, respectively. The corresponding permutation U is the form as follows:

$$U = \begin{bmatrix} 0 & 1 & 0 & 0 \\ 1 & 0 & 0 & 0 \\ 0 & 0 & 0 & 0 \\ 0 & 0 & 0 & 0 \end{bmatrix}. \tag{17}$$

Note that it is possible that there exists tie, for instance, there are two combinations of subcarriers with the same product and both these two combinations satisfy Constraints 2 and 3. For this case, arbitrary tie-breaking will be adopted which does not change the optimality of the algorithm.

5. The Scenario with Multiple Relays Multiple Users

In this section, we aim to extend the linear processing scheme proposed in the previous section to the scenario with multiple relays multiple users. We assume there are one BS, M relays, K users and N subcarriers. To make the computation feasible, in the first time slot, we denote the channel matrix between BS and all relays, the transmit power coefficient matrix and the transmission symbol vector as $H_{sr} = \text{diag}\{\underbrace{H_{sr}, H_{sr}, \ldots, H_{sr}}_{K}\} \in \mathbb{C}^{MNK \times MNK}$ $D_{sr} = \text{diag}\{\underbrace{D_{sr}, D_{sr}, \ldots, D_{sr}}_{K}\} \in \mathbb{C}^{MNK \times MNK}$ and $s = [\underbrace{s^\dagger, \ldots, s^\dagger}_{K}]^\dagger \in \mathbb{C}^{MNK \times 1}$, respectively. Then, the combined received signals of all relays can be given by $y_r = H_{br}D_{br}s + n_r$, where $n_r \in \mathbb{C}^{MNK \times 1}$. Similarly, the processing matrix of all relays is extended to $W = \text{diag}\{\underbrace{W, \ldots, W}_{K}\} \in \mathbb{C}^{MNK \times MNK}$, and the linear processing matrix $U = \text{diag}\{\underbrace{U, \ldots, U}_{K}\} \in \mathbb{C}^{MNK \times MNK}$.

Note that U is also a special permutation matrix, in which there are only N columns and N rows with nonzero entries.

We let $H_{rd}^k = \text{diag}\{H_{rd}^{1,k}, H_{rd}^{2,k}, \ldots, H_{rd}^{M,k}\}$ denote the combined channel gain matrix from all relays to user k, where $H_{rd}^{i,k} \in \mathbb{C}^{N \times N}$ is the channel gain from relay i to user k, and let $D_{rd}^k = \text{diag}\{D_{rd}^{1,k}, D_{rd}^{2,k}, \ldots, D_{rd}^{M,k}\}$ denote the power amplification matrix from all relays to user k, where $D_{rd}^{[i,k]} = \text{diag}\{d_{i,n,k}, \ldots, d_{i,N,k}\} \in \mathbb{C}^{N \times N}$ denote the power amplification matrix from relay i to user k. $d_{i,n,k}$ denotes the power amplification factor from relay i to user k over subcarrier n.

Next, let $H_{rd} = \text{diag}\{H_{rd}^1, H_{rd}^2, \ldots, H_{rd}^K\} \in \mathbb{C}^{MNK \times MNK}$ and $D_{rd} = \text{diag}\{D_{rd}^1, D_{rd}^2, \ldots, D_{rd}^K\} \in \mathbb{C}^{MNK \times MNK}$ denote the combined channel gain matrix from all relays to all users and the combined power amplification matrix from all relays to all users, respectively. The received signals of all users can be given by: $y_d = H_{rd}D_{rd}s + n_d$, where $n_d = [n_d^{[1]\dagger}, \ldots, n_d^{[K]\dagger}] \in \mathbb{C}^{MNK \times 1}$ represents the AWGN at users. Finally, the received signals at all users can be expressed as:

$$\begin{aligned} y_d &= H_{rd}W\left(H_{sr}D_{sr}s + n_r\right) + n_d \\ &= H_{rd}D_{rd}U\left(H_{sr}D_{sr}s + n_r\right) + n_d. \end{aligned} \tag{18}$$

The achievable rate in this scenario with multiple relays multiple users can be given by

$$R(U) = \frac{1}{2}\log\left|I + R_n^{-1}\widetilde{H}(U)\widetilde{H}^\dagger(U)\right|, \tag{19}$$

where $H(U)$ and R_n have the similar form as that in previous section.

So far, it is easy to prove that the Theorem 4 can be extended to the scenario with multiple relays multiple users. Thus, for the multiple relays multiple users network, when the power allocation is fixed, we can obtain the optimal permutation U which corresponds to the optimal subcarrier pairing, relay selection and channel assignment through the linear processing scheme proposed in this paper.

The computational complexity of the proposed linear processing scheme for the scenario with multiple relays multiple users is given by $\mathcal{O}((MN)^2 K \log(MN)^2 K + (MN)^2 K)$. In general, M and K are in the order of $\mathcal{O}(\sqrt{N})$. Therefore, the overall complexity of the proposed linear process scheme is $\mathcal{O}((7/2)N^3 \sqrt{N} \log N + N^{7/2})$. Note that the computational complexity of the proposed algorithm is less than most of the algorithms which solve the similar problems, such as network flow method.

Remark 5. Reference [20] studies the resource allocation of MIMO-OFDM relaying system and achieves excellent performance. However, it is worth nothing that [20] considers the power allocation and subcarrier assignment of relay system, and studies resource allocation problem from scalar optimization. In this work, we consider the linear processing scheme of AF relay system, which is a signal processing problem. We formulate the problem as a matrix optimization problem, and solve it using the method of matrix analysis.

6. Joint Optimization with Power Allocation

The relay selection and subcarrier pairing under fixed power allocation have been optimized in the previous section. In this section, we intend to tackle the joint resource allocation involving relay selection, subcarrier pairing, channel assignment as well as power allocation using the Lagrange dual domain method and the proposed linear processing scheme.

Firstly, we give a set of binary assignment variables corresponding to the permutation matrix U:

u_{mnik}: indicates whether subcarrier pairing (m, n) is assigned to relay i to help user k. Note that u_{mnik} is dependent on the permutation matrix U. In the aforementioned section, we have optimized the binary assignment variables u_{mnik} which indicates the subcarrier pairing, relay selection and channel-user assignment under fixed power allocation through optimizing the linear processing of the relays. In this section, we take into account to optimize the power allocation.

Since both F and T are diagonal matrices, we can readily convert (19) into scalar form as follows:

R

$$= \sum_{k=1}^{K} \sum_{i=1}^{M} \sum_{m=1}^{N} \sum_{n=1}^{N} u_{mnik} \log \left(\frac{\left| h_{i,n,k} d_{i,n,k} h_{b,r,n} \sqrt{P_{b,i,n}} \right|^2}{\sigma_d^2 + \sigma_r^2 \left| h_{i,n,k} d_{i,n,k} \right|^2} \right). \quad (20)$$

In accordance with the definition of D_r, $d_{i,n,k}$ can be represented as

$$d_{i,n,k}^2 = \frac{p_{i,n,k}}{\sigma_i^2 + h_{b,i,n}^2 P_{b,i,n}}, \quad (21)$$

where $p_{i,n,k}$ is the transmit power of relay i to user k over subcarrier n in the second time slot. Let $g_{b,i,n} = |h_{i,n,k}|^2/\sigma_r^2$, and $g_{i,n,k} = |h_{i,n,k}|^2/\sigma_d^2$. Substituting $d_{i,n,k}^2 = p_{i,n,k}/(\sigma_r^2 + h_{b,r,n}^2 P_{b,i,n})$, $g_{b,i,n} = |h_{i,n,k}|^2/\sigma_r^2$ and $g_{i,n,k} = |h_{i,n,k}|^2/\sigma_d^2$ into (20), and then (20) can be rewritten as:

R

$$= \sum_{k=1}^{K} \sum_{i=1}^{M} \sum_{m=1}^{N} \sum_{n=1}^{N} u_{mnik} \log \left(1 + \frac{g_{b,i,m} P_{b,i,m} g_{i,n,k} P_{i,n,k}}{1 + g_{b,i,m} P_{b,r,m} + g_{i,n,k} P_{i,n,k}} \right). \quad (22)$$

Our objective is to jointly optimize the subcarrier pairing, relay selection, channel assignment and power allocation to maximize the overall system throughput. Let $\mathbf{U} \triangleq [u_{mnik}]_{N \times N \times M \times K}$, $\mathbf{P} \triangleq \{[p_{b,i,m}]_{M \times N}, [p_{i,n,k}]_{M \times N \times K}\}$ and

$$R_{mnik} = \log \left(1 + \frac{g_{b,i,m} P_{b,r,m} g_{i,n,k} P_{i,n,k}}{1 + g_{b,i,m} P_{b,i,m} + g_{i,n,k} P_{i,n,k}} \right). \quad (23)$$

Then, the joint optimization problem can be formulated as:

$$\text{P1} \quad \max_{\mathbf{U,P}} \quad \sum_{k=1}^{K} \sum_{i=1}^{M} \sum_{m=1}^{N} \sum_{n=1}^{N} u_{mnik} R_{mnik} \quad (24)$$

$$\text{s.t.} \quad \sum_{k \in K} \sum_{i \in M} \sum_{n \in N} u_{mnik} = 1 \quad m = 1, 2, \ldots, N, \quad (25)$$

$$\sum_{k \in K} \sum_{i \in M} \sum_{m \in N} u_{mnik} = 1 \quad n = 1, 2, \ldots, N, \quad (26)$$

$$\sum_{i \in M} \sum_{m \in N} p_{b,i,m} \leq P_b, \quad (27)$$

$$\sum_{k \in K} \sum_{n \in N} p_{i,k,n} \leq P_i, \quad i = 1, 2, \ldots, M, \quad (28)$$

$$\mathbf{P} \succeq 0, \quad (29)$$

where constraint (25) and (26) mean each subcarrier can be assigned at most one node (user or relay) at each time slot to avoid interference. (27) and (28) mean the peak power constraints of BS and relays which are converted from (4) and (5), respectively.

Problem P1 is a mixed-integer nonlinear programming problem. Furthermore, P1 is nonconvex, which is difficult to

tackle in general. However, it has been shown by [13] that the duality gap is close to zero as the number of subcarriers goes to infinity. Therefore, the Lagrange dual domain method can be used to solve problem P1.

It is worth nothing that the logarithm function in (23) is not concave. To make it concave, we remove the "1" in the denominator. Then, the original logarithm function becomes:

$$R_{mnik} \approx \log \left(1 + \frac{g_{b,i,m} P_{b,i,m} g_{i,n,k} P_{i,n,k}}{g_{b,i,m} P_{b,i,m} + g_{i,n,k} P_{i,n,k}} \right). \quad (30)$$

It is clear that (30) is an upper bound of (23). The authors in [24] have shown that this upper bound is tight at moderate high SNR.

Next, we solve the problem P1 using Lagrange dual domain method. The Lagrangian function of P1 is:

$$L(\mathbf{U}, \mathbf{P}, \boldsymbol{\lambda}) = \sum_{k=1}^{K} \sum_{m=1}^{N} \sum_{n=1}^{N} \sum_{i=1}^{M} u_{mnik} R_{mnik}$$

$$+ \lambda_b \left(P_b - \sum_{r \in M} \sum_{m \in N} P_{b,i,m} \right) \quad (31)$$

$$+ \sum_{i \in M} \lambda_i \left(P_i - \sum_{k \in K} \sum_{n \in N} P_{i,k,n} \right),$$

where λ_b and λ_i, $i = 1, 2, \ldots, M$ are the Lagrangian multipliers associated with the constraint (27) and (28).

The dual function of P1 can be formulated as:

$$g(\boldsymbol{\lambda}) = \max_{\mathbf{U,P}} \quad L(\mathbf{U}, \mathbf{P}, \boldsymbol{\lambda}). \quad (32)$$

6.1. Optimizing U and P with Given λ. In this subsection, we optimize **U** and **P** with given λ to compute the dual function.

In the first, we optimize **P** with fixed **U**. The optimal **P** can be readily obtained using KKT conditions as follows:

$p_{b,i,m}$

$$= \frac{1}{1 + \sqrt{g_{b,i,m}/g_{i,k,n}}} \left[\frac{1}{\lambda_b} - \frac{\left(\sqrt{g_{b,i,m}} + \sqrt{g_{i,k,n}}\right)^2}{g_{b,i,m} g_{i,k,n}} \right]^+,$$

$p_{i,k,n}$

$$(33)$$

$$= \frac{1}{1 + \sqrt{g_{r,i,n}/g_{b,i,m}}} \left[\frac{1}{\lambda_i} - \frac{\left(\sqrt{g_{b,i,m}} + \sqrt{g_{i,k,n}}\right)^2}{g_{b,i,m} g_{i,k,n}} \right]^+,$$

where $[a]^+ = \max\{a, 0\}$.

In the previous section, we can obtain the optimal $\mathbf{U}^*$ with fixed power allocation at the BS and the power amplification at the relays using the proposed linear processing scheme. Substituting $\mathbf{U}^*$ into (24), we can obtaine the corresponding optimal **P**. Then, an iterative algorithm can be designed. The details will be shown in the following subsection.

6.2. Optimizing the Dual Function. In the aforementioned subsection, the dual function of the problem P1 has been

Initialize the Lagrange multiplier $\boldsymbol{\lambda}^{(0)}$ and power allocation $\mathbf{P}^{(0)}$
repeat
    set $t \leftarrow t+1$
    For given $\boldsymbol{\lambda}^{(t)}$ and $\mathbf{P}^{(t)}$, obtain the optimal $\mathbf{U}^{(t)}(\boldsymbol{\lambda})$ using the proposed linear process scheme. Then, according to the $\mathbf{U}^{(t)}(\boldsymbol{\lambda})$,
    obtain $\mathbf{P}^{(t+1)}(\boldsymbol{\lambda})$ using the Lagrange dual domain method
    Update $\boldsymbol{\lambda}$ through $\boldsymbol{\lambda}^{t+1} = [\boldsymbol{\lambda}^t - \Delta\lambda v^t]^+$, where v^t is the step size at the tth iteration.
until The convergence of $\min_t g(\boldsymbol{\lambda}^t)$

ALGORITHM 1

computed. Now, we minimize the dual problem using subgradient method:

$$\min \quad g(\boldsymbol{\lambda}) \tag{34}$$
$$\text{s.t.} \quad \boldsymbol{\lambda} \succeq 0.$$

Any subgradient-based method can be adopted to minimize $g(\boldsymbol{\lambda})$. Then, a subgradient at the point $\boldsymbol{\lambda}$ can be given by

$$\Delta\lambda_b = p_b - \sum_{r \in M}\sum_{m \in N} p_{b,r,m},$$
$$\Delta\lambda_i = p_i - \sum_{k \in K}\sum_{n \in N} p_{i,k,n}, \quad \forall i \in M. \tag{35}$$

Now, we propose an iterative algorithm to optimize the variables $\mathbf{U}$ and $\mathbf{P}$. Firstly, Lagrangian multiplier $\boldsymbol{\lambda}$ and power allocation $\mathbf{P}$ are initialized, and then $\mathbf{U}(\boldsymbol{\lambda})$ can be obtained by the proposed linear processing scheme. In accordance with $\mathbf{U}(\boldsymbol{\lambda})$, we can obtain corresponding power allocation $\mathbf{P}(\boldsymbol{\lambda})$. If $\mathbf{P}(\boldsymbol{\lambda})$ satisfies constraints (27) and (28), the solution $\mathbf{U}(\boldsymbol{\lambda})$ and $\mathbf{P}(\boldsymbol{\lambda})$ are global optimal solution. otherwise, iterative algorithm would continue to refine the solution. The details of the iterative algorithm is given in Algorithm 1.

7. Simulation Results

In this section, we evaluate the performance of the proposed algorithm using simulations. We assume that the BS is located in the center of the cell, and all users are placed randomly, and the relays are uniformly distributed on the centered circle. Cost231-Hata model is adopted as the path loss model, and the small scale fading is modeled as Rayleigh fading. The simulation parameters are summarized in Table 1.

7.1. Comparing with Benchmarks. In this subsection, we compare the performance of the proposed iterative algorithm with that of some common benchmarks as follows:

7.1.1. No Pairing. In this scheme, the subcarrier pairing is not considered, that is, the subcarrier in the first hop is the same with that in the second hop. Thus, we should only optimize the channel assignment, relay selection and power allocation.

7.1.2. No PA. In this scheme, we assume each subcarrier is allocated uniform power. Therefore, the original problem

TABLE 1: Simulation Parameters.

simulation parameter	value
Cell radius	1 Km
Inner boundary radius	0.6 Km
Subcarrier bandwidth	15 KHz
Number of relays	4
Noise power spectral density	−174 dBm/Hz
Path loss coefficient	4
Peak power of BS	30 dBm
Peak power of relay	25 dBm

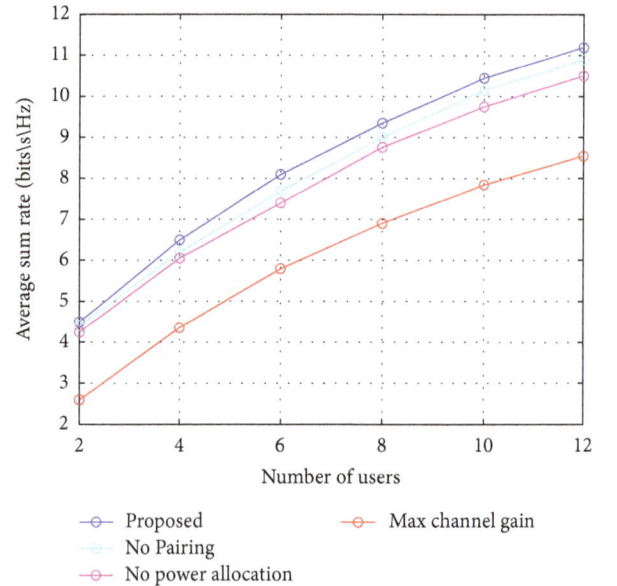

FIGURE 2: Operator's profits versus different number of users.

becomes a combinatorial problem since the original problem only consists of the subcarrier pairing and relay selection.

7.1.3. Max Channel Gain. In this scheme, there is no subcarrier pairing and power allocation. That is, the subcarrier in the first hop and in the second hop are the same. The channel is assigned according to the channel gain over the second hop.

Figure 2 shows the system throughput versus the number of users. The number of subcarriers and relays are set to $N = 16$, $M = 4$, respectively. The number of users changes from 2 to 12. It can be seen from Figure 2 that the system

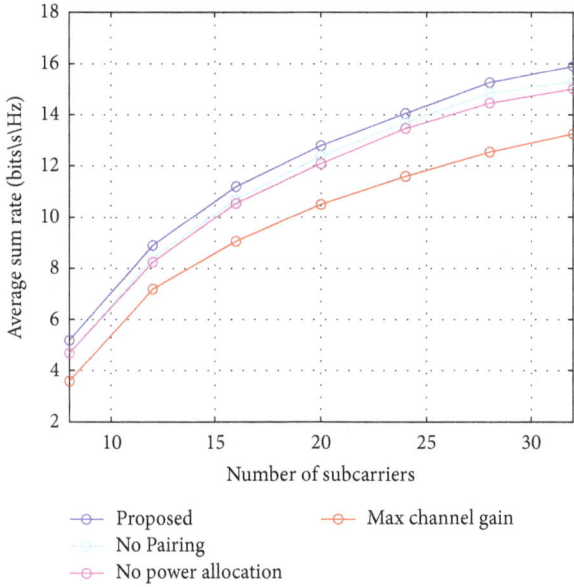

FIGURE 3: Operator's profits versus different number of users.

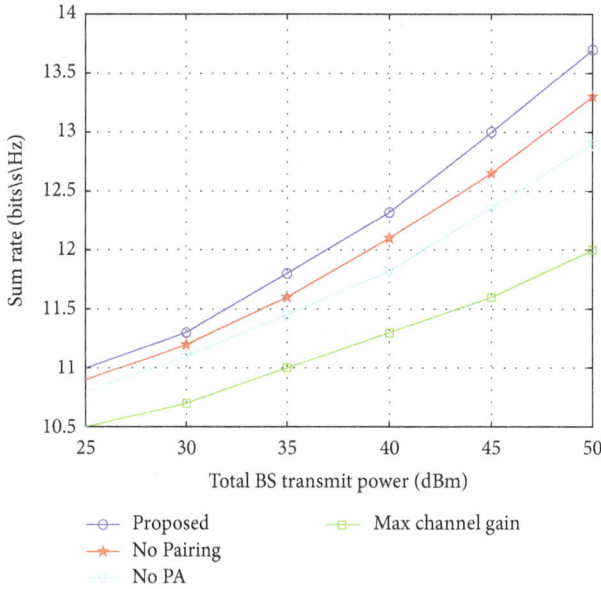

FIGURE 4: Operator's profits versus different number of users.

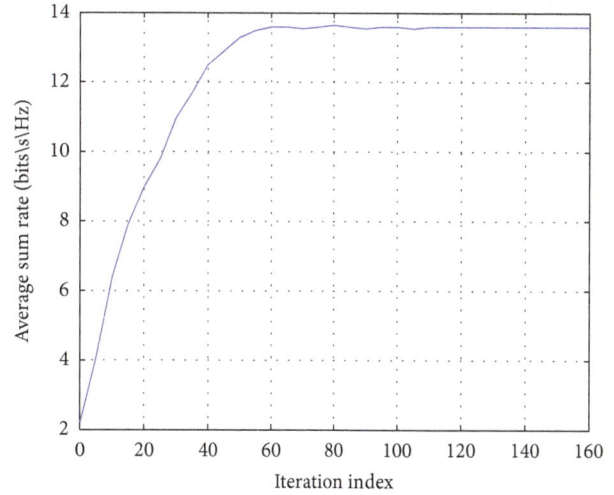

FIGURE 5: Operator's profits versus different number of users.

are set to $N = 16$, $M = 2$ and $K = 4$, respectively. Figure 3 shows that the proposed algorithm achieves more system throughput than these benchmark schemes. The result illustrates that it is necessary to jointly optimize the subcarrier pairing, relays selection and power allocation, and the proposed algorithm is effective.

7.2. Convergence Performance and System Capacity. In this subsection, we evaluate the convergence performance of the proposed algorithm and the system capacity versus the number of subcarriers.

Figure 5 shows the system throughput of the proposed algorithm versus the iteration indices. From Figure 5, it can be seen that, since the proposed linear processing scheme is polynomial time algorithm, the proposed algorithm have satisfactory convergence performance, and it converges within 50 iterations.

Figure 6 depicts the system throughput of the proposed algorithm versus the number of subcarriers with different total transmit power. The result shows that the system throughput can be improved significantly with the increase of the system bandwidth and SNR.

8. Conclusion

This paper investigates the joint resource allocation problem in multiple relays multiple users MIMO AF relaying network. The joint resource allocation problem involves subcarrier pairing, relay selection, channel assignment and power allocation which is difficult to tackle due to the combinatorial nature of the subcarrier pairing, relay selection and channel assignment. The subcarrier pairing, relay selection and channel assignment can be regarded as a linear processing of the relays which can be represented by a permutation matrix. We exploit the linear processing scheme for the scenario with multiple relays multiple users, and then propose an optimal linear processing design with fixed power gain to find the optimal permutation matrix which corresponds to

throughput increases with the increase of the number of users since the subcarriers have higher possibility to be allocated to the users who have better channel gain. Meanwhile, the proposed algorithm achieve 15 percent system throughput improvement than the *No Pairing* scheme.

Figure 3 illustrates the system throughput versus the number of subcarriers. It can be shown that the system throughput increases with the increase of the number of users. The number of subcarriers changes from 8 to 32. This result can be easily explained as that the system throughput is proportional to the system bandwidth.

Figure 4 depicts the system throughput versus the total transmit power. The number of subcarriers, relays and users

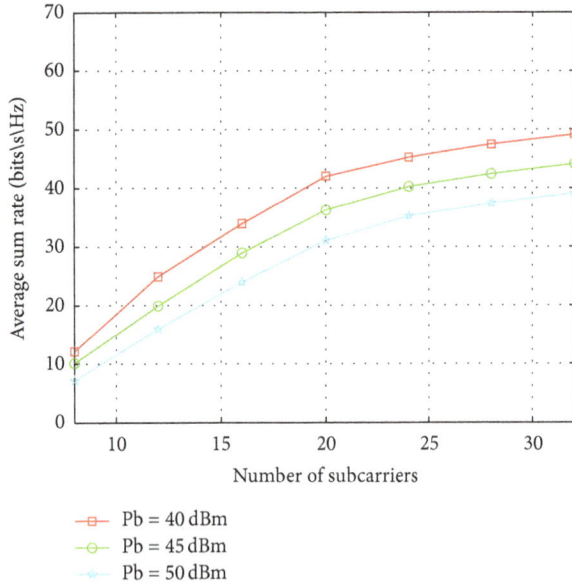

FIGURE 6: Operator's profits versus different number of users.

the optimal subcarrier pairing, relay selection and channel assignment. Through this linear processing design, we can obtain the optimal subcarrier pairing, relay selection and channel assignment with fixed power allocation. Then, the Lagrange dual domain method is adopted to solve the power allocation. Finally, an iterative algorithm is proposed to solve the combinatorial optimization problem and power allocation alternately.

The massive MIMO technology is one of the key topics of 5G wireless networks [28]. An issue with this technique is to eliminate the pilot population. In the future works, we will pay more attention to the resource allocation and pilot population elimination for the relay-enhanced massive MIMO system.

Appendix

Proof. In accordance with the property of the determinant $|AB| = |A||B|$, the objective function (9) can be rewritten as:

$$\frac{1}{2} \log \left| I + (PUQ)^{\dagger} PUQ \right|$$

$$= \left| Q^{\dagger} \right| \left| \left(Q^{\dagger} Q \right)^{-1} + U^{\dagger} P^{\dagger} UP \right| |Q| \qquad (A.1)$$

$$= \left| Q^{\dagger} Q \right| \left| \left(Q^{\dagger} Q \right)^{-1} + U^{\dagger} P^{\dagger} UP \right|.$$

Since $Q^{\dagger} Q$ is not a function of U, we only need to consider the second determinant:

$$\max_{U} \quad \left| \left(Q^{\dagger} Q \right)^{-1} + U^{\dagger} P^{\dagger} UP \right|. \qquad (A.2)$$

By the property of determinant

$$|A + B| \leq \prod_{n=1}^{N} \left(\lambda_n (A) + \lambda_{N+1-n} (B) \right), \qquad (A.3)$$

where $\lambda_n(A)$ and $\lambda_n(B)$ are the eigenvalues of A and B, respectively, sorted in ascending order, the equality is reached when A and B are both diagonal with the diagonal entries being inverse-order matched. Then, we have

$$\left| \left(Q^{\dagger} Q \right)^{-1} + U^{\dagger} P^{\dagger} UP \right| \leq \prod_{n=1}^{N} \left(\frac{1}{\left| q_{(n)} \right|^2} + \left| p_{N+1-n} \right|^2 \right). \qquad (A.4)$$

The equality is reached when U^* is the permutation matrix such that the entries of the ordered sequences $\{|p_{(i)}|\}$ and $\{|q_{(i)}|\}$ are one-to-one matched.   □

Conflicts of Interest

The authors declare that there is no conflict of interest regarding the publication of this paper.

References

[1] S. Wang and H. Ji, "Distributed power allocation scheme for multi-relay shared-bandwidth (MRSB) wireless cooperative communication," *IEEE Communications Letters*, vol. 16, no. 8, pp. 1263–1265, 2012.

[2] H. Al-Tous and I. Barhumi, "Joint power and bandwidth allocation for amplify-and-forward cooperative communications using stackelberg game," *IEEE Transactions on Vehicular Technology*, vol. 62, no. 4, pp. 1678–1691, 2013.

[3] F. Héliot, "Low-complexity energy-efficient joint resource allocation for two-hop MIMO-AF systems," *IEEE Transactions on Wireless Communications*, vol. 13, no. 6, pp. 3088–3099, 2014.

[4] M. Herdin, "A chunk based OFDM amplify-and-forward relaying scheme for 4G mobile radio systems," in *Proceedings of the 2006 IEEE International Conference on Communications, ICC 2006*, pp. 4507–4512, Turkey, July 2006.

[5] W. Dang and J. Huang, "Subcarrier-pair based resource allocation for cooperative multi-relay OFDM systems," *IEEE Trans, Wireless Commun*, vol. 9, no. 5, pp. 1640–1649, 2010.

[6] M. Dong, M. Hajiaghayi, and B. Liang, "Optimal fixed gain linear processing for amplify-and-forward multichannel relaying," *IEEE Transactions on Signal Processing*, vol. 60, no. 11, pp. 6108–6114, 2012.

[7] M. Hajiaghayi, M. Dong, and B. Liang, "Jointly optimal channel and power assignment for dual-hop multi-channel multi-user relaying," *IEEE Journal on Selected Areas in Communications*, vol. 30, no. 9, pp. 1806–1814, 2012.

[8] Y. Liu and W. Chen, "Limited-feedback-based adaptive power allocation and subcarrier pairing for OFDM DF relay networks with diversity," *IEEE Transactions on Vehicular Technology*, vol. 61, no. 6, pp. 2559–2571, 2012.

[9] T. Riihonen, S. Werner, and R. Wichman, "Hybrid full-duplex/half-duplex relaying with transmit power adaptation," *IEEE Transactions on Wireless Communications*, vol. 10, no. 9, pp. 3074–3085, 2011.

[10] Z. Mo, W. Su, S. Batalama, and J. D. Matyjas, "Cooperative communication protocol designs based on optimum power and time allocation," *IEEE Transactions on Wireless Communications*, vol. 13, no. 8, pp. 4283–4296, 2014.

[11] Y. Liu, W. Chen, J. Zhang, and Z. Luo, "Power allocation for the fading relay channel with limited feedback," in *Proceedings of the 2010 IEEE International Conference on Communications*, May 2010.

[12] N. Aboutorab, W. Hardjawana, and B. Vucetic, "A new resource allocation technique in MU-MIMO relay networks," *IEEE Transactions on Vehicular Technology*, vol. 60, no. 7, pp. 3485–3490, 2011.

[13] M. Tao and Y. Liu, "A network flow approach to throughput maximization in cooperative OFDMA networks," *IEEE Transactions on Wireless Communications*, vol. 12, no. 3, pp. 1138–1148, 2013.

[14] X. Zhang, X. Tao, Y. Li, N. Ge, and J. Lu, "On relay selection and subcarrier assignment for multiuser cooperative OFDMA networks with QoS guarantees," *IEEE Transactions on Vehicular Technology*, vol. 63, no. 9, pp. 4704–4717, 2014.

[15] L. Sanguinetti and A. A. D'Amico, "Power allocation in two-hop amplify-and-forward MIMO relay systems with QoS requirements," *IEEE Transactions on Signal Processing*, vol. 60, no. 5, pp. 2494–2507, 2012.

[16] T.-S. Chang, K.-T. Feng, J.-S. Lin, and L.-C. Wang, "Green resource allocation schemes for relay-enhanced MIMO-OFDM networks," *IEEE Transactions on Vehicular Technology*, vol. 62, no. 9, pp. 4539–4554, 2013.

[17] J. Liu, N. B. Shroff, and H. D. Sherali, "Optimal power allocation in multi-relay MIMO cooperative networks: theory and algorithms," *IEEE Journal on Selected Areas in Communications*, vol. 30, no. 2, pp. 331–340, 2012.

[18] Y. Liu and M. Tao, "An auction approach to distributed power allocation for multiuser cooperative networks," *IEEE Transactions on Wireless Communications*, vol. 12, no. 1, pp. 237–247, 2013.

[19] E. Monroy, S. Choi, and B. Jabbari, "Linear precoding with resource allocation for MIMO relay channels," *IEEE Transactions on Wireless Communications*, vol. 12, no. 11, pp. 5704–5716, 2013.

[20] D. W. K. Ng, E. S. Lo, and R. Schober, "Dynamic resource allocation in MIMO-OFDMA systems with full-duplex and hybrid relaying," *IEEE Transactions on Communications*, vol. 60, no. 5, pp. 1291–1304, 2012.

[21] I. Hammerstrom and A. Wittneben, "Power allocation schemes for amplify-and-forward MIMO-OFDM relay links," *IEEE Transactions on Wireless Communications*, vol. 6, no. 8, pp. 2798–2802, 2007.

[22] C. Li, F. Sun, J. M. Cioffi, and L. Yang, "Energy efficient MIMO relay transmissions via joint power allocations," *IEEE Transactions on Circuits and Systems II: Express Briefs*, vol. 61, no. 7, pp. 531–535, 2014.

[23] X. Hong, Y. Jie, C.-X. Wang, J. Shi, and X. Ge, "Energy-spectral efficiency trade-off in virtual MIMO cellular systems," *IEEE Journal on Selected Areas in Communications*, vol. 31, no. 10, pp. 2128–2140, 2013.

[24] R. Arab Loodaricheh, S. Mallick, and V. K. Bhargava, "Energy-efficient resource allocation for OFDMA cellular networks with user cooperation and QOS provisioning," *IEEE Transactions on Wireless Communications*, vol. 13, no. 11, pp. 6132–6146, 2014.

[25] Q. Dong and J. Li, "Structured beamforming designs for spectral efficiency and energy efficiency in a three-node amplify-and-forward relay network," *IET Communications*, vol. 11, no. 8, pp. 1207–1215, 2017.

[26] X. Xu, J. Bao, H. Cao, Y.-D. Yao, and S. Hu, "Energy-efficiency-based optimal relay selection scheme with a BER constraint in cooperative cognitive radio networks," *IEEE Transactions on Vehicular Technology*, vol. 65, no. 1, pp. 191–203, 2016.

[27] J. Chen, H. Chen, H. Zhang, and F. Zhao, "Spectral-Energy Efficiency Tradeoff in Relay-Aided Massive MIMO Cellular Networks With Pilot Contamination," *IEEE Access*, vol. 4, pp. 5234–5242, 2016.

[28] V. W. Wong, R. Schober, D. W. Ng, and L. Wang, *Key Technologies for 5G Wireless Systems*, Cambridge University Press, Cambridge, 2017.

Permissions

The contributors of this book come from diverse backgrounds, making this book a truly international effort. This book will bring forth new frontiers with its revolutionizing research information and detailed analysis of the nascent developments around the world.

We would like to thank all the contributing authors for lending their expertise to make the book truly unique. They have played a crucial role in the development of this book. Without their invaluable contributions this book wouldn't have been possible. They have made vital efforts to compile up to date information on the varied aspects of this subject to make this book a valuable addition to the collection of many professionals and students.

This book was conceptualized with the vision of imparting up-to-date information and advanced data in this field. To ensure the same, a matchless editorial board was set up. Every individual on the board went through rigorous rounds of assessment to prove their worth. After which they invested a large part of their time researching and compiling the most relevant data for our readers.

The editorial board has been involved in producing this book since its inception. They have spent rigorous hours researching and exploring the diverse topics which have resulted in the successful publishing of this book. They have passed on their knowledge of decades through this book. To expedite this challenging task, the publisher supported the team at every step. A small team of assistant editors was also appointed to further simplify the editing procedure and attain best results for the readers.

Apart from the editorial board, the designing team has also invested a significant amount of their time in understanding the subject and creating the most relevant covers. They scrutinized every image to scout for the most suitable representation of the subject and create an appropriate cover for the book.

The publishing team has been an ardent support to the editorial, designing and production team. Their endless efforts to recruit the best for this project, has resulted in the accomplishment of this book. They are a veteran in the field of academics and their pool of knowledge is as vast as their experience in printing. Their expertise and guidance has proved useful at every step. Their uncompromising quality standards have made this book an exceptional effort. Their encouragement from time to time has been an inspiration for everyone.

The publisher and the editorial board hope that this book will prove to be a valuable piece of knowledge for researchers, students, practitioners and scholars across the globe.

List of Contributors

Qin Qin and Yong-qiang He
College of Computer, Henan Institute of Engineering, Zhengzhou 450007, China

Li-ming Nie
School of Software Technology, Dalian University of Technology, Dalian 116621, China

Shoufa Chen
School of Information and Electronic Engineering, Zhejiang University of Science and Technology, Hang Zhou 310023, China

Zhongpeng Wang
School of Information and Electronic Engineering, Zhejiang University of Science and Technology, Hang Zhou 310023, China
State Key Laboratory of MillimeterWaves, Southeast University, Nanjing 210096, China

Yishan He, Yufan Cheng, Gang Wu, Binhong Dong and Shaoqian Li
National Key Laboratory of Science and Technology on Communications, University of Electronic Science and Technology of China, Chengdu 611731, China

Shuming Chen, Yang Guo, Shenggang Chen and Hu Chen
College of Computer, National University of Defense Technology, Changsha, Hunan 410073, China

Xiaowen Chen
College of Computer, National University of Defense Technology, Changsha, Hunan 410073, China
Department of Electronic Systems, KTH-Royal Institute of Technology, Kista, 16440 Stockholm, Sweden

Zhonghai Lu
Department of Electronic Systems, KTH-Royal Institute of Technology, Kista, 16440 Stockholm, Sweden

Axel Jantsch
Institute of Computer Technology, Vienna University of Technology, 1040 Vienna, Austria

Jiang Wu
School of Information, Zhejiang Sci-Tech University, Hangzhou 310023, China

Zhongpeng Wang
School of Information and Electronic Engineering, Zhejiang University of Science and Technology, Hangzhou 310023, China

State Key Laboratory of Millimeter Waves, Southeast University, Nanjing 210096, China

Miltiadis Moralis-Pegios, Pelagia Alexandridou and Christos Koukourlis
Telecommunications Systems Laboratory, Electrical and Computer Engineering Department, Democritus University ofThrace, 67100 Xanthi, Greece

Yanfei Jia
College of Information and Communication Engineering, Harbin Engineering University, Heilongjiang 150001, China
College of Eletrical and Information Engineering, Beihua University, Jilin 132012, China

Xiaodong Yang
College of Information and Communication Engineering, Harbin Engineering University, Heilongjiang 150001, China
Collaborative Research Center, Meisi University, Tokyo 1918506, Japan

Bin Ge, Kai Wang and Bao Zhao
School of Computer Science and Engineering, Anhui University of Science & Technology, Huainan, Anhui 232001, China

Jianghong Han
School of Computer and Information, Hefei University of Technology, Hefei 230009, China

Yiying Zhang and Kun Liang
College of Computer Science and Information Engineering, Tianjin University of Science & Technology, Tianjin, China

Yeshen He, Yannian Wu, Xin Hu and Lili Sun
China Gridcom Co., Ltd, Shenzhen, Guangdong, China

Hasna Kilani and Rabah Attia
SERCOM Laboratory, Tunisia Polytechnic School, Carthage University, 2078 La Marsa, Tunisia

Mohamed Tlich
CodinTek Company, El Ghazala Technological Park, 2088 Ariana, Tunisia

Xin Xu, Wanhua Zhu, Xiaojuan Zhang and Guangyou Fang
Key Laboratory of Electromagnetic Radiation and Detection Technology, Chinese Academy of Sciences, North 4th Ring RoadWest, Haidian District, Beijing 100190, China

Chao Huang and Dunge Liu
Key Laboratory of Electromagnetic Radiation and Detection Technology, Chinese Academy of Sciences, North 4th Ring RoadWest, Haidian District, Beijing 100190, China
Graduate University of Chinese Academy of Sciences, No. 19A, Yuquan Road, Beijing 100049, China

Jiezhuo Zhong and Wenlong Feng
College of Information Science and Technology, Hainan University, Haikou, Hainan, China

Wei Wu
College of Information Science and Technology, Hainan University, Haikou, Hainan, China
Institute of Deep-Sea Science and Engineering, Chinese Academy of Sciences, Sanya, Hainan, China

Chunjie Cao
College of Information Science and Technology, Hainan University, Haikou, Hainan, China
State Key Laboratory of Marine Resource Utilization in the South China Sea, Hainan University, Haikou, Hainan, China

Chin E. Lin and Ying-Chi Huang
Department of Aeronautics and Astronautics, National Cheng Kung University, Tainan, Taiwan

Shujuan Wang and Guiru Cheng
Changchun University of Technology, Changchun 130012, China

Long He
Sinopharm A-THINK Pharmaceutical Co. Ltd., Changchun 130012, China

Manh Cong Tran and Yasuhiro Nakamura
Department of Computer Science, National Defense Academy, 1-10-20 Hashirimizu, Yokosuka, Kanagawa 239-0811, Japan

Gang Zhang, Niting Cui and Tianqi Zhang
School of Communication and Information Engineering, Chongqing University of Posts and Telecommunications, Chongqing 400065, China

Shingo Yoshizawa
Kitami Institute of Technology, Kitami, Japan

Takashi Saito and Yusaku Mabuchi
Mitsubishi Electric TOKKI Systems Corporation, Kamakura, Japan

Tomoya Tsukui and Shinichi Sawada
IHI Corporation, Yokohama, Japan

Fang Liu and Yongxin Feng
School of Information Science and Engineering, Shenyang Ligong University, Shenyang 110159, China

Ali Khan, Qifu Tyler Sun and Zahid Mahmood
School of Computer and Communication Engineering, University of Science and Technology Beijing, Beijing, China

Ata Ullah Ghafoor
Department of Computer Science, National University of Modern Languages, Islamabad, Pakistan

Lei Wang
School of Computer, Nanjing University of Posts and Telecommunications, Nanjing 210023, China

Qing Wang
School of Tongda, Nanjing University of Posts and Telecommunications, Yangzhou 225127, China

Shingo Yoshizawa and Hiroshi Tanimoto
Department of Electrical and Electronic Engineering, Kitami Institute of Technology, Kitami, Japan

Takashi Saito
Mitsubishi Electric TOKKI Systems Corporation, Kanagawa, Japan

Yongxin Feng
Communication and Network Institute, Shenyang Ligong University, Shenyang, Liaoning, China

Fang Liu, Xudong Yao and Xiaoyu Zhang
School of Information Science and Engineering, Shenyang Ligong University, Shenyang, Liaoning, China

Huimin Duan, Hui Huang and Cuihua Li
Department of Electronic and Electrical Engineering, Hefei University, Hefei, China

Qiang Wang, Tiejun Chen and Tingting Lan
Yulin Normal University, Yulin, China

Index